Yearbook on Space Policy

For further volumes:
http://www.springer.com/series/8166

Yearbook on Space Policy

Edited by the

European Space Policy Institute

Director: Peter Hulsroj

Cenan Al-Ekabi • Blandina Baranes •
Peter Hulsroj • Arne Lahcen
Editors

Yearbook on Space Policy 2011/2012

Space in Times of Financial Crisis

Springer

Editors
Cenan Al-Ekabi
Blandina Baranes
Peter Hulsroj
Arne Lahcen
European Space Policy Institute
Vienna, Austria

ISSN 1866-8305
ISBN 978-3-7091-1991-4 ISBN 978-3-7091-1649-4 (eBook)
DOI 10.1007/978-3-7091-1649-4
Springer Wien Heidelberg New York Dordrecht London

Softcover reprint of the hardcover 1st edition 2014

Cover illustration: Artist's impression of Herschel, ESA – C. Carreau.

Printed on acid-free paper

Springer is part of Springer Science+Business Media (www.springer.com)

Preface Yearbook 2011–2012

'Space in Times of Financial Crisis'

The financial crisis and the subsequent economic crisis have yet again shown how interconnected our world has become. The space sector – in spite of its somewhat particular and uncommon structure – is not entirely immune to the merciless dynamics that are affecting the functioning of the overall economy. Fortunately, the space sector has been relatively protected from the most volatile caprices that typically characterise commercial markets however. This is due to the fact that, despite increasing trends of commercialisation and privatisation of space activities, the lion's share of activities and funding is still in one way or another managed by governments, on the national level or on intergovernmental basis. Moreover, space is very much research and development driven and because of its development requirements operates on a mid- to long-term time horizon. As such, investment in space activities during times of economic turmoil is administrated as a way of countercyclical economic stimulation à la Keynes. These features have created a time lag that has long shielded the space sector.

Recently, it seems that the dynamics might be somewhat changing, however. Financial austerity is forcing some governments to revise and reconsider their approach vis-à-vis investment in space activities. This is especially true for the governments of the countries that feel the effects of the crisis the most. As opposed to some examples in the regular economy, the question for space is actually not so much one of life or death, given that many governments remain capable and willing to fund space activities. Rather, the question is how the financial crisis will affect a sector which has been, and still is, a showpiece of cutting-edge human ingenuity, fruitful international cooperation and scientific and technological progress. It is for these reasons that the thematic title of this Yearbook highlights the implications of the financial crisis on the space sector.

As in previous editions, this Yearbook is comprised of three parts. The first part seeks to set out a comprehensive overview of the economic, political, technological and institutional trends that affect space activities. It is prepared in-house in the

European Space Policy Institute (ESPI), and while its perspective is European, it also provides a comparative analysis of space efforts around the world.

The second part of the Yearbook includes seven external contributions that bring together the views of various professionals in the space field. Because of their divergent backgrounds, this part of the Yearbook reveals some interesting dynamics on, inter alia, the implications of the crisis for space industry, its effects on political decision-making, international cooperation and the position of young professionals in the sector. The overall picture shows that the implications of the financial crisis are ambiguous to say the least. While it clearly poses considerable challenges for some specific actors and branches in the space sector, others are very likely to encounter opportunities as the diverse space ecosystem moves towards a novel state of equilibrium.

An important milestone in the preparation of the Yearbook was again the ESPI Autumn Conference, where authors met for an exchange on the content of their contributions. Having taken place in Vienna in September 2012, it provided a forum for constructive exchange and coordination of the contributions.

The third part of the Yearbook carries forward the character of the Yearbook as an archive of space activities. Again prepared in-house by ESPI, a bibliography, chronology and data about institutions are provided where readers of the now six volumes of the Yearbook can identify statistical development and evolutions.

In closing, we would like to thank the contributors of the articles in Part II for their engagement in this publication, as well as the ESPI staff that has been instrumental for its production.

Vienna, Austria

Cenan Al-Ekabi
Blandina Baranes
Peter Hulsroj
Arne Lahcen

Acronyms

ACE	Advanced Composition Explorer
AEB	Brazilian Space Agency
AHP	Analytic Hierarchy Process
AIA	Atmospheric Imaging Assembly
ALHAT	Autonomous Landing and Hazard Avoidance Technology
AMS	Alpha Magnetic Spectrometer
AO	Announcement of Opportunity
APL	Johns Hopkins University Applied Physics Laboratory
APRSAF	Asia-Pacific Regional Space Agency Forum
APSCO	Asia-Pacific Space Cooperation Organization
Ariane 5 ME	Ariane 5 Midlife Evolution
ARTEMIS	Acceleration, Reconnection, Turbulence and Electrodynamics of Moon's Interaction with the Sun
ARTEMIS	Advanced Responsive Tactically Effective Military Imaging Spectrometer
ARTIST	Austrian Radionavigation Technology and Integrated Satnav Services and Products Testbed
ASAP	Austrian Space Applications Programme
ASAT	Anti-satellite Missile Test
ASI	Agenzia Spaziale Italiana (Italian Space Agency)
ATCS	Active Thermal Control System
ATV	Automated Transfer Vehicle
BMD	Ballistic Missile Defence
BSP	Basic Plan for Space
CAGR	Compound Annual Growth Rate
Casbaa	Cable and Satellite Broadcasting Association of Asia
CAST	China Academy of Space Technology
CCL	Commerce Control List
CD	Conference on Disarmament
CDI	Call for Declarations of Interest
CDTI	Centre for the Development of Industrial Technology

CEO	Chief Executive Officer
CFAS	Federal Commission for Space Affairs
CHF	Swiss franc
CHIRP	Commercially Hosted Infrared Payload
CM12	European Space Agency Council at Ministerial Level 2012
CMB	Cosmic Microwave Background
CMEs	Coronal Mass Ejections
CNES	Centre National d'Etudes Spatiales (French Space Agency)
CODA	Central Office for Delay Analysis
CONCORD	European Conference on Corporate Research and Development
CONEA	National Space Activities Commission of Argentina
COP 10	10th Conference of the Parties
COROT	Convention, Rotation and Planetary Transits
COTS	Commercial Orbital Transportation Services
CSA	Canadian Space Agency
CSO	Czech Space Office
CST-100	Commercial Space Transportation-100
CVA	Communauté des Villes Ariane
DARPA	Defense Advanced Research Project Agency
DARS	Digital Audio Radio Service
DBS	Direct Broadcast Services
Dextre	Special Purpose Dexterous Manipulator
DLR	German Aerospace Center
DoD	Department of Defense
DRDO	Defence Research and Development Organisation
DSCOVR	Deep Space Climate Observatory
DTH	Direct to Home
DTU	Technical University of Denmark
DWD	Deutsche Wetterdienst (German Meteorological Service)
DWSS	Defense Weather Satellite System
EAC	European Astronaut Centre
EADS	European Aeronautic Defence and Space Company
EAP	Environmental Action Programme
EARSC	European Association of Remote Sensing Companies
EBITDA	Earnings Before Interest, Taxes, Depreciation and Amortization
EC	European Commission
ECMWF	European Center for Medium Range Weather Forecasts
EDA	European Defence Agency
EDRS	European Data Relay System
EFQM	European Foundation for Quality Management
EJSM	Europa Jupiter System Mission
ELC	Express Logistics Carrier
ELISA	Electronic Intelligence Satellite
EO	Earth Observation

EPOXI	Extrasolar Planet Observations and Characterisation/Deep Impact Extended Investigation
EPS	Eumetsat Polar System
ERA	European Research Area
ESA	European Space Agency
ESF	European Science Foundation
ESM	European Support Module
ESOC	European Space Operations Center
ESPI	European Space Policy Institute
ESTEC	European Space Research and Technology Center
EU	European Union
EUMETSAT	European Organisation for the Exploitation of Meteorological Satellites
EUTELSAT	European Telecommunications Satellite Organization
EVE	EUV Variability Experiment
FAA	Federal Aviation Administration
FCC	U.S. Federal Communications Commission
FCT	Portuguese Foundation for Science and Technology
FFG	Forschungsförderungsgesellschaft (Austrian Research Promotion Agency)
FP7	Seventh Framework Programme
FSO	Forecast-Sensitivity-to-Observation
FSS	Fixed Satellite Services
GDP	Gross Domestic Product
GENIE	Guidance Embedded Navigator Integration Environment
GEO	Geostationary Orbit
GEO	Group on Earth Observations
GEOSS	Group of Earth Observation System-of-Systems Cooperation Platform
GER	Global Exploration Roadmap
GERD	Gross Domestic Expenditure on Research and Development
GES	Global Exploration Strategy
GFCS	Global Framework for Climate Services
GME	Global Forecast Model
GMES	Global Monitoring for Environment and Security
GMT	Greenwich Mean Time
GNI	Gross National Income
GNSS	Global Navigation Satellite System
GOCE	Gravity Field and Steady-State Ocean Circulation Explorer
GPM	Global Precipitation Measurement
GPS	Global Positioning System
GRAIL	Gravity Recovery and Interior Laboratory
GSLV	Geosynchronous Satellite Launch Vehicle
GSRT	General Secretariat for Research and Technology

GTO	Geosynchronous Transfer Orbits
HFI	High Frequency Instrument
HMI	Helioseismic and Magnetic Imager
HST	Hubble Space Telescope
HTV	H-II Transfer Vehicle
IASI	Interféromètre Atmosphérique de Sondage dans l'Infrarouge
ICBM	Inter-continental Ballistic Missile
ICG	International Committee on Global Navigation Satellite Systems
IEA	International Energy Agency
IGS	Information Gathering Satellite
IKAR	Interdepartmental Coordination Committee for Space
ILN	International Lunar Network
IMF	International Monetary Fund
IMO	International Maritime Organisation
IMPEL	Implementation and Enforcement of Environmental Law
INPE	National Institute for Space Research
Intelsat	International Telecommunications Satellite Consortium
IOV	In-Orbit Validation
IPCC	Intergovernmental Panel on Climate Change
IRGC	Islamic Revolution Guards Corps
ISECG	International Space Exploration Coordination Group
ISPS	Innovative Space Propulsion Systems
ISRO	Indian Space Research Organisation
ISS	International Space Station
ISU	International Space University
ITAR	International Traffic in Arms Regulations
ITU	International Telecommunication Union
JADE	Jovian Auroral Distributions Experiment
JAXA	Japan Aerospace Exploration Agency
JEDI	Jupiter Energetic Particle Detector Instrument
JFK	John Fitzgerald Kennedy
JIRAM	Jovian Infrared Auroral Mapper
JWST	James Webb Space Telescope
KAI	Korea Aerospace Industries
KARI	Korea Aerospace Research Institute
KSLV	Korea Space Launch Vehicle
LADEE	Lunar Atmosphere and Dust Environment Explorer
LEO	Low Earth Orbit
LFI	Low Frequency Instrument
LRO	Lunar Reconnaissance Orbiter
LTRO	Long-Term Refinancing Operation
LWS	Living with a Star
MAG	Magnetometer
MAVEN	Mars Atmosphere and Volatile Evolution

MCTES	Ministry of Science, Technology and Higher Education
MDA	Missile Defense Agency
MDA	McDonald Dettwiler and Associates
MEP	Member of the European Parliament
MER	Mars Exploration Rover
MESSENGER	Mercury Surface, Space Environment, Geochemistry and Ranging
METoP	Polar Orbiting Meteorological Satellites
MEXT	Ministry of Education, Culture, Sports, Science and Technology
MFF	Multiannual Financial Framework
MHS	Microwave Humidity Sounder
MIRI	Mid-infrared Instrument
MMO	Mercury Magnetospheric Orbiter
MPCV	Multi-purpose Crew Vehicle
MPLM	Multi-purpose Logistic Module
MPO	Mercury Planetary Orbiter
MRO	Mars Reconnaissance Orbiter
MSL	Mars Science Laboratory
MSS	Mobile Servicing System
MWR	Microwave Radiometer
NASA	National Aeronautics and Space Administration
NASRDA	National Space Research and Development Agency of Nigeria
NATO	North Atlantic Treaty Organization
NGA	National Geospatial-Intelligence Agency
NOAA	National Oceanic and Atmospheric Administration
NOFBX	High Performance Nitrous Oxide Blended Monopropellant
NRC	National Research Council
NRO	National Reconnaissance Office
NSC	Norwegian Space Centre
NSI	National Space Institute of Denmark
NSO	The Netherlands Space Office
NTIA	National Telecommunications and Information Administration
NTU	Nanyang Technological University
NWP	Numerical Weather Prediction
OECD	Organisation for Economic Co-operation and Development
OSE	Observing System Experiments
OSP	Orbital/Suborbital Program
PARP	Polish Agency for Enterprise Development
PHAs	Potentially Hazardous Asteroids
PLA	People's Liberation Army
PM	Pump Module
PMM	Permanent Multipurpose Module
PPP	Purchasing Power Parity
PPP	Public Private Partnership

PROBA	Project for Onboard Autonomy
PSLV	Polar Satellite Launch Vehicle
PTSS	Precision Tracking Space System
PWR	Pratt & Whitney Rocketdyne
QZSS	Quasi Zenith Satellite System
RCM	Radar Constellation Mission
RHESSI	Reuven Ramaty High Energy Solar Spectroscopic Imager
ROEM	Renseignement d'Origine Electromagnétique
ROSA	Romanian Space Agency
RRM	Robotic Refuelling Mission
RXTE	Rossi X-ray Timing Explorer
SGAC	Space Generation Advisory Council
SAR	Synthetic Aperture Radar
SDO	Solar Dynamics Observatory
SESAR	Single European Sky Air Traffic Management Research
SIA	Satellite Industry Association
SIS	Space Infrastructure Service
SLS	Space Launch System
SM	Standard Missile
SMEX	NASA Small Explorer
SNSB	Swedish National Space Board
SOHO	Solar and Heliospheric Observatory
SpaceX	Space Exploration Technologies Corporation
SPOT	Satellite pour l'Observation de la Terre
SRON	Dutch Space Research Organization
SSC	Swedish Space Corporation
SST	SpaceShipTwo
SSTA	Singapore Space & Technology Association
SSTL	Surrey Satellite Technology Limited
STEREO	Solar Terrestrial Relations Observatory
STFC	Science and Technology Facilities Council
TCBM	Transparency and Confidence Building Measures
TEKES	Finnish Funding Agency for Technology and Innovation
THEMIS	Time History of Events and Macroscale Interactions During Substorms
TRL	Technology Readiness Level
TÜBITAK/UZAY	Scientific and Technological Research Council of Turkey
UAS	Unmanned Aerial Systems
UK	United Kingdom
UN	United Nations
UN COPUOS	United Nations Committee on the Peaceful Uses of Outer Space
UN FCCC/COP	United Nations Framework Convention on Climate Change/Conference of Parties

UN OOSA	United Nations Office of Outer Space Affairs
UN REDD	United Nations Programme on Reducing Emissions from Deforestation and Forest Degradation in Developing Countries
UN SAP	United Nations Programme on Space Applications
UN SPIDER	United Nations Platform for Space-Based Information for Disaster Management and Emergency Response
UNCTAD	United Nations Conference on Trade and Development
UNGA	United Nations General Assembly
UNGIWG	United Nations Geographic Information Working Group
UNIDIR	United Nations Institute for Disarmament Research
UNISPACE	United Nations Conference on the Exploration and Peaceful Uses of Outer Space
UNSDI	United Nations Spatial Data Infrastructure
USAF	United States Air Force
USAT	Ultra Small Aperture Terminals
USML	U.S. Munitions List
VSAT	Very Small Aperture Terminals
WISE	Wide-Field Infrared Survey Explorer
WMO	World Meteorological Organisation
WRC	World Radiocommunication Conference

Contents

Part I
The Year in Space 2011/2012

Chapter 1
European Space Activities in the Global Context

Cenan Al-Ekabi

1.1 Global Political and Economic Trends

1.1.1 Global Economic Outlook

In the reporting period covering the second half of 2011 and the first half of 2012, the world economy seemed to be on a successful path out of the financial and economic crisis. The booming economies of China, India, Brazil, Russia in particular, and some other emerging and developing countries were providing impetus to global economic recovery. Due to strong export orientation, industrialised countries, such as Germany, had benefitted from demand from these countries. Thus Germany had record growth rates and declining unemployment.

Nevertheless, there were developments in the global economy that gave cause for concern. For many countries the financial and economic crisis of the previous 3 years had begun to acquire a more permanent character:

- Capital movements are rather detached from investments in the real economy;
- National and international financial markets are above all secondary markets;
- Financial markets are still highly speculative;
- Financial markets are volatile due to overreaction and underreaction in financial and foreign exchange markets, as well as abrupt and voluminous international and domestic capital movements;
- High reliance on liquidity interventions; however, with IMF, central banks and governments having limited room for backup at this time.[1]

[1] Kamppeter, Werner. "International Financial Crises in Comparison - Lessons for Managing the Current Crisis." May 2011. Friedrich-Ebert-Stiftung – International Policy Analysis. 11 May 2012 http://library.fes.de/pdf-files/id/ipa/08081.pdf.

C. Al-Ekabi (✉)
ESPI (European Space Policy Institute), Vienna, Austria
e-mail: Cenan.Al-Ekabi@espi.or.at

C. Al-Ekabi et al. (eds.), *Yearbook on Space Policy 2011/2012*, Yearbook on Space Policy, DOI 10.1007/978-3-7091-1649-4_1, © Springer-Verlag Wien 2014

The combined effects amassed over time on a global scale have created new concerns and negative consequences. Not only has global economic activity weakened, but the economic pace has also become more uneven, exacerbating inequalities on a global, regional and societal scale. These growing imbalances in the global distribution of wealth and trade could potentially increase geopolitical tensions and encourage politically radical and economically protectionist reflexes. However, financial confidence in the United States began increasing, and the glacial European response to the Euro crisis had possibly finally reached a turning point, despite recent events in Greece. Still, the world faces a multitude of unresolved structural economic and financial deficiencies. Unforeseen geopolitical events, like the Arab spring, and natural disasters (such as the great earthquake of Japan) have multiplied the shocks to an already unsettled global financial and economic community.

In the past 3 years mainly public funds have been used to secure the European financial system and buffer the Euro crisis, in turn triggering austerity measures in almost all countries but the United States. The newly elected President of France, François Hollande, planned to reopen talks on the European Union's hard-won fiscal austerity pact in order to refocus the EU's economic policy on growth. The realism of reopening the austerity pact might be questionable, but there was growing recognition also within the European Commission that austerity alone will not solve the situation, and that active investment in growth must operate in parallel. In this sense, Europe will likely start to move in the direction of the approach of the United States, where growth has always been the priority.

Considering the aforementioned conditions, one must be cautious regarding the global economy's short term prospects. In developed economies the pace of economic recovery continues to be sluggish at best, with real economic growth reaching only 1.5 % in 2011 and not expected to increase beyond 2 % in 2012. On the other side, emerging economies have demonstrated a much more robust expansion of approximately 6 %, which was expected to continue in 2012 as well. As fiscal problems in developed economies persist, however, the prospects for emerging economies begin to look more uncertain too.[2] This is particularly true for countries that are more heavily dependent on foreign demand for their products and services. The current global outlook is based on a series of assumptions, such as the containment of the Euro crisis; continued and successful fiscal consolidation in the U.S. without obstruction of economic recovery, especially on the demand side; the ability of markets in advanced economies to absorb the shock of the gradual withdrawal of quantitative easing measures previously applied; the absence of further major geopolitical shocks; and that global market volatility and risk aversion will not become more pronounced.

Concerns about banking sector losses and fiscal sustainability have lowered credit confidence and widened the sovereign spreads for a large number of Euro area countries, which towards the end of 2011 reached highs not seen since the

[2] International Monetary Fund. World Economic Outlook 2011. Washington DC: IMF, 2011.

launch of the Economic and Monetary Union. As a result of this deterioration, bank funding came to a near stop across the entire eurozone towards the end of 2011, prompting the European Central Bank to initiate debt re-monetarisation measures through its Long-Term Refinancing Operation (LTRO). In general, bank lending conditions deteriorated across most advanced economies. More importantly, currency market volatility increased significantly, witnessing a sharp appreciation of the Japanese Yen and an equally violent depreciation of the currencies of several emerging countries.

The uneven nature and geographical distribution of post-crisis economic expansion is a major reason behind the significant risks to global economic activity. In fact, two simultaneous balancing acts should take place. On the one hand, private demand should take over again from public demand. Worryingly enough, advanced economies seem to be lagging behind on this front. On the other hand, an effort should be made to rebalance global trade volumes and international payments to pre-crisis levels. This would imply that economies with large external surpluses should increase their domestic orientation, while countries with trade deficits should do the opposite or radically increase their productivity. At present, this transition is still underway, and much of its eventual success depends on the ability of policy makers in crisis-hit economies to resist the temptation of choosing accommodating monetary policies to mend balance sheets.

In many advanced economies, notably those with external deficits, the drop in house prices is expected to continue in the longer term. Furthermore, structural reforms are still needed to boost production, including measures to reform the labour and production markets and strengthen their resistance to the adverse effects of population ageing.

Youth unemployment, with extremes in Greece and Spain, but high in almost all industrialised countries, is both a short and a long-term challenge with a possible fundamental societal impact on the wealth and stability paradigm of these countries.

1.1.2 Political Developments

1.1.2.1 Geopolitics

The year 2011 ushered in a wave of political demonstrations and unrest within the Arab world. Dubbed the Arab Spring, the consequences of this mass social-political upheaval continued to show throughout the year with profound influence on the geo-political climate. Following the initial uprisings in Tunisia, inciting similar action in neighbouring states, the Arab Spring spread with varying results across all of North Africa, and to most of the Middle East countries. Syria had been experiencing its own long-lasting turmoil, and the crisis there had other regions promoting conflicting interests. While Western Europe and the United States backed the uprising, Russia sided with traditional government interests, and

China adopted a policy of non-interference, politically motivated by Tibet and Taiwan. As a natural consequence, the UN Security Council was stuck between two stools, despite very active and constructive diplomacy by some members, such as France and the United Kingdom. A sort of 'Great Game' was being replayed in the Middle East as a result of the Arab spring, with neighbouring states taking traditional positions based on historical and religious affiliations, along with internal influences. Iran is a prime example, expounding its religious and geopolitical interests, and seizing the opportunity to deflect attention from its own nuclear affairs.

The Arab Spring situation illustrates the vulnerability of old regimes faced with modern attitudes and the supra-regional interconnectedness among neighbouring states stretching from Morocco to Asia. Whereas the Arab Spring sparked other regions to assess their current situation, Palestine and Israel have struggled with their own autonomy issues since World War II, yet have still been profoundly affected by the uprisings. The problems with Iran both directly and indirectly influenced the situation in Iraq, with a chain reaction affecting the situation in Afghanistan, and complicating the already highly critical matters in Pakistan, which thereafter touched India, and introduced their local nuclear weapons issues into the equation, along with the problems with North Korea. From there both China (as a neighbour of Pakistan and North Korea), and Japan were affected – with an eventual trickledown effect that reached the United States and Western Europe.

Pakistan struggled with overcoming the results of the disastrous flooding that affected it, while also showing signs of being marginalized, which is highly alarming for the nuclear state, and for the already uneasy relations with its neighbour India.

And North Korea was in the throes of dealing with the death of Kim Jong-Il, and the installation of his inexperienced son as successor. The succession is likely to give rise to internal power struggles creating difficulties not only for China, but for all regional powers in addition to all countries concerned about proliferation of nuclear arms. However, should current power structures crumble, the result while unappealing in the short term might also provide an opportunity for normalisation.

The overall context was heavily influenced by the start of the election process in the United States. These elections meant that political compromise across party lines was nearly impossible, as illustrated by the debt ceiling debacle. Real progress on the 2013 budget would have to wait for the November 2012 elections.

The overall context was also heavily influenced by the continued assertiveness of China on the global stage. China is undergoing a very different kind of leadership election and eventual transition. Notwithstanding the Bo Xilai situation, tensions in China were barely noticeable with the likely future president, Xi Jinping, visiting the United States in a symbolic gesture of comity. Still, China continued its aggressive investment policies in Africa, and continued to leverage its strength in space geopolitically, as well as domestically. The rendezvous and docking of two Chinese spacecraft was a triumph, setting the scene for a full-blown space station. Less noticeable was Russian assistance behind the scenes; quickly absorbed and internalised by China.

Russia, having finished its presidential elections, continued with Vladimir Putin at the helm. Interestingly, Putin's absolute grip on power – typical of the last 12 years – was increasingly questioned in the run-up to the election and even thereafter; with the lingering effects of the Arab Spring noticeable even in Russia.

The proactive civic spirit injected into the political process thanks to the Arab Spring and social media was an underlying theme during the reporting period, the Occupy Wall Street movement with its many off-springs is but a further example.

1.1.2.2 Environment

Space applications have an important role in the monitoring and protection of the environment. Space assets are uniquely positioned to offer a global perspective on climate change. Moreover they are helping to better manage disaster situations around the world. They often represent a common multinational platform for collecting relevant meteorological and environmental data. These characteristics make them ideal promoters of international understanding and cooperation in this field. Satellite based systems, for example, have recently been used to gather information about the situation of the melting of the ice-caps; to scan the rising global sea level and to map regions mostly affected by global warming. Remote sensing technologies can also be used to monitor deforestation and land use, and are important for better utilization of fresh water sources. There is no doubt that space technologies will play an important role in human and environmental security in the future, hence technical development of their capabilities is necessary.

Despite the financial crisis, climate change remains one of the commonly recognized agenda topics within the global political debate. Following the complicated acceptance and application of the 1997 Kyoto Protocol, which was set to expire in 2012, and the "Copenhagen Accord" of 2009 establishing voluntary emissions cuts,[3] the new round of negotiations over the follow-up international agreement continued in 2011. From 28 November to 11 December 2011, the 17th Conference of Parties to the UN Framework Convention on Climate Change (UN FCCC/COP) took place in Durban, South Africa. After days of difficult negotiations between delegates, an agreement on the preparation of a new legally binding treaty to address global warming and its solutions was reached in the final period. The treaty is supposed to be prepared by 2015 and enter into force by 2020.[4] With this effort, the international community showed willingness to move away from the long lasting deadlock and towards real political solutions. Traditional differences between the positions of developed and developing countries were

---

[3] "United Nations Climate Change Conference kicks off in Copenhagen." 7 Dec. 2009. United Nations Development Programme 20 Feb. 2012 http://content.undp.org/go/newsroom/2009/december/historic-united-nations-climate-change-conference-kicks-off-in-copenhagen.en.

[4] "Durban Climate Change Conference – November/December 2011." United Nations Framework Convention on Climate Change 20 Feb. 2012 http://unfccc.int/meetings/durban_nov_2011/meeting/6245.php.

significant and were stressed again during the conference. Nonetheless, emerging economies such as India and China and the representatives of major current CO_2 emitters (except the U.S.), took leading roles during the final negotiations.

The Durban conference saw progress in the creation of the Green Climate Fund by adoption of its management framework. In the future, this fund will gather and distribute approx. €75.5 billion to help developing countries handle the side-effects of global warming and climate change.[5] The other important outcome was the progress on the UN Programme on Reducing Emissions from Deforestation and Forest Degradation in Developing Countries (UN REDD). Deforestation and soil degradation are well known, not only as the negative effect of global climate change, but also as a significant contributor to it.

In sum, recent progress on global climate issues shows that the gap between leading industrialised countries and countries in the development process is slowly changing. It seems that developing and emerging countries are beginning to accept greater involvement in climate change avoidance. Despite this progress, one event immediately following the Durban conference might in the future undermine the role of industrialised countries in climate change negotiations. The reaction of Canada, which a day after the conference in Durban decided to formally withdraw from the Kyoto protocol to save an estimated $14 billion in penalties, illustrated that real progress in the international community can be counteracted by unilateral action.[6]

Environmental and sustainable development issues are still important for both internal and external EU policies. At the EU level, several important initiatives regarding environmental issues were undertaken during the Polish Presidency in the second half of 2011.[7] Thus the EU Council stressed the urgency of establishing an ambitious international regime to solve global climate changes and called for agreement on a global and comprehensive legally-binding framework to keep the future increase of global temperature below 2 °C.

Furthermore, the European Council adopted a decision establishing the position of the EU for the tenth Conference of the Parties (COP 10) to the Basel Convention on the Control of Trans-boundary Movements of Hazardous Wastes and their Disposal. In parallel, the European Commission stressed the urgent need to effectively prevent, combat and significantly reduce illegal waste shipments. According to EU officials, this should be achieved through improved use and coordination of existing tools, such as the European Union Network for the Implementation and Enforcement of Environmental Law (IMPEL), and inter-agency cooperation.[8]

---

[5] Black, Richard. "Climate Talks End With Late Deal." 11 Dec. 2011. British Broadcasting Corporation 20 Feb. 2012 http://www.bbc.co.uk/news/science-environment-16124670.

[6] "Canada Pulls Out of Kyoto Protocol." 12 Dec. 2011. CBC News 21 Feb. 2012 http://www.cbc.ca/news/politics/story/2011/12/12/pol-kent-kyoto-pullout.html.

[7] Council of the European Union. The 318th Council Meeting: Environment. Press Release 15321/11 final of 10 Oct. 2011. Brussels: European Union.

[8] Council of the European Union. Preparation of the Tenth Meeting of the Conference of the Parties to the Basel Convention on the Control of Transboundary Movements of Hazardous

During the Polish presidency, the EU Council also adopted conclusions on the assessment of the EU's sixth Environmental Action Programme (EAP) 2002–2012. The conclusions proposed by the European Commission are seen as comprehensive and forward-looking, and take into account existing initiatives such as the Europe 2020 Strategy; the EU positions on the UN Conference on Sustainable Development (Rio +20); the post-2010 Biodiversity Strategy; and more. The sixth EAP should allow for the consolidation and completion of legislation in almost all areas of EU environmental policy.

Regarding the 2012 Rio +20 UN Conference the Polish Presidency significantly assisted in establishing the EU's general positions, mostly stressing the use of green energy in the context of sustainable development and poverty eradication, and the establishment of an institutional framework for sustainable development.[9]

1.1.2.3 Energy

Despite the fact that economic recovery in 2011 and 2012 remained sluggish and uneven, and with future economic prospects still uncertain, global primary energy demand marked a remarkable 5 % increase on the preceding year of 2010, thus pushing CO_2 emissions to a new high. One of the key planks during the period in review was an increased effort in many countries to improve their energy efficiency. In spite of this, global energy use increased for the second consecutive year. Furthermore, events such as the Fukushima Daiichi nuclear power plant disaster, the resultant nuclear energy scepticism, and continued turmoil in parts of the Middle East and North Africa have cast doubts on the reliability of energy supply. But also the persisting global economic crisis has considerably lowered market confidence in sovereign financial performance, further weakening government resolve to implement comprehensive energy policies and achieve global climate change objectives.

Another recurring trend during the review period was the increasingly strong role of emerging countries in determining energy market behaviour. Emerging countries are projected to account for 90 % of global population growth, 70 % of global GDP increase and 90 % of energy demand growth over the period 2010–2035.[10] China in particular expected to further strengthen its position as the world's largest energy consumer: by 2035, it is expected to consume almost 70 % more energy than the United States. Even in this case, however, Chinese per capita consumption would remain far lower than that of the United States. At the same time, energy consumption growth rates in other emerging countries, such as India, Indonesia, and Brazil are expected to grow even faster than in China. An important

---

Wastes and Their Disposal, Cartagena de Indias, Colombia, 17–21 Oct. 2011. Council Conclusion 15393/11 final of 11 October 2011. Brussels: European Union.

[9] Council of the European Union. Rio + 20: Towards Achieving Sustainable Development by Greening the Economy and Improving Governance. Council Conclusion 15388/11.

[10] International Energy Agency. World Energy Outlook 2011. IAE: Paris, 2011.

side effect of this trend is that significant investments in energy production and distribution will have to be made to keep pace with growing demand. By some estimates, as much as $32 trillion will have to be invested in related infrastructure by 2035, of which $20 trillion would be for fossil fuel exploitation.[11] On the supply side, market volatility and geopolitical events in oil-rich regions in 2011 and 2012 demonstrated the fragility of fossil fuel reliance. In the short term, the economic slow-down provoked by the crisis contributed to lowering pressure on oil prices. Long-term consumption trends remain unchanged, however, prices are expected to rise significantly once global economic growth returns. All of the projected net increase in oil demand is expected to come from the transport sector of emerging countries.

Thus the age of fossil fuels is not over, in spite of a slight anticipated relative decline in their demand, i.e. despite the increase in energy consumption, the share of fossil fuels in the global primary energy consumption mix is projected to fall only slightly from 81 % in 2010 to approximately 75 % in 2035. Only natural gas is expected to increase its relative share within the fossil fuel mix. At the same time, the problem of decreasing the environmental footprint of fossil fuel energy consumption remains unsolved. As a matter of fact, 4/5 of the total energy-related CO_2 emissions permissible by 2035 are already produced today by current capital stock (power plants, buildings, factories, etc.). If no corrective measures are taken, energy-related infrastructure is projected to generate all of the 2035 allowed CO_2 emissions nearly two decades too early, i.e. by 2017.[12]

1.1.2.4 Resources

Space applications and Earth monitoring technologies play an important role in the area of resource management. They provide better control and support of the utilisation of scarce natural resources. Likewise, satellite based technologies can perform indispensable tasks for accommodating international trade, e.g. by streamlining global business transactions and payments. Global navigation satellite systems are already an integral part of transportation and utilisation of natural, agricultural and industrial resources. And the use of meteorological and imaging satellites is making agricultural output bigger and more reliable. For many developing countries the rationale for investment in space is improvement of the management of their agricultural and natural resources.

In 2011 and 2012, commodity prices and consumer price inflation receded, yet risks remained. The observed tendency was attributed to weaker global demand as a result of the economic crisis. Still, in this period, adverse weather in many regions affected several agriculture markets, as well as coal and metals production. The political situation, mainly in North Africa and the Middle East with the Arab

[11] Id.

[12] Id.

Spring, which resulted in political and economical destabilisation, and the international sanctions against Iran, resulted in a loss of oil supply. Furthermore, currency fluctuations affected domestic prices of commodities, sometimes increasing demand. The long period of dollar weakness, in particular, contributed to upward pressure on commodity prices.[13]

In 2011, the growth of international trade was expected to return to a single-digit figure, in the range of 7–8 %. However, the revival of trade has been uneven among countries and geographic regions. For instance, in industrialised countries trade has yet to bounce back to a level above its pre-crisis levels. These countries recovered part of their previous trade losses between mid-2009 and mid-2010, but there has been no further growth since. Another example is the situation in the economies in transition. Their trade failed to reach its pre-crisis level by the end of 2010. In sharp contrast, the volume of both imports and exports in most groups of developing and emerging countries already exceeded their 2008 peak in the course of 2010. In this case, the countries in the East Asia region took the leading position. In countries that produce durable and capital goods, such as China and Japan, exports increased in volume by almost 30 %. In developing countries that export mainly primary commodities, the volume of exports was relatively stable.

It was significant during the period that oil and mineral exporters experienced significant gains. Oil prices maintained their price level, primarily because of supply insecurity related to geopolitical events. In general, geopolitical risks were expected to remain high, causing oil prices to ease only marginally during the second half of 2012. This development occurred in a period of further economic stagnation and crisis in the global economy, thus multiplying cyclical pressures. To compare the situation with previous years, according to the figures of the United Nations Conference on Trade and Development (UNCTAD) crude oil market prices fluctuated within a $70–80 band during the first three quarters of, 2010. They then surged in the last quarter of the year, to reach a monthly average of $116.3 during the fourth month of 2011.[14] On average, global oil demand rose by 3.2 % in 2010. While the demand in non-OECD countries grew by 5.5 %, OECD countries witnessed an increase of only 1.1 %. The highest demand for crude oil came from China, with demand growing by 12.3 % in 2010. At the same time, oil supply increased by only 2.1 % in 2010, further increasing the gap between supply and demand.[15]

On the other hand, non-oil commodity prices continued their decline, thanks to improved supply conditions and a steep decline in global demand. In fact, non-oil commodity prices were projected to have fallen by 14 % by the end of 2012. In the near term, prices are expected to drop even further. During the previous years, uncertainty and instability have been the major distinguishing feature of commodity

---

[13] The International Bank for Reconstruction and Development. Global Economic Prospects 2011. Washington DC: World Bank, 2011. 51–65.

[14] Id. at 14.

[15] Id. at 14.

markets. This is also reflected in the greater volatility of commodity prices. According to an UNCTAD report, between the two periods 2002–2005 and 2006–2011, the simple measure of volatility increased by a factor of 3.8 for food commodities and vegetable oil seeds and oils; by 2.7 for agricultural raw materials; and by 1.6 for minerals and metals and crude petroleum. In the markets for metals and minerals prices rose steeply in the second half of 2010, and peaked during the first months of 2011. Gold and silver in particular, benefited from uncertainties about the global economy. During the second half of 2011 the prices for gold reached a historical maximum and floated between $1,690 per ounce and $1,714 per ounce. The reason for these high prices was clearly the safe haven status of gold and silver.[16]

The overall trend of stabilising or lower commodity prices would explain the relative decline of consumer price inflation during the period. In advanced economies, inflation was expected to be low, especially when compared to the effect of last year's higher commodity prices. Inflation in these economies was projected to fall to about 1.5 % in the course of 2012, down from about 2.75 % in 2011. In emerging and developing economies, pressures were also expected to drop, as both growth and food price inflation slow. Overall, consumer prices in these economies were projected to decelerate, with inflation of around 6.25 % during 2012, down from over 7.25 % in 2011.[17]

1.1.2.5 Knowledge

There is no doubt that sustained education and knowledge improvement is one of the necessary conditions for successful space activities, as well as for the full exploitation of their societal benefits. In general, space technology and development, drawing on multiple scientific disciplines, is one of the most difficult and challenging fields in scientific and technical research. Therefore, coherent and sustainable strategies aimed at improving higher education and supporting technical and scientific activities are particularly relevant and necessary for space sector activities.

For Europe, as a leading actor in the field of scientific and technological R&D related to space, maintaining and expanding its large pool of highly skilled and specialised scientists and professionals should be a constant priority. Recent evidence suggests that Europe's leading position and excellence in this field could be disputed in the medium term, both by developed and emerging economies. In order for Europe to continue to be among the leaders in the global race for knowledge and excellence in space R&D, current levels of financial spending and political commitment in this area will have to be increased, and not only simply maintained. For

---

[16] United Nations Conference on Trade and Development. Trade and Development Report, 2011. Geneva: UNCTD, 2011. 15.

[17] International Energy Agency. World Energy Outlook 2011. Paris: IAE, 2011.

example, today only a mere 26 % of the European working age population holds a higher educational degree compared to 41 % in the U.S., 44 % in Japan and 50 % in Canada.[18] Furthermore, this relative lack of skilled workers is most likely to be accentuated as demand for their services is expected to increase.

According to a 2010 report of the European Centre for the Development of Vocational Training, the European skills forecast indicates that by 2020 around 35 % of jobs in the EU will require higher education qualifications.[19] Based on this finding, for Europe to maintain its competitiveness and current position on the global scientific stage a 30 % increase in higher education graduates will have to be achieved by 2020. Encouragingly, the population of university and higher education students in Europe has been constantly increasing over recent years. Today, there are approximately 4,000 universities and other kinds of higher education institutions in Europe, with more than 19 million students. Nevertheless, this quantitative increase has not been accompanied by qualitative improvements in governance structures and proposed academic curricula, or by increases in funding. Although increasing in size, Europe's higher education system has not yet achieved an academic curricula distribution that will train scientists and professionals with the right kind of skills to support economic growth and scientific excellence in new technologies. This is especially true for the space sector, which has a relatively limited human resources supply and demand chain.[20]

According to a European Commission report, the potential of European higher education institutions to fulfil their role in society and contribute to Europe's prosperity is not fully exploited. EU officials pointed out that greater capacity for research and development could fuel innovation across all sectors of the economy, improving competitiveness and fostering job creation. The same report highlights the potential technological spin offs of innovation and their capacity to revitalise more traditional economic sectors and rural areas, multiplying their broader societal impact.[21] In this context, the role of space applications should be highlighted, as for example in the case of using communications satellites to improve broadband Internet connectivity in remote areas. Finally, the new strategy proposed by the European Commission identifies priority areas where EU countries need to do more to achieve shared education objectives and describes how the EU can support the modernisation of their national higher education policies. According to the EC proposal, EU-level initiatives will include a multi-dimensional university ranking which will better inform students about university courses and support competition between universities.

---

[18] European Commission-Eurostat. Europe in Figures. Eurostat yearbook 2011. Luxembourg: Publications Office of the European Union, 2011. 212–218.

[19] European Centre for the Development of Vocational Training. Skills Supply and Demand in Europe: Medium-Term Forecast up to 2020. CEDEFOP, 2010.

[20] Id.

[21] European Commission. Communication from the Commission. Europe 2020: A Strategy for Smart, Sustainable and Inclusive Growth. COM (2010) 2020 final of 3 Mar. 2010. Brussels: European Union.

Sustained and efficient investments in higher education are important prerequisites for quality improvement. For example, in 2008 the average level of direct spending on higher education in the EU, both public and private, was approximately 1.3 % of the European GDP. The majority of expenditure on higher education usually comes from the public sector and it has been constantly increasing. Private expenditure on education is less pronounced in the EU, thus showcasing the traditional role of the public purse in public endeavours. By comparison, U.S. total private and public investment for the same period amounted to 2.7 % of GDP.

The EU budget dedicated to funding programmes in education and training is about €8 billion for the period 2007–2013. In addition to this, the EU will spend an additional €4.7 billion on training, mobility and career development for researchers. For the 2014–2020 budget period the European Commission emphasises the priority of "smart growth" through knowledge creation, consequently allocating substantial funds to skills and knowledge creation. According to the budget proposal, a single funding programme for education, training and youth will receive €15.2 billion, which means an increase of 73 % compared to the 2007–2013 period. The Commission has also proposed a 46 % rise in research funding under the planned Horizon 2020 strategy that will bring it up to €80 billion.[22]

Considering the aforementioned shortcomings, it was not a surprise that initiatives in the scientific research and education fields were among the top priorities of the 2011 Polish EU Council Presidency. For example, increasing continuous learning, as well as educational and professional mobility, and modernising higher education infrastructure and curricula have been some of the key initiatives in this respect. The implementation of these priorities followed the broad lines of the Europe 2020 Strategy and its flagship initiative "Youth on the Move", for the creation of new skills and new jobs. In this context, the Polish presidency launched a comprehensive debate on learning and mobility issues in the framework of EU educational programmes, taking into account its international dimension and also including EU neighbours. These actions were further supported by the European Commission, which set the goal of doubling the number of EU grant recipients for studying and training abroad at 800,000 students by 2020. In addition to this, around €72.5 billion is to be spent on education and training across Europe's regions until 2013, and this pace is expected to be maintained in the future.[23] Finally, the EC plans to emphasise quality support in teaching and scientific research, as well as improving governance practices and transparency in higher education establishments.[24]

[22] "An EU Strategy for Modernising Higher Education – Questions and Answers." 20 Sept. 2011. European Commission – Press Release. 1 Mar 2012 http://europa.eu/rapid/pressReleasesAction.do?reference=MEMO/11/615&format=HTML&aged=0&language=EN&guiLanguage=fr.

[23] "Extra Funds for Education, Youth and Creativity Will Boost Jobs, says Commission." 11 June 2011. European Commission – Press Release. 29 Feb 2012 http://europa.eu/rapid/pressReleasesAction.do?reference=IP/11/857&format=HTML&aged=0&language=EN&guiLanguage=fr.

[24] Programme of the Polish Presidency of the Council of the European Union. 1 July – 31 December 2011. Warsaw: Poland Ministry of Foreign Affairs, 2010. 32.

1.1.2.6 Mobility

Maritime commerce accounts for the bulk of global trade, whereas air traffic carries most of the world's passenger traffic. Space assets are indispensable to both, as they provide meteorological, navigation and communication services that make sea and air transport safer and cheaper. Maritime navigation and mobile communications are two of the upcoming spheres with significant financial interests in the development of new generations of satellite-based applications.

In 2011 and 2012 the transport sector continued to suffer from the effects of the global financial crisis. As the principal facilitator of global commodity flows, the transport sector continued its steep decline in demand due to the crisis, and the subsequent slowdown in the demand for raw materials. Furthermore, the continued financial crisis and the tightening of lending by the banking sector also adversely affected transport, restraining necessary credit flows that facilitate international commerce transactions. Indeed, the collapse of confidence that has accompanied the sovereign debt crisis in Europe has had significant repercussions in the transport sector, with several banks refusing to issue letters of credit or accepting certain forms of collateral for loans.

According to the UNCTAD 2011 maritime transport report, imports to Asia grew at a slightly slower pace than exports, an average of 15.2 %. The differences between developed and emerging countries were noticeable. Imports in developing countries expanded at a faster rate of 18.7 % than the exports rate of 16.6 %. This was particularly driven by growth in import volumes of developing countries in Asia. Nevertheless, transition economies also recorded growth in import volumes. They reached an average of 17.8 %, which means a rate faster than the rate of exports. Finally, positive growth was also recorded in import volumes of developed countries. These reached an average of 16.5 %, led by the positive performances of the United States and the European Union. Considering the earthquake disaster in Japan, the World Trade Organization expected Japan's export volumes to drop by 0.5–0.6 % and its imports to increase by 0.4–1.3 %.[25] But already in May 2011, Japan's exports began to rebound from the major disruption caused by the crisis situation in March. Nevertheless, export growth decelerated later in the year. One of the reasons was softening global demand during this period. Finally, shortly after the earthquake disaster, imports rose notably, pushed mainly by higher demand for food. At the beginning of June 2011, exports rose to pre-crisis levels and beyond.

Maritime transport is the most commonly used form of transport for international trade, representing the bulk of global trade transport (90 %). Consequently, it suffered the greatest blow from the continued crisis. Indeed, the 2008 financial crisis and its repercussions terminated one of the longest uninterrupted growth periods in recorded maritime history. The timing of the crisis was particularly

[25] United Nations Conference on Trade and Development. Review of Maritime Transport 2011. Geneva: UNCTD, 2011. 5.

adverse, as ship owners had enjoyed the most profitable financial results of all time before the crisis, and had an unprecedented number of vessels on order. As a result, during the reporting period a great number of orders continued to be cancelled. During the last 12 months, decreased maritime activity has led to a wave of ship order cancellations and an unprecedented level of distress demolitions (projected to have reached 15–18 % of world fleet capacity in 2010). If these estimates materialize, all sectors of the maritime industry will suffer from considerable unemployment. Another medium-term consequence of the financial crisis with effects for sea trade could be the appearance of protectionist measures that would further hinder world trade.

Another challenge for the maritime industry in the past 12 months has been the increased number of piracy incidents, especially off the Somalia coast. Although the international military presence in the region has somewhat increased security, piracy incidents have persisted. This surge in piracy in the Gulf of Aden has generated considerable costs especially for Europe as 80 % of shipments that pass through the area are coming from or going to this continent. Re-routing shipments around the Cape of Good Hope alone is estimated to create over $7.5 billion of additional shipping costs annually with associated increased energy consumption. Coordinated measures were undertaken by the International Maritime Organisation (IMO) to face the deteriorating security environment in the area and provide security for the ships navigating in these waters. In this context, it authorised ship owners to voluntarily arm their vessels at their own expense, notably by hiring armed guards when passing through the region.[26]

1.1.3 Main Science and Technology Indicators Relevant for Space Activities

The space sector requires the continued input of cutting-edge technologies, facilitated by a centralised public sector that occasionally bets against unfavourable financial market conditions. In order to compete on a world scale while mitigating uncertain market conditions, a web of private and public technology development schemes exist that interact with and complement each other at various levels and to different extents in the space industry and associated business operations. This interdependence likewise exists at the European Union level where knowledge and innovation are the key drivers of future economic and societal development in a broader context. This is also reflected in the continued development of the European Research Area (ERA), which consists of activities, programmes and policies that are designed and operated at regional, national, and European levels, and in

[26] White Papers of the Government of China. "China's Space Activities in 2011." Beijing 29 Dec. 2011. 6 Mar. 2012 http://www.china.org.cn/government/whitepaper/node_7145648.htm.

which the space sector occupies an important place. Technological development and innovation is in fact a quantifiable activity, based on its input and output. The term 'input' describes investment in the resources necessary to conduct research and development, e.g. financial investment and scientific personnel; 'output' is the tangible result of these inputs, consequently measuring the added value of innovation in the process of transforming resources into concrete scientific knowledge and technological inventions.[27]

1.1.3.1 Science and Technology Inputs

Science and technology inputs at the European level are constantly measured against the benefits they make available. In 2011 and 2012, amid the global economic crisis, R&D investment underwent ambitious policy reforms with the aim of increasing efficiency and effectiveness and developing stronger R&D integration in broader industrial and macro-economic policies. This reporting period is measured by gross domestic expenditure on R&D (GERD), a statistical tool showing nominal changes in those expenditures.

Between 2007 and 2011, overall GERD over GDP increased from 1.85 % to 2.03 %; between 2000 and 2007, the GERD had hovered at around 1.85 %.[28] While the EU's 2011 GDP increased to €12.638 trillion in current price terms, its 2011 GERD was €256.55 billion; that amounted to a 4.6 % increase in GERD from the previous year.[29] Moreover, GERD itself increased from 2 % to 2.03 % between 2010 and 2011. In contrast, this performance stayed below Japan's and the United States' indicators during the same period, as is the long term trend in this field. As in 2010, Japan's 2011 GERD equalled 3.3 % over its GDP, amounting to about $144.1 billion (about €111 billion); and the US saw an increase in its GERD from 2.3 % in 2010 to 2.7 % in the following year, amounting to $405.3 billion (about €313 billion).[30] This highlights that despite the financial crisis, investment in innovation still has the potential for growth, and may be seen as a partial remedy to further decline. On the national level, the top three European countries with the highest

[27] Godin, Benoit. "Science, Accounting and Statistics: the Input–output Framework." Canadian Social Sciences and Humanities Research Council (2005): 5 http://www.csiic.ca/PDF/Godin_31.pdf.

[28] "Gross domestic expenditure on R&D (GERD) - % of GDP." Eurostat 22 Apr. 2013 http://epp.eurostat.ec.europa.eu/tgm/table.do?tab=table&plugin=1&language=en&pcode=t2020_20; *see also* "R&D expenditure." Oct. 2012. European Commission – eurostat 22 Apr. 2013 http://epp.eurostat.ec.europa.eu/statistics_explained/index.php/R_%26_D_expenditure.

[29] C.f. "National accounts – GDP." Oct. 2012. European Commission – eurostat 22 Apr. 2013 http://epp.eurostat.ec.europa.eu/statistics_explained/index.php/National_accounts_%E2%80%93_GDP.

[30] Grueber, Martin and Tim Studt. "2011 Global R&D Funding Forecast: Stability Returns to R&D Funding." 15 Dec. 2010. R&D 22 Apr. 2013 http://www.rdmag.com/articles/2010/12/2011-global-r-d-funding-forecast-stability-returns-r-d-funding.

investment levels of GDP on innovation kept their positions in 2011, with Finland (3.78 %), Sweden (3.37 %) and Denmark (3.09 %), while in 2011 ten European Member States reported expenditures under 1 %, based mainly in Southern and Eastern Europe.[31]

A further breakdown of statistical data helps to underscore the impact of private R&D investment in Europe, when paired alongside public funded projects. As in the previous year, countries with higher GERD levels also showed a generally high level of innovation investment within the private business sector; the top three again being Finland (2.67 %), Sweden (2.34 %), and Denmark (2.09 %), in addition to Austria and Slovenia. As with previous years, countries with larger GDPs (such as France and Germany) tended to field equally large government led investments in R&D, mostly through higher education research.[32]

In 2011, private sector funded R&D accounted for nearly 62.1 % of the EU's total GERD, as compared with figures for Japan at 75.6 % and the U.S. at 70.4 % in 2009. As shown above, the proportion of Government expenditures in this field also varied significantly from one European region to the other, with the highest levels observed in Central and Eastern European countries.[33] The mixed approaches to R&D funding in Europe pose an expected challenge when planning appropriate policies at the European level.

1.1.3.2 Science and Technology Outputs

In contrast to R&D input, the outputs achieved from investments in innovation are harder to measure, due to the complex market dynamics within the private sector. Innovation offers a competitive edge to companies that are willing to invest and develop R&D projects. On the other hand, statistics such as the number of patents lodged by country, industrial sector and individual companies are merely indicative of output, as their effect on financial performance is also heavily dependent on external market competition and other factors. Based on the European Commission's 2012 EU Industrial R&D Investment Scoreboard, in 2011 approximately 1,500 companies each invested more than $35 million in R&D worldwide. Total investment in R&D by leading companies amounted to €510.7 billion; i.e. European companies constitute 28.3 % (€144.6 billion) of that figure while the US at 34.9 % and Japan at 21.8 %, constitute a total of 85.1 % of the companies surveyed. Companies realise they can gain a cutting edge by investing heavily in R&D when operating in highly competitive environments, as this action appears to

[31] "Research and Innovation performance in EU Member States and Associated countries – 2013" European Commission – Innovation Union 22 Apr. 2013: 5 http://ec.europa.eu/research/innovation-union/pdf/state-of-the-union/2012/innovation_union_progress_at_country_level_2013.pdf.

[32] "Research and development expenditure, by sectors of performance. - % of GDP." Eurostat 22 Apr. 2013 http://epp.eurostat.ec.europa.eu/tgm/refreshTableAction.do?tab=table&plugin=1&pcode=tsc00001&language=en.

[33] Id.

have a linear link between investment and market performance. This trend might be the case with top performing companies, but is highly dependent on the nature of the market, and would likely not hold true in every sector. What is clear however is that in 2011, the EU experienced growth that was above the pre-crisis levels at 8.9 %, compared to 8.1 % in 2008, and −2.6 % in 2009.[34]

The uneven benefits of corporate innovation on a European scale are further demonstrated in the European Commission's bi-annual competitiveness assessment conferences, CONCORD. During the October 2011 meeting, the existence of concrete market benefits for innovative companies was confirmed; however, their globally non-linear nature makes their marginal efficiency difficult to measure. It would seem that innovation-fuelled corporate performance greatly depends on the market sector's level of demand for cutting-edge technology. From this point of view, it is clear that the space sector is highly influenced by R&D. In general, the more high-tech a sector is, the greater and more robust the investment returns that can be expected from innovation. By contrast, innovation in low-tech industry segments focuses primarily on improving production efficiency rather than directly creating new products or markets. This situation, in turn, creates the need for carefully planned and finely tuned innovation policies on a European scale that would be able to meet a great variety of cases. As mentioned earlier, ambitious European policy reforms with the aim of increasing efficiency and effectiveness had already begun in 2011; that said, at a European level any horizontal measures taken to improve competitiveness through innovation would have to be accompanied by appropriate custom-made policies for the different market segments, corporate sizes and geographical regions of Europe.[35]

1.2 Worldwide Space Policies and Strategies

1.2.1 The United Nations System

Various institutions within or associated with the United Nations are relevant for space policy. In this section, the UN General Assembly, its Committees, and other UN bodies and organs are discussed regarding space activities.

---

[34] European Commission. "The 2012 EU Industrial R&D Investment Scoreboard." EU R&D Scoreboard 23 Apr. 2013 http://iri.jrc.ec.europa.eu/docs/scoreboard/2012/SB2012.pdf; *see also* European Commission. "EU Industrial R&D Investment Scoreboard 2012." 6 Dec. 2012. Europa – Press Release RAPID 23 Apr. 2013 http://europa.eu/rapid/press-release_MEMO-12-948_en.htm.

[35] European Commission Joint Research Centre, Institute for Prospective Technological Studies. CONCORD-2011 – 3rd European Conference on Corporate R&D: The dynamics of Europe's industrial structure and the growth of innovative firms – Summary Report. Oct. 2011. Seville, Spain. JRC Scientific and Technical Reports, 2012.

1.2.1.1 United Nations General Assembly

During its 2011–2012 session, the United Nations General Assembly (UNGA) passed three Resolutions pertaining to the use and exploration of Outer Space. The first document, the 'International Day of Human Space Flight' Resolution 65/271 adopted on 21 June 2011, sets the annual date of 12 April as the International Day of Human Space Flight to celebrate the beginning of human exploration of the Universe.

On 12 January 2012, UNGA adopted Resolution A/RES/66/27 on the 'Prevention of an arms race in outer space.' The resolution, similar to A/RES/65/44 adopted in the previous year, reaffirmed the importance of transparency and confidence building measures (TCBM) to avoid an arms race in space and, whilst recognizing that the outer space legal regime does not guarantee the prevention of an arms race in outer space, acknowledged that the regime plays a significant role in the prevention of an arms race in that environment, and that there is a need to consolidate and reinforce the regime and enhance its effectiveness. Moreover, it again recognized the importance of strict compliance with existing bilateral and multilateral agreements. The Conference on Disarmament (CD) was seen as the relevant multilateral disarmament forum for further discussions. The Resolution also called for the establishment of an Ad Hoc Committee on the Prevention of an Arms Race in Outer Space within the CD. In general, it also recognised the complementary nature of multilateral and bilateral efforts in this area.[36]

On the same day, UNGA adopted its annual Resolution on "International cooperation in the peaceful uses of outer space" (66/71). This Resolution was also very similar to the one tabled in 2010, noting all the central aspects and challenges of the peaceful use of outer space and recalled the crucial importance of international cooperation to tackle all relevant issues. Furthermore, it provided an overview of the necessary steps that should be taken in this regard, including conferences, sessions of relevant entities and noteworthy progress in implementing corresponding programmes.[37]

1.2.1.2 UNGA Committees

UNGA has several committees that address space policy and associated matters. Several of those committees are discussed below.

[36] United Nations General Assembly. Resolution adopted by the General Assembly – Prevention of an arms race – Sixty-Sixth Session. UN Doc. A/RES/66/27 of 12 January 2012. Vienna: United Nations http://www.un.org/ga/search/view_doc.asp?symbol= A/RES/66/27.

[37] United Nations General Assembly. Resolution adopted by the General Assembly – International cooperation in the peaceful uses of outer space – Sixty-Sixth Session. UN Doc. A/RES/66/71 of 12 January 2012. Vienna: United Nations http://www.oosa.unvienna.org/pdf/gares/A_RES_66_071E.pdf.

Disarmament and International Security Committee

The Disarmament and International Security Committee works in close cooperation with the United Nations Disarmament Commission and the Geneva-based Conference on Disarmament to deal with disarmament, global challenges and threats to peace that affect the international community and seeks solutions to the challenges in the international security regime.[38] In December 2011, UNGA relied on the recommendations of the Disarmament and International Security Committee in its Plenary to adopt 47 resolutions and five decisions to revitalize the lagging multilateral disarmament process.[39]

Committee on the Peaceful Uses of Outer Space (COPUOS)

UN COPUOS activities in this 2011/2012 reporting period included the Committee's plenary session on 1–10 June 2011 and the annual sessions of its technical and its legal subcommittees on 6–17 February 2012 and 19–30 March 2012 respectively. At the 2011 plenary session, the Committee considered how to contribute to the implementation of the recommendations of the Third United Nations Conference on the Exploration and Peaceful Uses of Outer Space (UNISPACE III); it also endorsed the findings and recommendations of its subcommittees from their preceding sessions, and considered the current status of spin-off benefits of space technology; space and society, space and water, space and climate change; the use of space technology in the United Nations system; and the future role of the committee.[40]

At the 2012 scientific and technical subcommittee meetings, discussions continued on a broad range of space related issues, continuing with matters relating to remote sensing of the Earth by satellite (i.e. applications for developing countries and monitoring of the Earth's environment); the issue of space debris and the mitigation measures already implemented by some States; the further development of the space-system-based disaster management support programme (i.e. UN SPIDER); recent developments in global navigation satellite systems; the use of nuclear power sources in outer space; possible dangers emanating from near-Earth objects; an international space weather initiative to understand and forecast

[38] "Disarmament and International Security." General Assembly of the United Nation 12 Apr. 2013 http://www.un.org/en/ga/first/index.shtml.

[39] "General Assembly, Gravely Concerned about Status of UN Disarmament Machinery: Especially in Conference on Disarmament, Invites States to Explore Options." 2 Dec. 2011. United Nations General Assembly 12 Apr. 2013 http://www.un.org/News/Press/docs//2011/ga11182.doc.htm.

[40] United Nations General Assembly. Report of the Committee on the Peaceful Uses of Outer Space on its Fifty-Fourth-Session, Held in New York from 1–10 June 2011. UN Doc. A/66/20 of 20 June 2011. New York: United Nations http://www.oosa.unvienna.org/pdf/gadocs/A_66_20E.pdf.

near-Earth space weather and to communicate that knowledge to interested parties; along with additional space safety topics. The subcommittee received and considered information provided by the Member States on their activities in all these fields.[41]

The proceedings of the legal subcommittee focussed on the status and application of the five United Nations treaties on outer space, in addition to exchanging information on the activities of international governmental and non-governmental organizations related to space law. The issues of the definition and delimitation of outer space were approached, with various points of view presented; certain States identified the legal uncertainty that currently exists, while others expressed their preference for a functional approach to delimitation, and some States expressed reticence in changing the current framework until such time as there was a demonstrated need and a practical basis for developing a definition or delimitation of outer space. Moreover, the States discussed the character and utilization of the geostationary orbit (e.g., including consideration of ways and means to ensure the rational and equitable use of the geostationary orbit without prejudice to the role of the International Telecommunication Union), in addition to conducting a review and possible revision of the Principles Relevant to the Use of Nuclear Power Sources in Outer Space, the draft protocol on matters specific to space assets to the Convention on International Interests in Mobile Equipment, and capacity-building in space law.[42]

1.2.1.3 Other UN Bodies and Organs Monitoring Outer Space Activities

Beyond the UN General Assembly and its Committees, there are other UN bodies, programmes and organs related to space activities. The following discusses the ITU (a specialised agency of the UN), UN-SPIDER; the UN Programme on Space Applications, the International Committee on Global Navigation Satellite Systems (ICG), the United Nations Spatial Data Infrastructure (UNSDI), the Conference on Disarmament (CD) and UNIDIR.

---

[41] United Nations General Assembly. Report of the Scientific and Technical Subcommittee on its Forty-Ninth Session, Held in Vienna from 6 to 17 February 2012. UN Doc. A/AC.105/1001 of 28 February 2012. Vienna: United Nations http://www.oosa.unvienna.org/pdf/reports/ac105/AC105_1001E.pdf.

[42] United Nations General Assembly. Report of the Legal Subcommittee on its Fifty-First Session, Held in Vienna from 19 to 30 March 2012. UN Doc. A/AC.105/1003 of 10 April 2012. Vienna: United Nations http://www.oosa.unvienna.org/pdf/reports/ac105/AC105_1003E.pdf.

International Telecommunication Union (ITU)

The International Telecommunication Union (ITU) held its annual Council at its headquarters in Geneva from 11 to 21 October 2011. The ITU's biennial World Radiocommunication Seminar (WRS) was held in Geneva on 6–10 December 2010. During those meetings, a significant number of space related issues were advanced, including radio spectrum allocation and orbital slots procedures, emergency communications standardisation, and the coordination of satellite networks; also addressed in the Council meeting were the strategic plan and budget of the ITU and a review of the implementation of the objectives of the World Summit on the Information Society. At the Council meeting, the key issue regarding information and communication technology development was how to create the new digital highways needed to accommodate the massive increase in data traffic.[43] These meetings provided a forum to exchange views on the associated technical, procedural and operational aspects of satellite communications and, particularly, technical issues related to technological advancements. The next WRS was scheduled in Geneva on 3–7 December 2012.[44]

UN-SPIDER

The United Nations Platform for Space-based Information for Disaster Management and Emergency Response (UN-SPIDER) organized several workshops and regional meetings in this reporting period. The platform was set up by UNGA in 2006 with the aim of providing universal access to all types of space-based information and services relevant to disaster management support. In this context, the International Charter on Space and Major Disasters was activated several times by the UN Office of Outer Space Affairs (UN OOSA) at the request of other UN entities. Key geopolitical events that activated the Charter included the volcano eruption affecting Chile and Argentina[45]; and floods in South Korea and in Japan in 2011[46]; and the volcano eruption that affected Guatemala in 2012.[47]

---

[43] "ITU Council meets to review strategies for ICT development – Increased data traffic demands high speed digital highways" 12 Oct. 2011. ITU 12 Apr. 2013 http://www.itu.int/net/pressoffice/press_releases/2011/CM15.aspx.

[44] "ITU Meetings and Conferences." ITU 12 Apr. 2013 http://www.itu.int/events/eventdetails.asp?eventid=9841.

[45] "UN-SPIDER June 2011 Updates." June 2011. United Nations Office for Outer Space Affairs 12 Apr. 2013 http://www.un-spider.org/sites/default/files/06%202011%20UN-SPIDER%20Updates.pdf.

[46] "UN-SPIDER July 2011 Updates." July 2011. United Nations Office for Outer Space Affairs 12 Apr. 2013 http://www.un-spider.org/sites/default/files/07%202011%20UN-SPIDER%20Updates.pdf.

[47] "UN-SPIDER May 2012 Updates." May 2012. United Nations Office for Outer Space Affairs 12 Apr. 2013 http://www.un-spider.org/sites/default/files/UN-SPIDER%20Updates%20May%202012.pdf.

UN Programme on Space Applications (SAP)

The UN Programme on Space Applications (SAP) is concerned with cooperation in space science and technology. In the reporting period between 2011 and 2012, several activities were carried out under SAPs auspices dealing with topics such as: the contribution of tele-epidemiology (i.e. disease analysis affecting large populations) to public health actions in the context of climate change adaptation; Use of Space Technology for Human Health Improvement; Space for Human and Environmental Security; International Space Weather Initiative, etc.[48]

International Committee on Global Navigation Satellite Systems (ICG)

The aim of the International Committee on Global Navigation Satellite Systems (ICG) is to promote cooperation in matters of satellite navigation. The UN OOSA serves as the Executive Secretariat of the ICG and the associated Providers' Forum. The sixth meeting of the ICG took place in Tokyo, Japan, on 5–9 September 2011. This time organized by the Government of Japan, the meeting drew participants from industry, government, non-governmental officials and academia to review and discuss developments in global navigation systems.[49]

Throughout the sessions of its Plenary, presentations were made to update the ICG Members, Associate Members, and Observers on matters of interest such as: Agriculture, Fishery and Information Technology Construction (Precise Positioning); Geographic Information System and Disaster Mitigation; and Intelligent Transportation System and Location Based Services. Four working groups also met during the Plenary, covering compatibility and interoperability; enhancement of performance of GNSS services; information dissemination and capacity building; and reference frames, timing and applications.[50] Within the Plenary, the ICG endorsed the decisions and recommendations of the working groups with regard to the implementation of the actions contained in its work plan; and took note of the reference frames, timing and applications working group's recommendation that it would be important to involve an official representative of the European Union in its discussions and deliberations.[51] The next ICG meeting will take place in Beijing, China.

---

[48] "United Nations Programme on Space Applications Activities Schedule: 2011." United Nations Office for Outer Space Affairs 12 Apr. 2013 http://www.oosa.unvienna.org/oosa/en/SAP/sched/2011.html.

[49] "ICG Meetings." United Nations Office for Outer Space Affairs 12 Apr. 2013 http://www.oosa.unvienna.org/oosa/en/SAP/gnss/icg/meetings.html.

[50] Id.

[51] United Nations General Assembly. Sixth Meeting of the International Committee on Global Navigation Satellite Systems – Note by the Secretariat. UN Doc. A/AC 105/1000 of 10 Nov. 2011. Vienna: United Nations http://www.oosa.unvienna.org/pdf/icg/2011/icg-6/report/49STSC-1000_E.pdf.

United Nations Spatial Data Infrastructure (UNSDI)

The United Nations Geographic Information Working Group (UNGIWG) held its twelfth annual meeting in Vienna, Austria, on 28–30 March 2012. The UNGIWG is a network of UN cartography and geospatial information management science professionals that lay the foundations for the United Nations Spatial Data Infrastructure (UNSDI). The UNSDI contributes substantively to the mission of the United Nations by functioning as a comprehensive, decentralised geospatial information network to facilitate decision-making mechanisms between member states, regional organizations and partners that advance social, economic development, environmental and humanitarian agendas.[52]

Conference on Disarmament (CD)

The Conference on Disarmament (CD) is the key multilateral disarmament and arms control negotiating forum within the international community. The stalemate in its work regarding space security has continued. Its 2012 session ended on 14 September 2012, with the next year's session to be conducted in three parts: first part on 21 January–29 March, second on 13 May–28 June, and third on 29 July–13 September 2013. In the course of the 2011 and 2012 sessions, the cessation of the nuclear arms race and nuclear disarmament, the prevention of nuclear war including all related matters, and prevention of an arms race in outer space are some of the topics that were addressed.[53]

United Nations Institute for Disarmament Research (UNIDIR)

The United Nations Institute for Disarmament Research (UNIDIR) conducts some projects that cover, both directly and indirectly, issues of space security. It seeks to review former proposals and to propose new options for breaking the deadlock in space weaponisation matters at the Conference on Disarmament (CD).[54]

---

[52] "12th UNGIWG Plenary Meeting." UNGIWG 13 Apr. 2013 http://www.ungiwg.org/meetings/12th-ungiwg-plenary-meeting-0.

[53] "Disarmament – Annual Reports" UNOG – The United Nations Office at Geneva 13 Apr. 2013 http://www.unog.ch/80256EE600585943/(httpPages)/58FCFA0DC9975CC9C1256F570056A178?OpenDocument.

[54] United Nations General Assembly. United Nations Institute for Disarmament Research – Note by the Secretary General, on its Sixty-Seventh Session. UN Doc. A/67/169 of 23 July 2012. http://www.unidir.org/files/Reports/Rep_2012_EN.pdf.

1.2.2 The Group on Earth Observation

The Group of Earth Observation (GEO) system-of-systems cooperation platform (GEOSS) is a voluntary partnership of governments and international organisations that facilitate the interlay of current and future Earth observing systems, while also providing coordinated and sustained data worldwide as a decision support tool for its users. In 2011 and 2012, GEO sought to further improve its governance mechanisms and operational capabilities. Further discussion on recent developments can be found in subsequent parts of this Yearbook.

1.2.3 Europe

1.2.3.1 European Space Agency

The second half of 2011 and the first half of 2012 constituted a degree of transition for the European Space Agency. On the one hand, with the Agency's budgetary cycle nearing its end, and a Ministerial Council coming up, the focus had been on the implementation of the strategic directions decided in 2008, such as developing space applications missions, increasing cooperation with the European Union, and utilising the European contribution to the ISS to its full potential. On the other hand, with preparations for the next budgetary cycle well underway, developments in ESA programmes also offered a glimpse of the mission areas that will come into focus shortly, such as the future of the ISS and Ariane 5. In the same context, the period in review also witnessed the emergence of trends in ESA policies as a response to the rapidly changing geopolitical and economic conditions of our time. These include an acute emphasis on international cooperation, as well as on the use of more flexible and purpose-specific funding mechanisms, capable of maximising investment returns, both on operational and industrial levels.

An example of this trend has been ESA's contribution to the development of public private partnerships (PPPs) in European space programmes. In 2011 and 2012 this mutually beneficial interaction between the public and private sectors slowly expanded on a European, as well as national level. In fact, on a national level, all major space faring countries are continuing to explore this sort of advantage, with the United Kingdom, Germany and France taking the lead. Especially in the United Kingdom, almost two thirds of all European PPPs are anchored. On a regional level, and despite the European Commission's reservations (mostly due to the negative prior experience on Galileo), the European Investment Bank has provided no less than €225 million in PPPs funding. It was ESA, however, that initiated the most important such partnerships on an institutional level, especially through its participation in satellite communications projects. In this context, ESA has implemented a PPP-type financial arrangement for a data-relay satellite system over Europe (EDRS) that would have commercial, civil-government and military uses.[55]

Economic and fiscal instability in Europe has also clearly impacted on the development of other important future ESA projects, such as the development of an Ariane 5 upgrade. Funding required to complete the development of a new upper stage for the rocket was expected to rise to approximately €1 billion. However, budgetary constraints have prompted ESA Member States to scrutinise the programme's funding and request that it be at least partially compensated by a reduction of the launcher's operational costs, which are also partially borne by ESA. In this context, significant efforts were undertaken to broaden the programme's supplier base as much as possible, within the limits of the Agency's fair geographic return principle. As in the case of PPPs, emphasis was placed on improving the cost to benefit ratio of the investment, mainly through increasing the programme's development flexibility and adjustability to changing conditions.[56]

Contrary to the uncertainty regarding the future of Ariane 5, 2011 and 2012 witnessed the successful development conclusion and inaugural launch of the European Soyuz and Vega rockets. The former is the European version of the Russian launcher, modified for operations from equatorial French Guiana. Further improvements included new safety, telemetry and tracking systems that would qualify the rocket for launch from the European spaceport. Thanks to the latter's geographical position, the European Soyuz has an increased lift capacity of over 3 t, making it suitable for placing light communications satellites in orbit. On the other hand, the rocket's development met with a 3 year delay and a 36 % budget increase, mainly related to the completion of ground launch infrastructure. Nonetheless, the launcher carried the first two Galileo satellites to orbit with its inaugural flight in October 2011.[57]

A third recurring trend in ESA policies in 2011 and 2012 was, as mentioned, increased emphasis on future international cooperation, especially in the fields of space transportation and exploration. The retirement of the space shuttle during this period increased the potential for cooperation between Europe and the United States in terms of access to space vehicles. At the same time, it indirectly increased the value and relevance of ESA's Automated Transfer Vehicle (ATV). The combined use of the ATV and other similar spacecraft from the U.S., Japan and Russia to serve ISS supply needs could create opportunities in the long term for the creation of common transportation policies among all participating space actors. Due to their technological proximity and operationally complementary nature,

[55] De Selding, Peter B. "Europe Knocked for Balking at Public-Private Satellite Venture." Space News, 20 June 2011: 7.

[56] De Selding, Peter B. "ESA Industrial Policy Limits Ariane 5 Cost-saving Potential." Space News 27 June 2011: 6.

[57] De Selding, Peter B. "European Soyuz Must Pass Final Exam Before October Debut." Space News 4 July 2011: 5; *see also* De Selding, Peter B. "Insiders Hard-pressed to Say Why European Soyuz Was Delayed." Space News 4 July 2011: 5.

these spacecraft could also pave the way for future cooperation on a technology development level.[58]

Another example of the European Space Agency's efforts to position itself at the centre of international cooperation manifested itself in October 2011, when the Agency formally invited Russia to join the U.S.-European ExoMars space exploration mission. The Russian space agency Roscosmos was invited to participate as a full partner in the endeavour, and to contribute a third of its budget. Its potential contribution to the mission could include the provision of the launcher and of the orbiter's entry, descent and landing system. ESA's decision was partially dictated by a previous NASA budgetary tightening that prohibited it from committing to the originally set 2016 launch timeframe. In this situation, the Agency would either have to curtail the mission, or seek a third partner.[59] Yet by the end of the reporting period, ESA's ExoMars project underwent a significant reorganization of support, with Russia's Roscosmos replacing NASA as ESA's main partner in launching the telecommunications orbiter, two landers and a rover to Mars in 2016 and 2018. Following this new pairing, the ExoMars is expected to cost ESA around 1.2 billion euros; however, ESA has only managed to commit 850 million euros out of its €1 billion originally planned funding commitment. Budget constraints forced NASA to back out of the project where it originally had agreed to provide Atlas V rockets for the 2016 and 2018 launches, and to share in the cost of the 2018 rover.[60] Roscosmos then entered offering to provide Proton rockets for both ExoMars launch dates, and a nuclear power source to a newly required European lander; with ESA responsible for 100 % of the costs of the rover and 20 % of the work on the Russian-led lander for 2018.[61] Additional sources of funding that are being considered include: engaging in additional cooperation with Russia on Europe's Juice science mission to Jupiter in 2022; delaying construction on ESA's new headquarters building; channelling contributions from new ESA members directly to ExoMars; soliciting funds from Europe's space science decision-making body; and trying to gain additional funds from ExoMars' current sponsors.[62]

Budgetary tightening in the U.S. might also take its toll on other cooperation already initiated between ESA and NASA on space science and exploration missions. Important joint projects, such as Laplace and LISA missions could be jeopardised by short term cuts in NASA's budget. Thus NASA might not be able to meet its obligations as a true partner in some of these missions. On the upside, however, budgetary constraints across the Atlantic could further increase the

[58] Svitak, Amy. "U.S. And Europe Explore Common Space Transportation Needs." Aviation Week & Space Technology 27 June 2011: 41.

[59] De Selding, Peter B. "ESA Formally Invites Roscosmos To Join ExoMars Mission as Full-fledged Partner." Space News 17 Oct. 2011: 1.

[60] De Selding, Peter. "ESA Ruling Council OKs Funding for Mars Mission with Russia." SpaceNews 19 Mar. 2012: 1.

[61] Id.; *see also* De Selding, Peter. "ExoMars Missions Back on the Brink." SpaceNews 14 May 2012: 1.

[62] De Selding, Peter. "ExoMars Wins One-month Reprieve." SpaceNews 21 May 2012: 8.

incentive for ESA to engage in international cooperation with various partners in this mission area. Increased cooperation in this case would aim at harmonising operational, technological and scientific space science and exploration mission objectives. The greater the degree of that cooperation, the more ESA could hedge against future budgetary threats to these missions. In reality, cooperation would not only need to be broadened, but also deepened in order to meet this objective.[63]

Finally, apart from cooperating on an international level, in 2011 and 2012 ESA continued its efforts to strengthen its relations with other European institutional space stakeholders. An apparent manifestation of this tendency was an agreement reached with the European Defence Agency (EDA) to cooperate on the development of dual use satellite applications. Although this agreement did not go as far as jointly financing such programmes, it did, however, create the vehicle to share competencies in the interest of developing European dual use technologies and operational concepts. The first such project would be a demonstration mission for the control of Unmanned Aerial Systems (UAS) by satellite.[64]

1.2.3.2 European Union

In 2011 and 2012, the European Union continued its efforts to position itself as a leading global actor in space activities. Despite the unquestionable stimulus that the Lisbon Treaty ratification contributed to this effort, significant challenges remained regarding the future of the European space policy and programmes heralded in the treaty, especially from a budgetary point of view. The effects of the continued economic crisis discussed above may have played a significant part in this state of affairs. In general, developments during the reporting period present a mixed picture and included considerable breakthroughs, but also set-backs.

An example of the latter was developments regarding the GMES programme, and especially the efforts undertaken to resolve outstanding budgetary and governance issues among the European Commission, the European Space Agency (ESA) and Member States. This important complication for the programme's future appeared in July 2011, when the European Commission announced that it did not intend to finance GMES through its next 7 year framework programme (known as FP8). This appeared to be an effort on the part of the EC to tighten its research and development budget in view of the ongoing economic crisis. The proposal of the Commission was to create a funding mechanism outside the Multiannual Financial Framework (MFF), based on voluntary subscription based on GNI contributions from all Member States. This proposition might prove challenging, considering that

[63] Svitak, Amy. "NASA's Money Woes Thwart Joint Science Missions with ESA." Aviation Week & Space Technology 8 Aug. 2011: 24.

[64] "ESA, European Defense Agency Strengthen Ties." Space News 27 June 2011: 5.

the project's annual costs were projected to reach €834 million[65] and the general reluctance of the EU stakeholders to create funding mechanisms outside the MFF.

Immediately after this announcement, the European Commission's decision came under scrutiny on multiple fronts. For example, concerns were raised by industry officials and associations, such as the European Association of Remote Sensing Companies (EARSC). They claimed that the EC's decision would create uncertainty regarding its funding, consequently deterring the private sector's further engagement in it and jeopardising its future. Other concerns were voiced by Member State officials, who considered this an effort on behalf of the EC to actually raise the programme's long-term budget.[66]

The greatest concern about this decision, however, seems to have been expressed by ESA, the European Commission's partner in GMES technology development and programme management. ESA officials were reported to have addressed this concern to ESA Member States. They stated that by excluding GMES from FP8, the European Commission increased uncertainty over the programme's funding beyond 2013. The programme was expected to enter its operational deployment phase at that time, with the EC assuming the bulk of its funding after ESA has completed the development. Finally, the fact that the EC took this decision unilaterally, at a moment when the programme's ownership had not yet been decided, was a further source of concern.[67]

The decision to remove GMES funding from the European Commission's medium-term research and development budget came as a surprise, as the EC had allocated €104 million to the programme only a month prior to this decision. This payment had allowed ESA to book launching slots for the GMES satellites starting from 2015, according to the original timetable. With the funding of the operational phase still pending, the decision to finance its launch now seems premature.[68] Finally, the programme's funding mechanism revision came at a time when its development was on track and no significant delays or budget overruns were anticipated, raising questions as to whether the decision was based on broader policy and budgetary considerations, rather than on the project's own merit.[69]

Criticism of the European Commission's decision was voiced in the European Parliament as well, with 44 of its members calling for its withdrawal. In their petition, the MEPs warned of the risk of the programme's collapse as a result of budgetary uncertainty. Furthermore, they pointed out that in a time of economic

[65] De Selding, Peter B. "European Commission Broaches New Funding Scheme for GMES." Space News 25 July 2011: 6.

[66] Id.

[67] Id.

[68] De Selding, Peter B. "European Commission Doles out Down Payment for Initial GMES Launches." Space News, 13 June 2011: 10.

[69] De Selding, Peter B. "European Commission Broaches New Funding Scheme for GMES." Space News 25 July 2011: 6.

crisis, the chances Indeed, failure to finance GMES operations would nullify the nearly €1.6 billion investment already made in spacecraft development by ESA.[70]

In stark contrast to developments regarding GMES, the second half of 2011 saw the achievement of a milestone first launch for the Galileo GNSS. On 21 October 2011, two Galileo satellites were placed to orbit on the inaugural flight of the European version of the Soyuz rocket, launched from the European spaceport in French Guiana. The successful launch signified the opening of the system's operational deployment phase, with the scheduled launch of 14 satellites out of a total of 30 required to achieve full operational capability. A further 6–8 spacecraft were expected to be ordered in early 2012, depending on the level of the funds remaining from the programme's original €3.4 billion budget. Further steps would include the development of a customised fairing stage for the Ariane 5 launcher, capable of carrying four Galileo satellites to orbit. This would enable the use of both types of rockets and consequently speed up the satellite's launch sequence.[71]

On the other hand, having met with a 40 % budget overrun, the Galileo project will require significant further funding to become fully operational. In fact, the effort of reducing the programme's budget overrun and securing its future funding also made progress in 2011 and 2012. In June 2011 it was projected that the project's original €3.4 billion budget would only cover the deployment of roughly 20 spacecraft. In order to complete the 30-satellite constellation, an additional sum of up to €1.9 billion would be required. In the course of July 2011, European Commission officials pledged to reduce this figure by at least €500 million, at the request of European Member States. At the same time, they announced the EC's decision to include funding for this additional sum in the EU's eighth framework funding programme (FP8) for research, thus drawing a distinction from the GMES funding. Finally, the programme's development goals were concluded with the finalisation of the system's ground segment procurement contracts.[72]

A further sign of the European Commission's renewed interest and status as a global stakeholder in GNSS development and operations was revealed when EC officials protested to their U.S. counterparts against the development of LightSquared. In their communications, European authorities expressed their concern about possible interference with Galileo signals over the United States, especially for safety critical operations.[73]

---

[70] "European Commission Urged To Put GMES Back in Budget." Space News 12 Sept. 2011: 3.

[71] De Selding, Peter B. "Soyuz Lofts Two Galileo Satellites In Debut from European Spaceport." Space News 24 Oct. 2011: 1.

[72] De Selding, Peter B. "European Governments Trim Galileo System Cost Overruns." Space News 27 June 2011: 5.

[73] De Selding, Peter B. "European Commission Adds Voice to LightSquared Opposition." Space News 25 July 2011: 21.

1.2.3.3 EUMETSAT

One of the key developments regarding EUMETSAT programmes was the progress made in the development of the organisation's next generation geo-stationary weather satellites. In 2011 and 2012, EUMETSAT pursued its plans to develop a polar-orbiting constellation to succeed the current METOP meteorological satellites in use. Despite reluctance from some of its members on budgetary grounds, the agency aspired to begin early design and conceptual work on the spacecraft. The current METOP program consists of three identical satellites launched at 6-year intervals between 2006 and 2018, with operations running until 2023. ESA invested about €1.5 billion in METOP, while EUMETSAT's share was €2.4 billion.[74] The timely launch of preparatory work on the second generation programme is crucial in order to achieve continuity of service.[75]

On 24 February 2012, ESA entered into a $1.8 billion deal with Thales Alenia Space to build six satellites for its METEOSAT Third Generation system, aiming to provide meteorological services for 20 years starting in 2018. ESA is covering 62 % of the cost of the six satellites while EUMETSAT will cover the other 38 % of the contract and is also funding more than two-thirds of the entire MTG program (budgeted at about €2.4 billion).[76]

Data from China's polar-orbiting FY-3B meteorological satellite has been distributed by EUMETSAT since 24 January 2012 as part of the EumetCast broadcast network. EUMETSAT has been distributing microwave humidity and temperature data from the FY-3A spacecraft since late 2010. Users in the Asia-Pacific region will also be receiving the Chinese satellite data from the China Meteorological Administration's (CMA) own dissemination network, which resembles EUMETSAT's EumetCast and was to be operational later in 2012.[77]

1.2.3.4 National Governments

France

Several different issues emerged regarding France and its space activities during the reporting period. As a leading country in European space cooperation activities and one of the two biggest contributors to the ESA budget, the main discussion related mostly to launch issues. Of significance was the debate over the successor to the Ariane 5 launcher, the development of France's and Germany's positions on

[74] De Selding, Peter B. "Figures Demonstrate Value of Polar-Orbiting Weather Satellites." Space News 21 Nov. 2011: 6.

[75] De Selding, Peter B. "Four Eumetsat Members Block Approval of Polar-orbiting Satellite Constellation." Space News 4 July 2011: 1 and 14.

[76] De Selding, Peter B. "ESA Signs $1.8B Deal with Thales Alenia for Six Weather Sats." 24 Feb. 2012. Space News 15 May 2012 http://www.spacenews.com/earth_observation/120224-esa-deal-thales-weather-sats.html.

[77] "Eumetsat to Distribute Chinese Weather Sat Data." Space News 23 Jan. 2012: 3.

relations between ESA and the EU, and finally French support in the field of international cooperation to build the space capacity of other actors.

France has already invested €250 million in a public bond offering to finance early development work on the Ariane 5 Midlife Evolution (Ariane 5 ME), one of the proposed future successors to ESA's heavy lift Ariane 5 launcher. This vehicle would be of modular design, capable of launching 10,500 kg into geo-stationary orbit, and is planned to debut in 2018. France and Germany had agreed that a new re-startable upper stage of the current Ariane 5 rocket would be a top priority for the ESA's next multiyear budget and spending priorities, on which negotiations have to be concluded at the November ESA Ministerial Council meeting. Completing this stage will likely cost €1.5 billion – a considerable impact on France's national space policy related to keeping launch abilities in a sustainable frame. At the end of 2011, France was undecided about whether ESA governments should invest the required €1 billion in an upgrade from the ESA's next budget. ESA governments pay around €120 million per year to Arianespace to offset fixed costs. Within this future project evaluation period, France will need to take a position on the already initiated Ariane 5 ME development or the development of the next generation Ariane rocket. While the German government has continued to stress its support for the Ariane 5 ME upgrade, French officials are uncertain that, in a time of enormous pressure on its public budget, the upgrade of the current launcher is the way to go. This notwithstanding, the then French President said in November 2011 that he would maintain France's space budget despite the pressure on the public budget, when the investment on space activities amounted to approximately €530 million from public bonds.[78] In early 2012, France and Germany agreed to move the discussion over the Ariane 5 successor to a newly established working group to resolve differences over the project's future. The working group expected to come to a conclusion on the Ariane 5 by late June 2012.

On 22 March 2012, France released a space strategy report advocating the absorption of ESA by the European Union. The report also stressed the need for EU help in financing Europe's Guiana Space Centre spaceport in Kourou. Europe's Ariane 5 launcher was designed to thrive in global market conditions that did not factor in competition from China, India and the commercial launch providers in the United States. Now, European nations are urged to confront the fact that this model has become obsolete, manifesting the need for a broad update of priorities. The report calls on ESA to soften its geographic return rules along with urging the European Commission to apply a different standard to its competition requirement in areas like space, as there is only one viable European launch provider. And it further went on to express France's position on Galileo, emphasizing the need for the full complement of 30 satellites plus spares.[79]

---

[78] De Selding, Peter B. "Sarkozy Reaffirms Commitment to Space Spending." SpaceNews 28 Nov. 2011: 6.

[79] De Selding, Peter B. "French Strategy Paper Says ESA Should Fall Under EU Authority." Space News 2 Apr. 2012: 4.

In a distinct development, on 7 June 2011, France concluded a cooperation agreement with Azerbaijan for the promotion of satellite communication services. The agreement followed Azerbaijan's procurement of its first national communications satellite. Its launch was scheduled for the end of 2012, onboard an Ariane 5 rocket, for which Azerbaijan planned to secure a loan from the French export credit agency Coface.[80]

Germany

In 2011 and 2012 Germany continued its effort to position itself as the European space technology leader by taking the lead in key European space technology development projects, both in the framework of the EU and ESA. At the same time, it increased the visibility and public impact of its technological capabilities, both through its strong support for the International Space Station (ISS) and by initiating its own national programmes. The German space programme focuses on the following priorities: the completion of the Galileo and GMES constellations; the full exploitation of ISS for as long as possible; the development of new or improved launch vehicles; and the realization of space exploration missions.[81]

After Galileo and GMES, a key issue for Germany during the review period was to reach an understanding with its European partners on the development of future access to space systems. Negotiations were undertaken with France to agree on this issue prior to the ESA Ministerial Council of late 2012 that would decide on the matter. From the outset, the German position favoured developing an upgrade for Ariane 5 (ME), instead of building an entirely new rocket. Equipped with a new upper stage, Ariane 5 ME would be able to lift nearly 10.5 t to GEO (1 t more that the current version) and could make its debut flight in 2018, for a cost of up to €1.5 billion.

The still unresolved challenges in the development of German national space programmes through PPPs were evident in projects such as the country's use and further development of Radar Earth observation spacecraft. Although German Radar EO satellites (TerraSAR-X and TanDEM-X) were developed on public funding, their products were partially commercialised on a private sector basis. Despite the original plan that their commercial operators would take over the development of the next generation of such satellites, their weak commercial revenue quickly suggested that this would not be possible. This example demonstrated, yet again, the difficulty of funding, developing and operating Earth observation services on a purely commercial basis in the absence of a minimum guaranteed commercial income provided by long-term government product

---

[80] De Selding, Peter B. "France-Azerbaijan Accord Emphasizes Space Cooperation." 8 June 2011. Space News 25 Jan. 2012 http://www.spacenews.com/policy/110608-france-azerbaijan-space-cooperation.html.

[81] De Selding, Peter B. "Germany Reaffirms Commitment to $2B Ariane 5 Upgrade." Space News 31 Oct. 2011: 6.

purchases, as in the case of the United States. On the other hand, it has become clear that in the absence of an anchor tenancy of the satellites' operational use, the German space agency would have no choice but to subsidize the development of their successors, in the interest of maintaining service continuity.[82]

Italy

Cooperative PPPs in developing space capabilities were also highly relevant for Italy in 2011 and 2012, particularly in the field of satellite communications. In this context, the Italian space agency (ASI) resorted to cooperation with the private sector, both in procuring satellite services and promoting the commercialization of its own satcom spacecraft. These seemingly opposite approaches had in fact a common denominator: scaling down operating costs for satellite application missions, either by outsourcing services or commercializing already acquired assets. An example of the former was ASI's €50 million deal with Eutelsat to purchase the exclusive use of ten Ku-band transporters onboard one of its future satellites. An example of the latter was the agency's decision on 15 July 2011 to create the commercial company Asitel, in order to promote services based on the Athena-Fidus Ka-band satellite, scheduled for launch by 2014. The set up of Asitel and the advanced commercialization of Athena-Fidus services even prior to its launch have the potential to transform the satellite's operational use to a PPP scheme, if commercial clients are willing to purchase not only services, but indeed exclusive transponders for use onboard the spacecraft.[83]

The United Kingdom

In 2011 and early 2012, the United Kingdom (UK) continued to successfully develop its space activities and cooperation with other partners, related in part to its recently established UK Space Agency. Within the reporting period, the UK signed several cooperation agreements. On 14 July 2011, officials from the Russian space agency and UK space agency signed an agreement that outlines cooperation in several areas of space activity. They will establish a working group consisting of experts from both sides to review application issues to the GLONASS system, in addition to potential microgravity experiments on the ISS.[84] Thereafter, on 18 April 2012, British and Japanese officials signed an agreement on space research and

---

82 "Public Funding on the Table for Germany's Next Radar Satellite." Space News 11 July 2011: 12.

83 De Selding, Peter B. "Italian Space Agency Signs Bandwidth Deals with Eutelsat." Space News 11 July 2011: 6.

84 "UK Space Agency and Roscosmos Signed Agreement." 14 July 2011. UK Space Agency 6 May 2012 http://www.bis.gov.uk/ukspaceagency/news-and-events/2011/Jul/uk-space-agency-and-roscomos-sign-agreement.

other potential commercial opportunities. One of the key domains is cooperation on Earth observation technologies, including the NovaSAR programme (i.e. the Disaster Monitoring Constellation) run by the British firm Surrey Satellite Technology Limited (SSTL).[85] Prior to this agreement, on 29 November 2011, the British government announced that it would finance about half of the costs of this project. The investment, amounting to £21 million, is for launching the first SSTL built NovaSAR satellite (the total cost amounts to £45 million).[86]

In October 2011, the UK Space Agency launched a call for proposals to support space science research related to Mars exploration. The call, amounting to a total of £1.6 million, will be distributed in the form of research grants to promote planetary science research in the UK, with a particular focus on exploiting data from NASA's Mars Science Laboratory and (at that time) the Russian Phobos-Grunt mission.[87]

1.2.4 The United States

From a policy perspective, the second half of 2011 and the first half of 2012 was a transition period for United States' space activities. This period witnessed the implementation and further elaboration of the Obama Administration's prior ground breaking decisions to adopt a new strategic orientation for NASA and to implement a new National Space Policy. During this time, concrete actions were undertaken to realise the new strategic orientations, amid political controversy over their declared objectives and how best to accomplish them.

A focal point related has been the development of new access to space technologies. In an important speech on 6 July 2011, President Obama laid down the strategic rationale behind his policy. He defended his decision to invest in the development of new relevant technologies, instead of diverting resources to the exploitation of existing ones that date back to the Apollo era. He maintained that rather than keeping on doing the same thing, it would be better to invest in the search for breakthrough technologies that would revolutionise spaceflight. Furthermore, he confirmed that LEO human spaceflight missions would be completely outsourced to the private sector, while the NASA Human Spaceflight programme would be reoriented for a mission to an asteroid, rather than Mars, which would become an ultimate goal.[88]

Another issue area that preoccupied the Obama Administration was its plan to reform the export licensing procedures regarding space technology components.

---

[85] "UK and Japan Commit to Greater Collaboration on Space." 18 Apr. 2012. Department for Business Innovation & skills, UK Space Agency 6 May 2012 http://www.bis.gov.uk/ukspaceagency/news-and-events/2012/Apr/uk-and-japan-commit-to-greater-collaboration-on-space.

[86] De Selding, Peter B. "British Government Backs SSTL's Low-Cost Radar Satellite Project." Space News 5 Dec. 2011: 4.

[87] "UK Space Agency Solicits Mars Research Grant Bids." Space News 24 Oct. 2011: 8.

[88] "Obama: Spaceflight Need a Technological Breakthrough." Space News 11 July 2011: 8.

After nearly 2 years of consultations, in July 2011 government officials announced their proposal for a comprehensive new regulation on the subject. In a nutshell, the proposed reform would remove a great number of space related items from the Department of State's U.S. Munitions List (USML) and return it to the Commerce Department's Commerce Control List (CCL). The immediate result of this reform would be to significantly simplify and speed up export procedures, thus restoring U.S. commercial space companies' competitive situation in the international space market. At the same time, U.S. officials offered reassurances that the reform would protect the country's national security interests by only affecting non-military parts and components. Furthermore, Administration officials qualified the new export rule as a "common sense" approach. They maintained that the new rule would contain a more precise description of the USML items, which would then be reviewed on a case by case basis based on their actual nature and not on the broad generic definitions used thus far. According to the Administration, the implementation of the new paradigm would result in the migration of as much as 90 % of the items to the CCL list, 50 % of which were expected to be relieved of any export licensing obligation whatsoever.[89] Despite the Obama Administration's zest in promoting the reform, its implementation was hampered by political controversy and administrative delays. On the one hand, the USML review that was a prerequisite for the reform's presentation to legislative authorities was reported to be 18 months overdue. On the other hand, the Administration's plans to couple the export licensing progress overhaul with the creation of a single government authority to supervise it ran into bureaucratic and political turmoil.[90]

Earlier, in 2009, the Obama Administration reaffirmed its support for ITAR reform, calling for a review of the current regime.[91] Recently, a report from the U.S. State and Defence Departments concluded that many satellites and their components could be transferred to the Commerce Department's Commerce Control List. While these satellites and their components are currently listed under the U.S. Munitions List, administered by the Department of State, the "1248 report" (invoked by Section 1248 of the National Defense Authorization Act for 2010) recommends giving the U.S. President the authority to determine export jurisdiction. The move would aid national security by allowing the government to focus its controls on technologies and capabilities that are the most sensitive; however, the report still recommends strengthening the U.S. ban on shipping U.S. satellite technology to China for launch on Chinese rockets. Meanwhile, if the 1248 report recommendations are applied, most communications satellites and some

89 Brannen, Kate. "Obama's Export Licensing Regulation Reform Takes Step Forward." Space News 1 Aug. 2011: 13.

90 Leone, Dan. "White House Report on Export Reform Still Expected This Fall." 17 Oct. 2011. Space News 6 Feb. 2012 http://www.spacenews.com/policy/111017-report-export-reform-fall.html.

91 Klamper, Amy. "Official Reaffirms White House Support for ITAR Reform." 14 Sept. 2009. Space News 24 Apr. 2012 http://www.spacenews.com/policy/official-reaffirms-itar-reform.html.

low-performing remote sensing satellites could be treated as non-military technology for export purposes, without affecting national security.[92]

In 2011, an estimated 71,000 people were employed in the U.S. space industry. That number is lower than the previous year, and this flow is in the opposite direction to the European counterpart that experienced a significant increase in employment in 2010. U.S. manufacturers have a natural tendency of dependence on their domestic market.[93] In 2010, 72.2 % of the sales in the U.S. aerospace sector were to the domestic market, in 2011 that figure increased by a small margin to 72.8 %. To summarize, these figures are a reminder that the domestic market plays a crucial role in the U.S. aerospace industry, clearly showing that public expenditure must increase at a constant pace to cover the weaknesses in the export market and ensure increased sales on a yearly basis. This tendency is even more pronounced if one looks at the space domain only. Employment figures strongly support this observation, as the decreasing numbers of employees mirror the small changes in sales. The U.S. is still at its turning point after recent political decisions; the outcome will depend on whether the U.S. space industry has the will to reform the ITAR export control rules, and whether the decision to privatize a significant portion of U.S. space activity will positively affect the sector.[94]

1.2.5 Canada

Despite the fact that Canada increased its budget for the period 2011–2012 to a record high of $424.6 million, and that the Canadian government has identified space as a key national strategic capability, its overall budgetary situation must be considered as one of the main reasons why it might be intended to keep expenditure on space activities rather low in the long-term. On the one hand, Canada is an important partner in the ISS program through its Mobile Servicing System – through its cooperation on the NASA's James Webb Space Telescope – in addition to having a status similar to that of an associate member with ESA. On the other hand, the country still lacks a comprehensive, broadly endorsed political approach toward space activity. In the review period, Canada released its new Report on Plans and Priorities in reference to its space activity. Despite this, Canada may still be waiting for clarification as to where its space agency's (CSA) sustainable and strategic plan is headed. For now, a wait-and-see approach has been noticeable, and has already brought some criticism.[95] Space industry representatives and politicians

---

[92] Leone, Dan. "U.S. Report Supports Sweeping Reform of Satellite Export Rules." 18 Apr. 2012. Space News 23 May 2012 http://www.spacenews.com/policy/120418-report-backs-reform-sat-export.html.

[93] Aerospace industry association, "Aerospace Industry Sales by Customer 1997–2011.", 2011 http://www.aia-aerospace.org/economics/year_end_review_and_forecast.

[94] Boessenkool, Antonie. "DoD: U.S. Space Industry May Lose Edge." 25 May 2010. Defense News 24 Apr. 2012 http://www.defensenews.com/story.php?i=4641686.

[95] "Canada's Space Planning Should Emphasize Flexibility." Space News 27 June 2011: 18.

have expressed concern that CSA is still without guidance from the government in handling anticipated problems, resulting in a haphazard approach to budgeting. The lack of information and direction from the Canadian government on its commitment to a long-term space plan also makes it difficult for other countries to be involved in Canada's space activities.[96]

In March 2011, the Canadian government decided to commit to the ISS from 2015 to 2020. This decision also means that in addition to technical cooperation on this international project (e.g. the Special Purpose Dexterous Manipulator), Canadian astronauts will return to space.[97]

According to Canada's Report on Plans and Priorities, the budget increase in 2011 and 2012 was mainly for finalizing the Radar Constellation Mission (RCM) and the Polar Communication and Weather mission. However, budget cuts for the next period make the future of the RCM uncertain. This project is part of a constellation of radar-imaging satellites conducting maritime and Arctic surveillance. The satellite's developer, McDonald Dettwiler, has raised doubts whether the project will proceed, as the government's budget does not include the funds required to continue to the next project's phase. This phase involves the production of the first satellite for the constellation. Given the uncertainty of the project's funding, McDonald Dettwiler will probably restructure its work force, likely resulting in employment layoffs.[98]

1.2.6 Russia

In 2011 and 2012 Russia continued the modernisation of its space infrastructure according to the 10 year Federal Space Programme announced in 2005. The document, which provides main strategic orientations and guidelines for the Russian space programme for the 2015 horizon, is still relevant for the interpretation of the country's current ambitions in space.[99] Russia's Security Council produced one more space policy related paper in 2008, updating the security related project priorities.[100] The implementation of this programme so far has demonstrated

---

[96] Pugliese, David. "Canadian Space Spending to Recede Following this Year's Spike." Space News, 20 June 2011: 7.

[97] Rakobowchuk, Peter. "Canada to Support Space Station till 2020." 1 Mar. 2012. Herald News 8 May 2012 http://thechronicleherald.ca/canada/68715-canada-support-space-station-till-2020.

[98] Pugliese, Dave. "CSA Budget Cuts Put Radarsat Constellation Mission in Limbo." Space News 9 Apr. 2012: 5.

[99] "Major Provisions of the Federal Space Programme of the Russian Federation for 2006–2015." 22 Oct. 2005. Russian Federal Space Agency (Roscosmos) 30 Apr 2010 http://www.federalspace.ru/main.php?id=85.

[100] "The Outlines of the Russian Federation Policy in the Field of Space Activities for the Period Until 2020 and Further Perspective" Apr. 2008. Russian Federation Security Council 30 Apr. 2010 (Available in Russian text) http://www.scrf.gov.ru/documents/96.html.

considerable resilience and political commitment, despite the recent years' economic downturn on a global scale. In fact, a declared objective of Russian space activities is precisely to revitalise its economy and consolidate its space industrial basis. In addition to this, it seeks to generate technological and economic spinoffs from space science research and development, for the benefit of the entire society. Finally, it seeks to improve its defence power.

Looking back, space technologies were first to contribute to the country's economic recovery after the "lost decade" of the 1990s, increase external trade volume, and exploit Russian industry's competitive advantage in space. Further research and development areas currently include expanding space application utilisation, as well as implementing modern programme management and funding tools (such as public-private partnerships) in order to conduct scientific research and diffuse its results to society at large.

The issue of improving Russian space industry's competitiveness on a global scale was a key plank of the country's space policies in 2011 and 2012. According to Russian space agency data, the country's space industry had only managed to capture 3 % of international space business, despite the fact that it conducted 40 % of worldwide launches and manufactured 20 % of all spacecraft produced. This evidence suggests that Russia's space market penetration is very narrow, especially compared to the size and expertise of its industrial base. In order to remedy this problem, during the reporting period the Russian government continued efforts to restructure the country's space industry.[101]

One of the key aspects of this effort was to improve the industry's workforce skills especially on a managerial level, through the adoption of common best practices and a unified management culture. Another key plank of the restructuring efforts was to break up contractual work among several holding companies, effectively replacing the existing public quasi-monopoly with an open market competition environment. This step would also gradually introduce more private funds into the space sector. Bringing private funds into Roscosmos programmes was in fact identified as a significant objective of the restructuring in itself. It is believed that creating a competitive environment would benefit the quality and innovative nature of Russia's space products and services. However, with the government being by far the greatest employer and client of the space industry, real privatisation efforts might prove elusive, especially if the state were to maintain its role as the single most important funding source for space technologies.[102]

At the same time, Roscosmos initiated post-ISS planning for the period beyond 2020, when the ISS programme is scheduled to be terminated according to its current budgetary provisions. On this issue as well, Russian authorities seemed to carefully balance between adopting a cooperative approach and maintaining their

[101] Zaborskiy, Victor. "Revamping Russia's Space Industry: For Better or Worse?" Space News 1 Aug. 2011: 19.

[102] Id.

independent space capabilities. Regarding the former, in October 2011 the Russian space agency agreed with NASA to set up an expert-level group to investigate the possible options for ISS use after 2020. Similar contacts had also been initiated with the other participating space agencies. One of the possible scenarios under discussion envisaged the exploitation of ISS components for the benefit of future space exploration missions.[103] In addition to this, however, Roscosmos also initiated studies on what it would take to operate the ISS alone beyond 2020. With the future of ISS still undecided, this should be considered as a purely theoretical problem. Nevertheless, it does illustrate Russia's rising sense of self-confidence in the field of space technology and operations.[104]

1.2.7 Japan

Japan's space activities in the second half of 2011 and the first half of 2012 followed the broad strategic orientations of the country's basic plan for space (BSP). The BSP had previously established a 5 year roadmap for 2009–2013, along six basic pillars: the realisation of a "secure, pleasant and affluent society" utilising space; the enhancement of Japan's national security; the promotion of "space diplomacy"; the creation of a "vigorous future" by promoting space related research and development; the fostering of strategic space industries for the twenty-first century; and consideration of the environment.

The Basic Plan clearly demonstrates the strategic importance given to space for the future prosperity and security of the entire country. Furthermore, it identifies the future areas of interest for the country's space policy. These are space applications, security, international cooperation, scientific development, industrial development and environmental protection. It should be noted that security in its broader sense (military, diplomatic and economic) becomes the cornerstone of the new policy, as it is depicted in three of the six pillars. Other key policy objectives include achieving full autonomy in space technologies and increasing public-corporate synergies in space activities. Finally, space exploration (including independent manned flights) also receives special attention in the document.

From a programmatic point of view, one of the key developments in Japanese space policy during the review period was the decision to proceed with the development of the Quasi-Zenith Satellite System (QZSS), the country's future regional satellite navigation constellation. The Japanese Cabinet requested an initial ¥4.1 billion ($53.4 million) to begin manufacturing the first satellite. QZSS is scheduled to be fully operational by 2020, and it is designed to augment the

[103] Morring, Jr., Frank. "Spacefaring Nations Regroup For Push beyond LEO." Aviation Week & Space Technology 10 Oct. 2011: 46.

[104] Morring, Jr., Frank. "Keep On Trucking." Aviation Week & Space Technology 10 Oct. 2011: 26.

regional accuracy of the GPS signal. The system promises to provide Japanese authorities with a more accurate, secure and independent service. The future deployment of four spacecraft will enable 24 h regional coverage, for an estimated cost of ¥170 billion ($220 million). The deployment of the full seven satellite constellation will dramatically decrease its dependence on GPS for regional coverage, but for an additional cost of ¥90 billion ($116 million). The decision to bring the system to its full potential might have been influenced by the swift deployment of China's Beidou GNSS, which was expected to be finished by 2018. From a strategic point of view, increased competition between China and the United States in the region might pose a long-term threat to GPS signal availability and reliability. If such an eventuality were to occur, the timely development of the full QZSS constellation would help to mitigate its effects on Japanese citizens.[105]

From a space governance perspective, the second half of 2011 also introduced possible plans to break-up the Japan Aerospace Exploration Agency (JAXA) and concentrate all space activities under a smaller dedicated space agency directly under the Prime Minister's authority. Japan's newly appointed Prime Minister Yoshihiko Noda was thought to be in favour of this already existing proposal, which would effectively remove control of the Japanese space program from competing ministries and place it under centralised Cabinet-level management. Officials from JAXA's parent Ministry of Education, Culture, Sports, Science and Technology (MEXT) were reported to be against it – this opposition was founded on budgetary grounds, as the plan would move a third of the Ministry's budget to the Cabinet. However, by shifting the balance of power within the Cabinet, the appointment of the new Prime Minister had the potential to provide a forward push for the reorganisation plan.[106]

1.2.8 China

Chinese space policy evolves around the country's 5 years' economic development plans. The country's space programme is therefore meant to support its overall development objectives, while maintaining a comprehensive set of objectives for space activities. The main challenge for the Chinese programme is to achieve the right mix of national space capabilities and participation in international space cooperation. In December 2011, the Chinese government announced its latest five-year plan for 2011–2016, in the form of a government White Paper entitled

---

[105] Kallender-Umezu, Paul. "Japan Commits To Deploying Satellite Navigation System by 2020." Space News 17 Oct. 2011: 14.

[106] Jayaraman, K.S. "Japan's Latest Political Shake-up Stirs Hope for Space Agency Overhaul."12 Sept. 2011. Space News 25 Jan. 2012 http://www.spacenews.com/policy/110912-japan-space-agency-overhaul.html.

"China's Space Activities in 2011."[107] This document updates and extends the country's strategic and operational objectives in space, depicting the progress made since 2006 and laying down its short term plans, divided into four main activity areas: space transportation, satellite development, space applications and space science. In addition to this, it presents China's ambitions to play a pivotal role on the international scene and underpins the contribution of space activities to achieving this purpose.

China's principle policy objectives in space are all related to promoting the country's scientific, economic and social development, securing its national security and independence (in its broader sense) and improving its international influence by engaging in space cooperation. More specifically, it relates space activities to achieving the objectives of its scientific and technological innovation policies. At the same time, it stresses the importance of maintaining independent space capabilities, while carefully trying to balance this with an open approach to international cooperation and insistence on the peaceful exploitation of space. In a nutshell, the new Chinese space policy's principal axes are:

- Enhance space science and technology capabilities through innovation.
- Maintain technological and operational self-reliance across the board.
- Adopt an open and constructive attitude to international cooperation on the basis of mutual benefit.
- Oppose space weaponisation and protect the space environment.

All of the aforementioned policies converge to achieve the general national objectives of economic development, social progress and comprehensive national strength.

Apart from these general principles, China's new space policy recites in detail the country's achievements in space over the past 5 years and declares its programmatic intentions for the future. The level of operational and technological details provided is unprecedented for a Chinese document of this kind; combined with the clear description of future programmes, it implies a higher level of confidence and pride in the country's space capabilities than before. The Chinese 5-year space programme focuses on four mission areas: transportation, satellites, spaceflight and applications.

In the field of space transportation, it focuses on the development of three new launcher configurations by 2016, using more efficient engines and an entirely new upper stage. They include Long March 5 (with a 14 t to GEO lift capacity), Long March 6 (1 t to LEO), and Long March 7 (5.5 t to LEO). The realisation of this programme will provide China with comprehensive and flexible access to space capability, in line with the current and prospective space rocket development plans of all other major space faring nations. This fact underpins China's willingness to improve its space capabilities on a peer-to-peer basis. Of particular interest is Long

[107] White Papers of the Government of China. "China's Space Activities in 2011." Beijing 29 Dec. 2011. 6 Mar. 2012 http://www.china.org.cn/government/whitepaper/node_7145648.htm.

March 6, which is described as a "high-speed response launch vehicle". This lightweight launcher will provide China with an operationally responsive launch capability for the first time, with obvious national security and commercial applications.

In the field of satellite development, the new Chinese space policy describes a comprehensive programme embracing all fields of satellite and spacecraft use. It calls for the development of improved weather and communications satellites, as well as of an entirely new Earth observation and electromagnetic monitoring satellite series. The key plank of this programme will be the development of a satellite fleet capable of all-weather 24-h operations worldwide, which would imply making significant advancements in space borne SAR and high resolution optical instrument technologies. Finally, the new policy reiterates China's goal of fielding a regional satellite navigation capability by the end of 2012 and completing the deployment of its entire 35 satellite Beidou GNSS constellation by 2020.

In the area of orbital spacecraft development and human spaceflight, the new Chinese space policy is much more reserved in providing details, conscious perhaps of the higher technical challenges involved in these endeavours. It nevertheless draws an accurate picture of its short term steps, principally related to the full exploitation of the Tiangong-1 vehicle for operational testing and technological R&D purposes. To this end, it announced the launch of two more spaceflight missions, Shenzhou-9 and Shenzhou-10. In the long run, the new policy reiterates China's ambitions to independently develop all technologies necessary for human space flight and for maintaining a permanent human presence in orbit. Finally, the new policy announces the beginning of concept studies for a future Chinese human lunar landing programme.

Regarding space applications, the new Chinese space policy represents a significant push towards the full exploitation and commercialisation of the country's emerging satellite capabilities. Particular attention is paid to the utilisation of the Beidou GNSS and the new Earth observation and communications satellite fleet mentioned above. A broad market-oriented distribution of such services is described in the document, indicating China's ambition to establish itself in the global commercial satellite services market in the same way it has in the commercial space launch market. The approach adopted to meet this objective seems to be equally motivated by the need to expand the industrial and commercial scope of the Chinese space industry, as well as by the willingness to improve the downstreaming of satellite services and consequently increase societal benefits from them.

The implementation of the aforementioned programmes is to be assisted by a set of appropriate policy measures. These include plans for a broad restructuring of the space related industrial and R&D base, including the encouragement of sustainable scientific innovation, the creation of new research facilities, and the renewal of the country's skilled human capital relevant to space activities. The new industrial capabilities are to be coupled with a suitable funding and legal framework that will include the adoption of a national space law. It is foreseen that this new framework will also encourage the development of space entrepreneurship and market-oriented satellite utilisation schemes. It should be noted that all these policies are explicitly reserved for the field of satellite application, which seems to be the focal point of the new space policy's practical implementation, and the activity area expected to be most influenced by it.

Finally, the new Chinese space policy pays significant attention to the role of international space cooperation, both as a means of fulfilling the country's national objectives in space and as a tangible recognition of its rising space status. The policy document gives a very detailed account of China's principal bilateral and multilateral cooperation agreements on space activities. It places particular emphasis on bilateral cooperation with Russia, Europe (ESA, EUMETSAT and individual Member States), Brazil, Ukraine and an array of smaller emerging countries. Furthermore, it underpins China's increasing engagement in multilateral cooperation, especially in the context of the UN and the Asia-Pacific Space Cooperation Organization (APSCO). The principal guidelines of China's international engagement in space activities remain unchanged, advocating the promotion of "...inclusive space development on the basis of equality and mutual benefit, peaceful utilization and common development."[108] However, increased emphasis is given to regional cooperation, and the further development of APSCO.

1.2.9 India

By tradition, Indian policy aims at achieving social and economic development through space activities. The Indian space programme currently operates under the guidelines of the current 11th five year plan (2007–2012) which focuses on creating space applications capable of providing tangible products that improve life conditions in the country. Self-reliance and space services oriented projects are the corner stones of India's space policy. These include two operational space systems, one for satellite communications and television broadcasting services and one for Earth observation.

A key plank of India's space policy in 2011 and 2012 was the emphasis on international cooperation. Encouraged by the development of its space programme, India's space agency adopted a more extrovert approach to relationships with established space faring powers, such as the U.S., but also Europe, albeit on a bilateral basis with European Member States, rather than on an institutional level. The rapid development of the country's space capabilities has boosted its self confidence to engage in joint projects with such partners, based primarily on the low cost of its technologies, but also on their increasing reliability. In this case as well, priority was given to space application missions. An example of this trend can be seen in the successful launch of a joint Earth observation mission with the French space agency CNES. On 12 October 2011, the Megha-Tropiques satellite was successfully placed in LEO on board a PSLV launcher. The spacecraft was

[108] Id.

designed to operate for 3–5 years, for the purpose of studying tropical monsoons with the use of a microwave instrument. The Indian contribution to the mission was significant, including the provision of the launcher, spacecraft development and 60 % of the approximately $152 million budget. Despite the project's successful development, no follow-up mission has been yet envisaged. In addition to its main payload, the PSLV rocket also carried three Indian microsatellites to orbit, including a demonstrator commercial ship-monitoring spacecraft.[109]

Another key plank of Indian space policy during the reporting period, and a sign of the country's increased self-confidence and open approach to international space activities, was the steps taken to boost commercial space activity, especially in the Earth observation product market. This was exemplified by the Indian authorities' decision in July 2011 to lift existing restrictions on the sale of high resolution satellite imagery of the Indian subcontinent. Until then, ISRO policy distinguished between government and commercial clients of Earth observation products from its satellites. Consequently, ISRO announced that all imagery of up to 1 m spatial resolution would be available on a non-discriminatory basis for distribution. Data of better than 1 m resolution destined to non-governmental users would be screened prior to distribution, to protect national security interests. Private entities requesting such products would still need the endorsement of a government agency. More importantly, ISRO approved the open sale of such products to foreign entities as well, consequently opening up relevant Indian data to the international market. In relation to this, a High Resolution Image Clearance Committee was set up to approve foreign sales of Earth observation images, conditional on the signing of nondisclosure agreements with ISRO's Remote Sensing Centre.[110]

1.2.10 Brazil

Despite long-standing obstacles in Brazil's space and R&D development, and the lack of expertise in its workforce, Brazil is fast becoming an important space player of the future due to its geographic position close to the equator, and its rapidly growing economy. On March 2011, Brazilian officials made an announcement on an ambitious plan to triple the space agency's budget from 300 to 900 million Brazilian reals (from $185 million to $555 million); it will also streamline its management, boost private sector involvement, and educate and train more aerospace workers.[111] In recent years, Brazil has reached a point of recognition as a global economic leading country. By 2011, Brazil was the sixth largest country by

---

[109] De Selding, Peter. "ISRO Launches Joint French-Indian Satellite To Study Tropical Monsoons." Space News 17 Oct. 2011: 5.

[110] "India Eases Restriction on Satellite Imagery Sales." Space News 11 July 2011: 8.

[111] Messier, Douglas. "Will a New Space Power Rise along the Atlantic?" 15 Aug. 2011. The Space Review 8 May 2012 http://www.thespacereview.com/article/1904/1.

nominal GDP, and it was expected advance to the fifth position by the end of 2012. Moreover, Brazil is a member of the BRICs, consisting of Brazil Russia, China, and India. At the last BRIC conference, several partners stated that their economic cooperation should also include cooperation between their space sectors.[112] While all BRIC members are space faring nations, and cooperation already exists between Brazil and Russia (and Ukraine) in the development of launch capabilities, in the wake of delays and funding uncertainties, strengthening cooperation between Russia and China would also help Brazil in developing its own space programme and industry.

Brazil is already undertaking cooperative programs with Ukraine and Russia to build six Cyklon-4 rockets that could launch small satellites into low Earth orbit (LEO) and heavy geosynchronous communications satellites. The head of the Ukrainian Space Agency expressed concern that the project is underfunded, however, as $280 million has been spent and around $260 million is still required; Brazil has invested $50 million more than Ukraine in that project.[113]

The United States is still an important partner for Brazilian space activity. In October 2011, NASA and the Brazilian Space Agency (AEB) signed an agreement for a scientific and engineering feasibility study for potential cooperation in the Global Precipitation Measurement (GPM) mission project (this project is already established between NASA and JAXA). The two agencies also signed a cooperation agreement on an ozone study. Moreover, Brazil will play an important role in the dissemination of NOAA data for the South America region. Brazil's National Institute for Space research (INPE) is already cooperating and trying to expand the GeoNetCast program. These activities foster collaboration among scientific organisations on an international scale by improving the availability of scientific data for *inter alia*, global climate protection and environmental monitoring. Brazil has also expressed strong interest in cooperating on international projects for crisis management dealing with major natural or technological disasters. On 5 April 2012, Brazil and USA held the first space related security dialogue, affirming both nations' commitment to collaboration in working towards a sustainable, safe and secure space environment.[114]

---

[112] "Dmitry Medvedev: BRIC Cooperation has Great Potential." 13 Apr. 2012. The Voice of Russia 8 May 2012 http://english.ruvr.ru/2010/04/13/6382853.html.

[113] Messier, Douglas. "Will a New Space Power Rise along the Atlantic?" 15 Aug. 2011. The Space Review 8 May 2012 http://www.thespacereview.com/article/1904/1.

[114] "State Dept. Fact Sheet: U.S.-Brazil Space Cooperation." 9 Apr. 2012. U.S. Department of State, Office of the Spokesman 8 May 2012 http://iipdigital.usembassy.gov/st/english/texttrans/2012/04/201204063457.html#ixzz1uHwajBn0.

1.2.11 Emerging Space Powers

In 2011 and 2012, South Korea was one of the most active emerging space actors, especially in the field of space applications. In October 2011, the Korean Aerospace Research Agency announced its plans to launch two additional Earth observation spacecraft of the Kompsat satellite series by mid-2012. The first would be equipped with a Synthetic Aperture Radar instrument, and the second with an improved optical one. Both spacecraft would be locally designed and built by the Korean space agency and the country's aerospace industry (KAI). This campaign would bring the number of Korean-built satellites in orbit to a total of five, significantly improving both ground resolution and operational flexibility. Although the satellites were poised to provide imagery for civilian and scientific purposes, their improved operational characteristics would enable their military use as well. It is noteworthy that Korean authorities opted for open commercial market procedures in awarding the programme's contract, because the same commercially-oriented approach is expected to prevail in the distribution of the satellites' products.[115] In addition to this, Korea's increased Earth observation capabilities have fuelled the country's stronger participation in international cooperation schemes in this field. For example, in July 2011 the Korea Aerospace Research Agency became the newest member of the International Charter on Space and Major Disaster. The country pledged to provide data from the Komp-2 spacecraft in support of the Charter's emergency relief support operations.[116]

During the reporting period, Singapore emerged as another Southeast Asian country with significant space aspirations. Most international commercial space companies maintain offices in the country, encouraged by favourable tax laws. In October 2011 Singapore's government unveiled its plans to transform this international space-related presence to a more comprehensive space industry and business hub for the entire Southeast Asia region. In recent years, Singapore has developed a full-fledged space research industrial and administrative infrastructure to support its space activities. Furthermore, the country's space programme was created on a fully outsourced PPP basis from the outset. The foundations of this approach were laid down in 2006, with the creation of the Singapore Space & Technology Association (SSTA), a private entity. SSTA's board includes senior representatives from the Singapore Economic Development Board, the Defence Sciences Organization, the Civil Aviation Authority of Singapore, the National University of Singapore and Singapore Technologies (ST) Electronics. The latter is a joint venture company created on a PPP basis for the purpose of commercialising locally developed EO technologies and services, including data from the country's first Earth-observation satellite X-SAT, built by the Nanyang Technological University (NTU). SSTA is the centre of space-related activities in Singapore. It organises an

[115] Perrett, Bradley. "Seoul Pushes Ahead With Spacecraft Program." Aviation Week & Space Technology 24/31 Oct. 2011: 44.

[116] "South Korean Space Agency Joins Disaster Relief Pact." Space News 25 July 2011: 8.

annual Global Space & Technology Convention, fuels the Asia-Pacific Regional Space Agency Forum in cooperation with JAXA, and runs the Singapore Space Academy.[117]

In the Middle East region, Iran continued to establish itself as a strong regional space actor, through the implementation of a comprehensive space programme. The focus of Iran's space activities in 2011 and 2012 was on the space application's mission area. The country's relevant programme culminated on 15 June 2011 with the launch of the Rasat-1 Earth observation satellite, onboard a Safir rocket. Rasat-1 was Iran's second locally designed and built satellite, after the launch of the spacecraft Omid in March 2011. Although Rasat-1 is considered to be a technology validation mission, the spacecraft's EO capabilities might be suitable for operational use. According to Iranian officials, the mission's objectives were to establish contact with Earth stations, obtain orders from those stations, take Earth images and transmit them along with telemetry information back to its ground station.[118]

On 12 March 2012, the official representatives of the two biggest satellite fleet operators based in the Middle East, Nilesat and Arabsat, said in separate statements that intentional jamming of their signals has resulted in substantial business and revenue-losses during the Arab Spring. Nilesat representatives said the jamming started with Libya and then moved to Bahrain, Iraq, Syria, Iran and other places. Nilesat's five satellite transponders carrying more than 60 television channels were interrupted for varying lengths of time. The international regulatory radiofrequency body (ITU) was asked for an official reaction, thus calling on operators to react to current jamming issues, and prevent problems from occurring in the future.[119] On 7 December 2011, five international television and radio broadcasters (incl. the BBC, Voice of America, Deutche Welle, Audiovisuel Exterieur de la France and Radio Netherlands Worldwide) called on international regulators to address the issue of international satellite signal interference as an agenda item at its 2012 ITU conference. These operators all indicated that Iran was trying to jam their satellite signals.

The ITU's World Radiocommunication Conference (WRC-12) was held in Geneva early in 2012. Despite the increased number of issues since the last conference, some issues were given higher priority for future consideration by international regulatory officials. In the short term, there will be increased demand for access to satellite orbital slots, mainly by developing and emerging nations. The inevitable result of current trends in demand for orbital slots has intensified competition between developing and emerging economies, mainly in Africa, Asia and the Latin America region.

---

[117] Francis, Leithen. "Singapore Aims To Become Space Industry Powerhouse." Aviation Week & Space Technology 17 Oct. 2011: 50.

[118] "Iran Lofts Imaging Satellite In Orbit atop Safir Rocket." Space News, June 20 2011: 8.

[119] De Selding, Peter. "Jamming No Mere Nuisance for Middle East Satellite Operators." Space News 26 Mar. 2012: 6.

The WRC delegates agreed to extend the amount of time a satellite operator has to replace a failed satellite, i.e. from 2 to 3 years; it also formally set the agenda for strengthening the rules related to registering satellite systems.[120] Finally, the long dispute over Iran's Zohreh-1 satellite moved to the next stage after the 153 delegates decided to reinstate Iran's right to the previously licensed slot position.[121]

1.3 Worldwide Space Budgets and Revenues

In general, worldwide national space budgets continued to grow in 2011. World government expenditure for civil space programs amounted to $44.92 billion.[122] Total world governmental expenditure on space programmes amounted to $72.77 billion.[123] The rate of increase of these space budgets has accelerated since 2010. Whereas the Compound Annual Growth Rate (CAGR) had dipped to 2 % between 2009 and 2010, following a stronger rate of 9 % during the period between 2004 and 2009, the aggregate growth rate rose to 6 % between 2010 and 2011.[124] A more detailed analysis of institutional budgets is set out in the following section.

The 2012 Space Report provides a guide to the commercial revenue of space activities, listing the 2011 total revenue of commercial satellite services at about $110.53 billion including activities such as telecommunications, Earth observation and positioning services (this amount represents a 9 % increase from $101.73 billion in 2010). The revenue of space-related commercial infrastructure including manufacturing of spacecraft and in-space platforms, launch services as well as ground equipment is estimated to have reached around $106.46 billion (resulting in a corresponding increase of 14 % in launch capacity compared to the decrease in 2010). Overall, total commercial space revenue in 2011 was $216.99 billion.[125]

1.3.1 Overview of Institutional Space Budgets

From the Space Report, total institutional spending, including that of intergovernmental organisations, on space programs in 2011 can be estimated to have been

[120] De Selding, Peter. "Iran Decision, British About-Face among Surprises at Radio Frequency Conclave." Space News 27 Feb. 2012: 1 and 4.

[121] "Looking Beyond WRC-12." Space News 12 Mar. 2012: 26.

[122] The Space Report 2012. Colorado Springs: The Space Foundation, 2012: 58. The amount was calculated by subtracting the total military expenditure on space from the total world governmental expenditure on space (see infra footnote).

[123] Id. at 42.

[124] Id. at 32.

[125] Id.

approximately $72.77 billion, a figure which showed a nominal decrease of 6 % compared to 2010.[126] This space spending was comprised of $44.92 billion in civil expenditures (61.7 % of the total) and $27.85 billion in defence expenditures (38.3 %). On the other hand, the Euroconsult report lists 2011 civil space expenditures as amounting to $40.3 billion, whereas its estimates for government expenditures for defence space programmes amounted to about $30 billion.[127] Consequently, the ratio of defence expenditure relative to civil expenditure went down compared to the previous year, when the ratio was $37 billion in civil expenditures (or 52 % of the total) and $34 billion in defence expenditures (or 48 %).[128]

Based on the Space Report 2012, of the estimated $27.85 billion of defence related space expenditure worldwide, $26.46 billion was spent by the United States, representing a share of 95 % and indicating a significant percentage increase compared to the previous year. These funds came from *inter alia*, the Department of Defence (DoD), the National Reconnaissance Office (NRO), and the National Geospatial-Intelligence Agency (NGA). It should be noted that a degree of uncertainty exists regarding expenditures on defence space activities as not all relevant funding is made public. However, it is clear that the activity in the United States is a driving force in worldwide space activity, particularly in the defence area (Fig. 1.1).

The expenditure hierarchy among states stayed roughly similar to 2010; China was the exception, having advanced its position in 2011. Maintaining its strong lead position, the U.S. had the largest budget, directing $20.79 billion toward civil expenditure, and $26.46 billion toward defence expenditure.[129] The low estimate of Russia's budget must be put into perspective, as it does not factor in the intensive military activity entailing regular classified launches, or the scientific programmes. Attention should be given to China's national space budget of $3.08 billion which overtook France's budget of $2.27 billion in 2011.[130] In 2010, China was already poised to overtake France's 2010 budget of $2.5 billion with a budget of $2.4 billion.[131] India managed to maintain its position as having the seventh largest space budget, surpassing Italy's budget by a significant margin.

The European Space Agency, as an international organization now composed of 20 member states, had a 2011 budget of €3.994 billion ($5.80 billion), an increase of 6.7 % above its 2010 budget of €3.745 billion ($4.60 billion).[132] In 2011, as a joint

[126] Cf. id. at 32 and The Space Report 2011. Colorado Springs: The Space Foundation, 2011: 32; note: Figures in this section are based on the Space Report 2012 data (USA, Russia, Japan, China, and France), while all other values in Figs. 1.1 and 1.2 come from the Euroconsult Report 2012 – Profiles of Government Space Programs.

[127] Euroconsult Report 2012.

[128] Pagkratis, Spyros. Space Policies, Issues and Trends in 2010/2011. Vienna: European Space Policy Institute, 2011: 12.

[129] The Space Report 2012, 44.

[130] Cf. id. at 48, 51.

[131] Space Policies, Issues and Trends in 2010/2011, 13.

[132] "ESA Budget for 2011." ESA 15 May 2012 http://download.esa.int/docs/DG/ESA_2011_Budget_040111_rev2.ppt#256,1,Slide1.

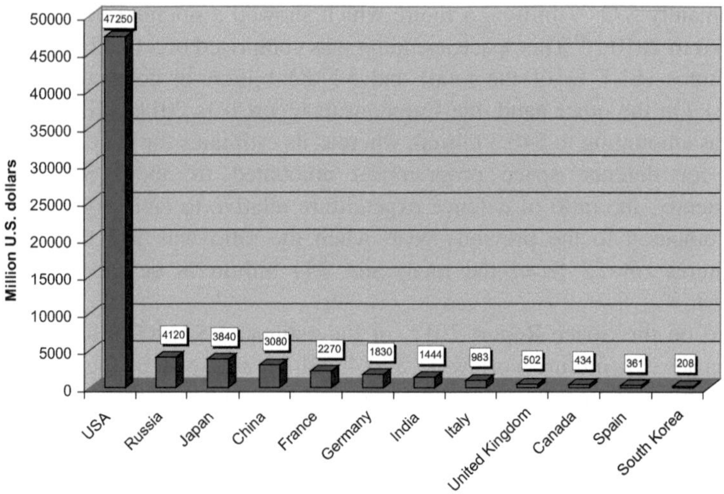

Fig. 1.1 Public space budgets of major space powers in 2011 (Based on Euroconsult and the Space Report 2012 data)

investment of the then 19 Member States, the five biggest contributors were France 18.8 %, Germany 17.9 %, Italy 9.5 %, U.K. 6.6 %, and Spain 5.1 %. The Space Report 2012 notes a slight decrease of 8.7 % in Japan's space budget amounting to ¥309.4 billion ($3.84 billion) when compared to ¥339 billion in 2010.[133] Japan's budget contributed to reducing the world concentration of space expenditure in the U.S., European countries, ESA and Russia, to 80.4 % in 2011 from 82 % in 2010.

When measuring the concrete effort of countries in the space sector it is necessary to put the figures into perspective in regard to GDP[134] (Fig. 1.2). However, considering the absolute numbers alone paints only a partial picture since comparisons between countries with different economic conditions (e.g. price or wage levels) can be misleading.

U.S. space budget figures underscore its strong engagement in the space field; however, its level of investment seems to have tapered off or even diminished slightly. As explained above, the Russian figures must be read with caution, yet Russia's space effort increased by 0.03 to be 0.22 % in 2011. France's ratio relating to space efforts as a percentage of GDP diminished to 0.08 %, with India surpassing it and Japan following behind at 0.09 % and 0.07 % respectively. The other leading space countries in Europe continued to invest 0.05 % or less of their GDP on space activities.

The U.S. continued to dominate per capita space expenditure at $151.59 (a decrease of 2.64 % from 2010), whereas France's per capita space budget was reduced to $35.86 (a decrease of 11.24 %) (Fig. 1.3). From the values listed in the

[133] The Space Report 2012, 55.

[134] The data used is the nominal GDP converted to current U.S. dollars using the official exchange rates as indicated by the International Monetary Fund.

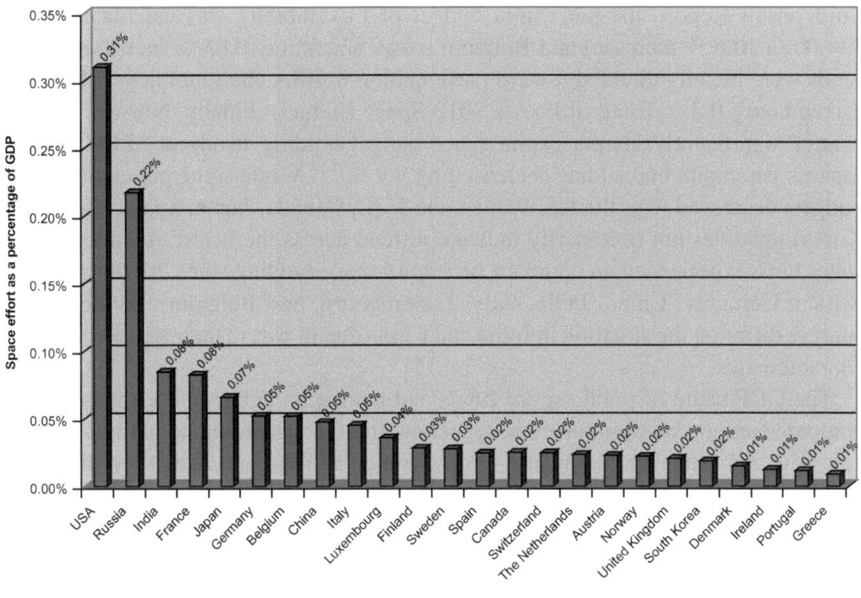

Fig. 1.2 Public space budgets (selection) as a share of nom. GDP in 2011 (Source: Euroconsult/ IMF)

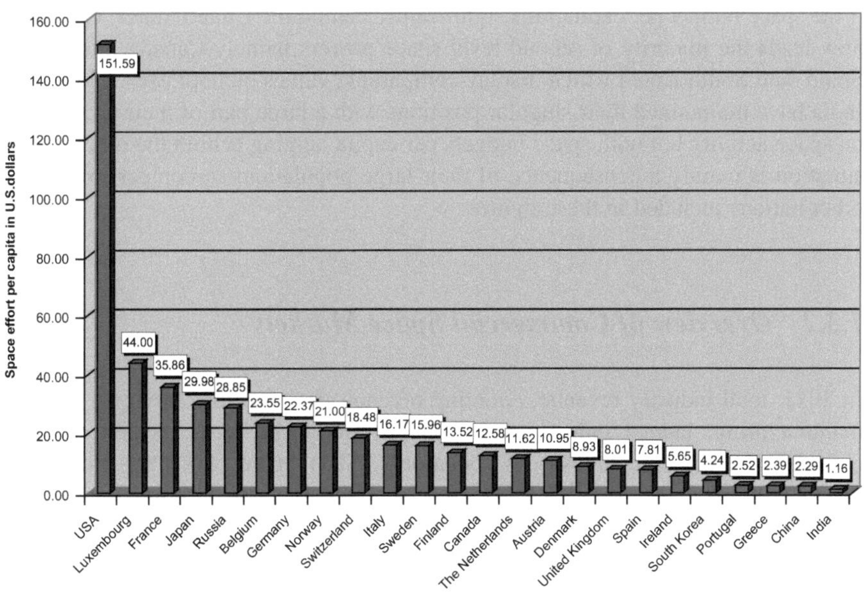

Fig. 1.3 Public space budgets per capita (selection) in 2011 (Source: Space Report 2012/ Euroconsult/UN World Population Prospects)

Euroconsult Report, the per capita budget of Luxembourg was calculated to be $44.00 (a 10.6 % increase) and Belgium's was $23.55 (a 10.05 % increase); these funds were heavily directed toward participation in ESA (Belgium now 4.1 % and Luxembourg 0.3 % listed in ESA's 2012 Space Budget). Finally, Norway did not change significantly, its per capita space budget holding steady at $21.00, while Japan's per capita budget has decreased by 9.9 %.[135] While some per capita space budgets decreased (e.g. the US. France, the U.K., Canada, Japan, Spain, and South Korea) this does not necessarily indicate a trend across the board; the majority of states have experienced an uptick in per capita space funding since 2010 (including Russia, Germany, China, India, Italy, Luxembourg, and Belgium). Authoritative sources differ on the situation in India and China due in part to their socio-economic characteristics.

The GDP share of public space funds and per capita public space funds can be compared among states involved in space funding. This is presented in Figs. 1.4 and 1.5, with the latter excluding the United States and Russia to display the other countries more clearly. The U.S. continues to excel by a significant margin in both the public space fund share of GDP and the space budget per capita. France still holds second position in space budget per capita, while Russia has been surpassed by Japan for third place. It should be noted that some caution is needed when considering the latter two due to the uncertainty of reported values. Additionally, Russia has maintained its second place position with respect to the space budget as a percentage of GDP, with a large amount of its GDP directed to space activity, even if the space budget per capita falls significantly behind the United States. Germany now leads the majority of second level space powers namely Canada, Italy, UK, Spain, and South Korea which display comparable values on both axes. China and India have maintained their singular positions with a large part of their GDP spent on space activity but with space budgets per capita lagging behind the others. This situation is mainly a consequence of their large populations in comparison to the other nations included in these figures.

1.3.2 Overview of Commercial Space Markets

In 2011, total industry revenue, counting revenue from satellite services, satellite manufacturing, launch industry, and ground equipment was expected to have exceeded the Satellite Industry Association's (SIA) figure of $168.1 billion for 2010.[136] In fact, the Space Report 2012 reported that total commercial space

[135] The Space Report 2012, 55.

[136] "State of the Satellite Industry Report". August 2011. Satellite Industry Association and Futron Corp 24 Apr. 2012 http://www.sia.org/PDF/2011_State_of_Satellite_Industry_Report_(August%202011).pdf.

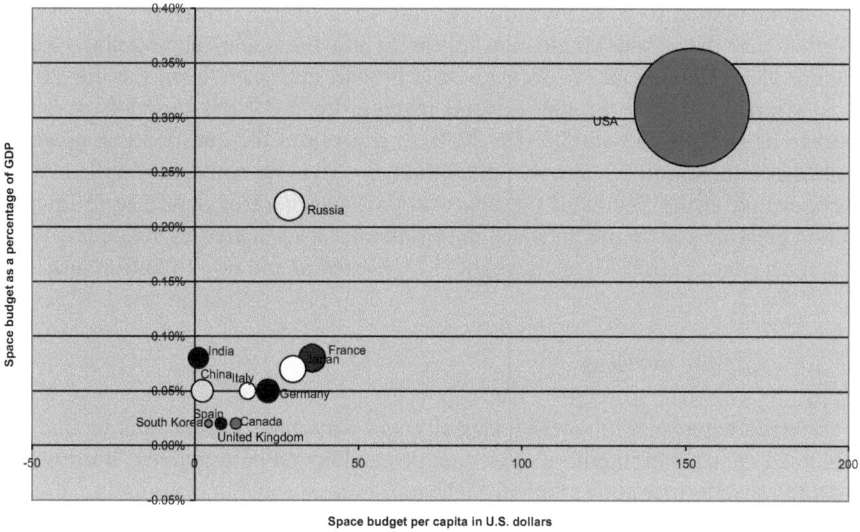

Fig. 1.4 Public space budgets as share of GDP mapped against space budgets per capita in 2011. The *bubble size* indicates the absolute space budget (Based on Space Report 2012 and Euroconsult data)

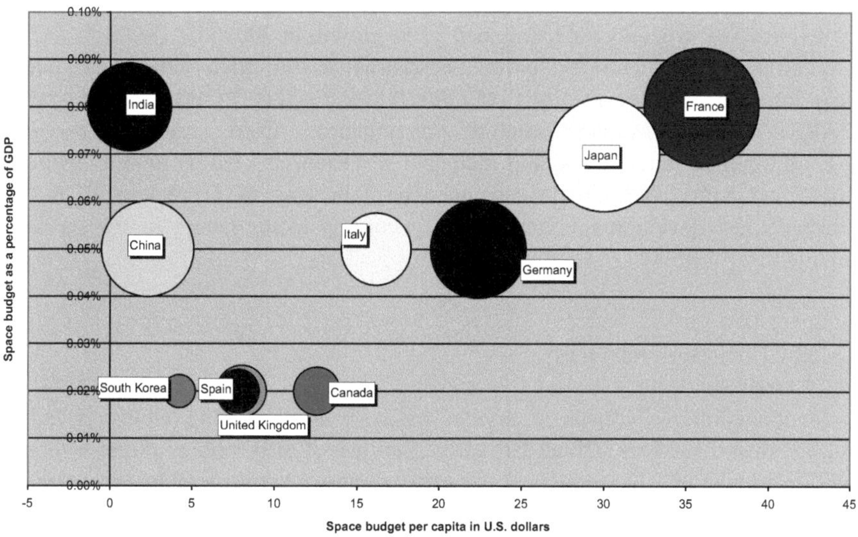

Fig. 1.5 Public space budgets as a share of GDP mapped against space budgets per capita in 2011, not including the U.S. and Russia. The *bubble size* indicates the absolute space budget (Based on Space Report 2012 and Euroconsult data)

revenue amounted to $216.99 billion for the year 2011.[137] However, it must be clarified that these sources do not appear to use the same methodology when making their assessments. In fact, revenue growth may have been low for 2011, as SIA reports that the growth of total revenue for 2010 fell short of the 2009 average of 11 %, eking out 5 % for 2010.[138] It is out of the question that revenue could have increased by almost $50 billion in 2011, as could be implied by a juxtaposition of the figures of the Space Report and those of SIA. The following section presents key figures and data on commercial space activities divided by field of activity, based primarily on available SIA figures in addition to Futron reports.

1.3.2.1 Satellite Services

A sustained expansion of satellite capacity and corporate revenue was fuelled by continued demand from emerging economies and developing regions. Worldwide satellite capacity rose above 900 36 MHZ transponder equivalents in 2011.[139] And despite the apparent slowing of growth within the satellite industry, satellite services maintained a consistent share portion among the other segments, including satellite manufacturing, launch services, and ground equipment; remaining at 60 % in 2010, the average since 2006. In 2010, satellite services exceeded $101.3 billion, due mostly to the increase in direct to home (DTH) satellite services. However, the rate of revenue growth for satellite services slightly decreased, experiencing 9 % growth in 2010, whereas it experienced 11 % growth in 2009.[140]

Satellite services can further be deconstructed into their component parts, including consumer services (e.g. satellite television (DTH/DBS), satellite radio (DARS), and consumer satellite broadband), fixed satellite services (e.g. transponder agreements, and managed services), as well as mobile services (voice and data), and remote sensing. The following is a breakdown of the industry's key developments and trends, according to the nature of the services provided.

1.3.2.2 Consumer Services

As mentioned above, consumer services are made up of satellite television, radio, and broadband services. Direct Broadcast Services (DBS) refer to direct-to-home satellite television. This section of the industry showed considerable development in 2010 and 2011, fuelled by the quantitative expansion in emerging markets and the qualitative increase in new technologies and services in developed ones.

---

[137] The Space Report 2012, 32.

[138] State of the Satellite Industry Report, 10.

[139] Satellite Telecommunications Report – 2011 Year-End Summary. Futron: 1.

[140] Id.

Demand in the latter manifested signs of recovery compared to the relatively flat revenue in 2010. In 2010, DBS revenue increased from $71.8 billion to $79.1 billion, an increase of 10.2 %. This trend was expected to continue into 2011 as the number of HDTV channels had been growing exponentially (e.g. by 42 % between 2009 and 2010).[141] By 2010, consumer satellite television services represented nearly 80 % of total satellite services revenue, with more than 70 % of the available HDTV channels servicing the Americas, and the remainder servicing Europe and Asia. Meanwhile, according to previous projections, the annual growth rate of DBS is expected to exceed that of the rest of the satellite services' sector as DBS will be replacing more traditional services such as video distribution.[142] By 2011, there were over 35 million satellite pay-TV subscribers in the U.S. Satellite radio revenue grew to $2.8 billion in 2010, from $2.5 billion the previous year (an increase of 12 %), while satellite broadband revenue grew by 10 %, from $1.0 billion in 2009 to $1.1 billion in 2010.[143] And the majority of worldwide broadband satellite revenue came from the U.S., generating 70 % thereof. All major satellite operators have increased their investments in new technologies and products, especially in developed markets.

1.3.2.3 Fixed Satellite Services

Fixed Satellite Services (FSS) refer to the use of spacecraft that utilise land terminals in fixed positions to broadcast. They include broadband internet, communications and network televisions and radio broadcasts. In 2010 and 2011, the FSS outlook remained buoyant, as operators continued to profit from previous investments in new capacity, as well as from the sustained demand for TV and broadband services.

While some reports anticipate a decrease in new satellite investment leading up to 2012–2013, the effect of the current boom in FSS is expected to continue throughout the decade. In 2010, industry-wide FSS revenue climbed 4.2 % in 2010, to $15 billion; while an increase, that amount was significantly reduced from the 10.8 % growth in 2009, amounting to $14.4 billion. The growth is explained by the insatiable demand for video and broadband, particularly from the Americas. The revenue generated by Eutelsat is a clear example of this upward trend with 2011 revenue climbing to €1.168 billion ($1.558 billion) and EBITDA to €926.4 million ($1.236 billion). These were increases of 11.2 % and 11.9 % from the previous year's revenue of €1.05 billion and €827.8 million respectively.[144]

---

[141] Id.

[142] 2010 Futron Forecast of Global Satellite Services Demand, 3.

[143] Id.

[144] Eutelsat Communications. Reverence Document 2010–2011 Including The Annual Financial Report. Paris 2011 http://www.eutelsat.com/investors/pdf/reference-document-1011.pdf.

To handle the increase in demand, commercial operators invested in technological upgrades that let them meet the needs of consumers while still making significant profits. For that purpose, a number of Ka-band satellite projects for broadband connectivity were announced in 2010 and 2011. However, transponder agreement revenue expanded only slightly between 2009 and 2010, with growth in 2010 masked partly due to a stronger U.S. dollar.[145]

Eutelsat began providing commercial broadband service with its Ka-Sat satellite on 31 May 2011.[146] This satellite is the first of a fleet of high throughput satellites under development by Astrium Satellites; and approximately 44,000 Surfbeam 2 terminals were already sent to the Netherlands in anticipation of the demand for broadband capacity. Ka-Sat distributes its bandwidth through 82 spot beams, each carrying 900 megabits per second of capacity. These beams are evenly distributed among points in Europe, and also with some points in Africa. However, the satellite lacks the onboard processing needed to maximize bandwidth and location flexibility. The coverage population is about 800 million; nearly three times that of its U.S. counterpart – ViaSat's WildBlue consumer broadband service. It is expected that European demand for Ka-Sat might be higher in suburbs than in rural areas, often the subject of government broadband stimulus programs.[147]

1.3.2.4 Remote Sensing

Remote sensing refers to commercial companies that provide optical and radar images to the open market; however, they are mostly used by government entities that have been increasingly outsourcing such capabilities over the past few years. With a considerable 37 % increase in 2009, commercial remote sensing revenue managed to reach the $1 billion benchmark in that year. However, remote sensing revenue experienced little growth in 2010, staying at that $1 billion value.[148] Government demand (particularly military demand) was the dominant factor boosting remote sensing services' providers' revenue; however the sector's private clients have been on the rise and contributed to the 2009 increase.

A total of 14 civilian remote sensing satellites were launched in 2011; an increase of 55.6 % compared to the nine remote sensing satellites launched the previous year (not including government-built, university-built, or research satellites).[149] Additionally, eight military surveillance satellites were launched in 2011; an increase of 28.6 % compared to seven the year before. While both civilian and military remote sensing satellites increased last year, it appears that the civil

[145] State of the Satellite Industry Report, 12.

[146] De Selding, Peter. "Ka-Sat Enters Services as European Broadband Market Heats Up." Space News 6 June 2011: 16.

[147] Id.

[148] State of the Satellite Industry Report, 13.

[149] Id. at 16.

industry is growing at twice the rate as military, indicating a shift in where government funding is directed, and suggesting that states are looking more toward PPPs in this field.

1.3.2.5 Mobile Satellite Services

Mobile satellite services offer both mobile data service and mobile voice service (including satellite phones). In 2010, mobile satellite services earned revenue of $2.3 billion, an increase of 4.5 % compared to $2.2 billion in the previous year. Within its segments, mobile voice services diminished by 9 %, whereas mobile data services revenue increased by 10 %. The latter segment comprises around 3/4 of all mobile satellite services revenue.[150]

In the U.S., opinions on the severity of LightSquared's interference with GPS signals shifted over the 2010–2011 period. LightSquared had developed plans to provide the U.S.'s first coast-to-coast hybrid wholesale wireless network. By June 2011, a 6-month study conducted by the Technical Working Group of the National Public Safety Telecommunications Council confirmed that interference to public-safety operations would occur.[151] This finding was especially damning since the U.S. Federal Communications Commission (FCC) gave conditional approval to LightSquared to deploy up to 40,000 base stations if it could demonstrate that it would not cause widespread interference with GPS. LightSquared is licensed to use two specific 10–MHZ blocks within the L-band frequency range; however GPS signals are also transmitted using the L-Band spectrum.[152] The ground-based transmissions by the former overpower the generally weak GPS signals from space. The consequence of such interference could result in aviation users losing GPS capabilities around densely populated areas with stations spaced out 400–800 m apart, and aircraft below 3,040 m in altitude could not rely on GPS in some states. Additionally, police cars could not acquire GPS signals within 182 m of a LightSquared tower that broadcasts at 15 kW.[153] Of the various mitigation options considered, the only viable option would be for LightSquared to acquire rights to another part of the electromagnetic spectrum.[154] Of note, LightSquared closed a $586 million additional loan to help fund the $7 billion project.[155]

---

[150] Id. at 13.

[151] De Selding, Peter. "LightSquared Suffers Setbacks on Two Fronts." Space News 20 June 2011: 5.

[152] Leone, Dan. "LightSquared Says Altered Plan Would Limit GPS Disruption." Space News 27 June 2011: 14.

[153] Brinton, Turner. "Reports: LightSquared Plan Poses Unacceptable Risk to GPS Service." Space News 13 June 2011: 1.

[154] Id. at 4.

[155] Warwick, Graham and Taverna, Michael A. "Tests Planned On Cell-Phone Jamming Of GPS." Aviation Week & Space Technology 28 Feb 2011: 48.

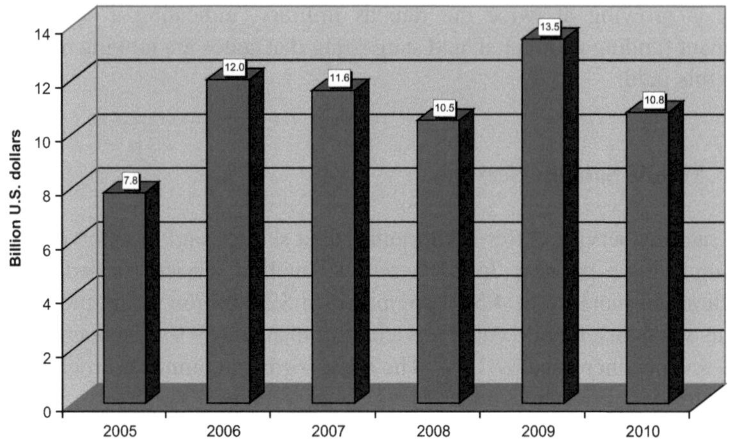

Fig. 1.6 World satellite manufacturing revenue (Source: SIA/Futron/Space Foundation)

1.3.2.6 Satellite Manufacturing

While satellite manufacturing revenue in 2011 is uncertain, it did not vary significantly from the previous levels in 2010 and 2009. The total revenue of satellite manufacturers that built satellites both for governmental and commercial launches was estimated to have reached $14.5 billion in 2010; in fact, however, it only reached $10.8 billion indicating a decrease of 20 % (instead of the 7 % increase) from the $13.5 billion in 2009.[156] Hence, Fig. 1.6 depicts a trend of decreased manufacturing revenue peppered with periods of increased revenue as in 2009 and 2006. It should be noted that despite the waning trend of revenue exhibited in Fig. 1.6, actual predictions forecast a regular increase of revenue to reach $15 billion in 2013.[157]

1.3.2.7 Commercial Space Launch

The year 2011 ushered in a total of 18 commercial launches (one commercial launch failed), with the objective of carrying 41 payloads into orbit, out of which 35 provided commercial services. While U.S. companies did not conduct any commercial launches, Russian companies again held the lion's share with a total of 10 launches composing 55.6 % of the total commercial launches. Moreover, there were four European launches at 22.2 %, and China had two commercial launches at 11.1 %, as did the multinational Sea Launch AG. The revenue from the 18 launches amounted to an estimated $1.9 billion, a decrease of 21.7 % or $526

[156] State of the Satellite Industry Report, 15.

[157] 2010 Futron Forecast of Global Satellite Services Demand: Executive Summary.

million from the previous year, and below 2009 revenue by $483 million. Due to the absence of U.S. commercial launches, where $307 million was generated in 2010, no U.S. commercial revenue was generated in 2011. European revenue was once again the highest, reaching $880 million (a 33.3 % decrease from last year), followed by Russian commercial launch revenue at approximately $707 million (a decrease of 14.4 %). Multinational revenue took the next position, generating approximately $200 million, and Chinese revenue amounted to around $140 million. The 21.7 % drop in industry revenue was a result of the decrease in net profits by all commercial launch service providers (not including Chinese or multinational commercial launches in 2011). However, these figures should only be considered as indicative of the sector as they do not depict current commercial launch contracts because contracts are typically prepaid 1–2 years prior to launch; instead they take into account the value of the past year's activity.[158]

Looking at European Launch activities in total, i.e. commercial and non-commercial, Arianespace expected a 10 % increase in revenue for 2011 compared with 2010, marking a slight profit after 2 years of losses.[159] In 2010, Arianespace earned €897 million ($1.197 billion), resulting in a decrease of €83 million ($110.7 million) in revenue. Thanks in part to the introduction of the Soyuz 2 in Kourou, Arianespace was able to avert loss for the third year in a row, making a small profit on revenue of €985 million ($1.275 billion). Four European launches were carried out by Arianespace onboard the Ariane 5 ECA launcher. Europe's Ariane 5 ES-ATV launcher made its second flight in February 2011, the maiden flight having been conducted in March 2008. This past launch carried the second ATV, Johannes Kepler, to the international space station, with three more missions planned in the coming years. The ATV-3, Edoardo Amaldi, was launched on 23 March 2012,[160] with upcoming ATV missions including the Albert Einstein in 2013, with the final ATV-5 scheduled in 2014.[161] Europe's spaceport in Kourou was outfitted to launch the Soyuz spacecraft in 2011. In that year, Arianespace launched two Soyuz 2 spacecraft, designated as Soyuz-ST. While marking a milestone in cooperation between Europe and Russia, this decision to develop the spaceport's launch infrastructure to launch the Soyuz benefitted both Europe and Russia, enhancing the competitiveness and flexibility of the launch systems available at the spaceport. Moreover, the Soyuz-ST launched the first pair of satellites for Europe's Galileo global navigation satellite system in October 2011; another

---

[158] Federal Aviation Administration. Commercial Space Transportation: 2011 Year in Review. Washington DC: FAA, Jan. 2012: 5.

[159] De Selding, Peter. "Arianespace Expects To Post 2011 Profit After 2 Years of Losses." 5 Jan. 2012. Space News 24 Apr. 2012 http://www.spacenews.com/launch/120105-arianespace-expects-profit.html.

[160] "ATV-3 launch." 23 Mar. 2012. ESA News 22 May 2012 http://www.esa.int/esaCP/SEM9UR2T00H_index_0.html.

[161] "Ariane 5 ES." ESA Launch Vehicles. 24 Apr. 2012 http://www.esa.int/esaMI/Launchers_Access_to_Space/SEM20W67ESD_0.html.

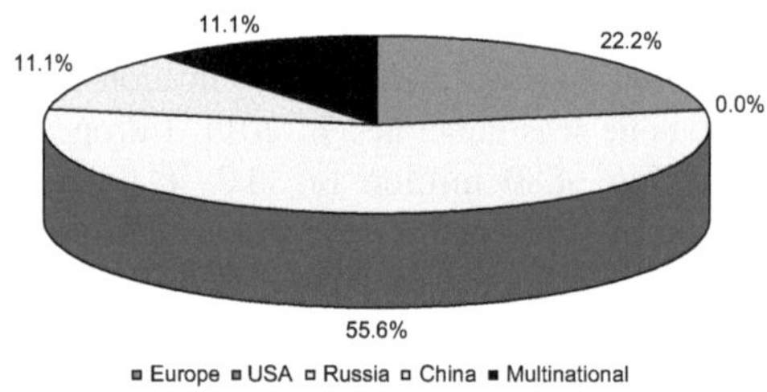

Fig. 1.7 Commercial launch activity by country in 2011 (Source: FAA)

hallmark of European and Russian commercial cooperation.[162] Europe's Vega Launcher made its debut launch on 13 February 2012. Vega can carry a 1,500 kg satellite into a 700-km orbit, This small satellite launcher will be priced commercially at around €32 million ($42 million) per launch, and is expected to be price-competitive with converted Russian ballistic missiles; if sufficient market demand exists, the price may drop to €22 million ($28.5 million) (Fig. 1.7).[163]

1.3.2.8 Ground Equipment

Ground equipment revenue include infrastructure elements, such as mobile terminals, gateways and control stations, and consumer equipment, such as very small aperture terminals (VSAT), ultra small aperture terminals (USAT), DTH broadcast dishes, satellite phones and digital audio radio satellite (DARS) equipment. Portable Navigation Devices (PND) form one of the sub-segments of end-user electronics incorporating GPS chip sets.

The PND market continued its expansion while still slowing down in comparison to 2008. The sharp increase of 34 % experienced in 2008 continued to fall below the 8 % growth reached in 2009 to the rate of 3 % in 2010 with a consolidated global turnover of $51.6 billion.[164] In 2010, ground equipment represented 31 % of world space business revenue maintaining its 3 % rate of growth in the global distribution of the previous year; it experienced slower growth in consumer equipment, while also experiencing accelerated growth in network equipment.[165]

The 2011 revenue profiles for the two companies leading the market parted ways. In 2010, TomTom announced its total revenue to be €1.521 billion ($2.016 billion), which decreased in 2011 to €1.273 billion ($1.648 billion)[166]; with total

[162] "One Soyuz Launcher, Two Galileo Satellites, Three Successes for Europe." 21 Oct. 2011. ESA – News. 24 Apr. 2012 http://www.esa.int/esaCP/SEM167GURTG_index_0.html.

[163] De Selding, Peter. "Vega Expected to be Price-competitive With Russian Rockets." 23 Jan. 2012. Space News 23 May 2012 http://www.spacenews.com/launch/012312-vega-expected-price-competitive-with-russian-rockets.html.

[164] State of Satellite Industry Report, 8.

[165] Id. at 9.

[166] "TomTom Annual Report and Accounts 2011." TomTom 24 Apr. 2012. http://files.shareholder.com/downloads/TOMTOM/1821069808x0x552704/24dcadbf-4781-4f88-81cc-84172c34e85e/Tomtom_JV2011.pdf.

revenue diminishing by 16.3 %. On the other hand, Garmin moderately rebounded from its 9.6 % drop in total revenue to $2.69 billion in 2010, rising to a total revenue of $2.76 billion (an increase of 2.6 %) in 2011.[167]

To better understand these figures, they can be divided into types of product and sale areas. Revenue in Garmin's automotive/mobile products in 2011 continued to decline to $1.591 billion from $1.669 billion in 2010; however, its 2011 segments related to marine, aviation and outdoor fitness either grew or stayed on par with 2010 resulting in steady growth. Meanwhile, TomTom increased its revenue in automotive systems by 31.3 % during the same period with global revenue of €235 million ($304.3 million) from €179 million ($237.20 million) in 2010. Geographically speaking, the growth of Garmin's sales in Asia slowed down compared to the 47 % increase in 2010; in 2011, revenue in Asia increased by only 12.5 % to $248.06 million from $220.49 million in 2010. Moreover, Garmin continued to lose revenue in the Americas, losing an additional 7.2 % in 2011, after losing 16.52 % in the prior year; i.e. 2011 revenue amounted to $1.527 billion in 2011 whereas it was $1.646 billion in 2010, and $1.972 billion in 2009. On the other hand, revenue in Europe increased by 19.5 % to $983 million in 2011, after holding steady at around $823 million in the two previous years.[168] TomTom was even less successful than its counterpart in North America with a decrease of 32.6 % in 2011, following the 7.54 % decrease in sales in 2010. Earning €256.6 million ($332.26 million) in 2011, down from €380.5 million ($504.22 million) in 2010, TomTom experienced a loss in revenue of €123.9 million. TomTom's revenue in Europe also decreased 12.4 %, from €1.07 billion ($1.42 billion) to €937.5 million ($1.21 billion). This loss in revenue followed an increase of 6.26 % in 2010. The rest of the world was the only sector that appeared to experience an increase in revenue with a growth rate of 12.3 %, earning €79.1 million ($102.42 million), up from €70.5 million ($93.42 million) in 2010; this increase in revenue is on top of the 12.9 % increase experienced in 2010.[169] While revenue growth in the Americas seems to be tapering off, Europe is in a state of flux between PND providers, with the most growth occurring in the Asia Pacific market (Table 1.1).

1.3.2.9 Insurance Sector

Insurance costs have steadily decreased as the space industry has continued to demonstrate increased hardware reliability, low accident rates and promising growth in recent years.[170] More competition results from a safer market, which

[167] "Garmin Annual Report 2011." 24 Apr. 2011. http://www8.garmin.com/aboutGarmin/invRelations/reports/2011_Annual_Report.pdf.

[168] "Garmin Ltd. Form 10-K for Period Ending 12/31/2011." 24 Apr. 2012 http://www8.garmin.com/aboutGarmin/invRelations/reports/xbrl/Fourth_Qtr_2011_view/index.htm.

[169] TomTom Annual Report and Accounts 2011, 54.

[170] De Selding, Peter. "Insurance Premiums Stay Flat Despite W3B Satellite Failure." 20 May 2011. Space News 24 Apr. 2012 http://www.spacenews.com/satellite_telecom/110520-insurance-premiums-flat.html.

Table 1.1 Understanding TomTom & Garmin variables

	Total revenue	2010	2011
TomTom		€1.521 billion ($2.016 billion)	€1.273 billion ($1.648 billion)
Garmin		$2.69 billion	$2.76 billion
	Products		
TomTom	Automotive systems	€179 million ($237.20 million)	€235 million ($304.3 million)
Garmin	Automotive/ mobile	$1.669 billion	$1.591 billion
	Geographical sales		
TomTom	Europe	€1.07 billion ($1.42 billion)	€937.5 million ($1.21 billion)
	North America	€380.5 million ($504.22 million)	€256.6 million ($332.26 million)
	Rest of world	€70.5 million ($93.42 million)	€79.1 million ($102.42 million)
Garmin	Europe	$822.84 million *($824.07 million in 2009)*	$983.00 million
	Americas	$1.646 billion *($1.972 billion in 2009)*	$1.527 billion
	Asia	$220.49 million *($149.92 million in 2009)*	$248.06 million

results in a change in consumer behaviour as well. Insurance is now procured 2 or 3 years before the launch of a spacecraft for the purpose of reducing the price. The coverage rates for space insurance have continued to decrease throughout the 2000s resulting in the withdrawal of actors from this market. This has led some insurers to be alarmed as to the fragile market that has resulted from the lower premiums and fewer market actors. With annual premiums for satellite insurance policies averaging between $800 and $900 million for the past several years, the double payload capacity of the Ariane 5 ECA has insurers nervous. In 2011, an Ariane 5 carried two satellites in a single launch, whose combined insurance was more than $700 million, over 80 % of the total premium revenue for the year. Had that launch failed, it would have taken 80 % of total premium revenue with one stroke.[171] Despite the risk, new underwriters keep entering the space insurance market expecting to make money on satellite coverage based on the recent track record. However, the decline in premiums, in addition to the potential fragility of the market has led some larger insurance brokers, e.g. London's Liberty Syndicate 4472 (associated with Lloyd's), to avoid launch coverage completely, focusing only

[171] De Selding, Peter. "Falling Satellite Insurance Premiums Put Market at Risk of Major Upheaval." 2 Mar. 2012. Space News 24 Apr. 2012 http://spacenews.com/satellite_telecom/120302-falling-sat-insurance-premiums-market-risk.html.

on the less risky in-orbit policies. With very narrow profit margins based on low rates, a loss of more than $1 billion in a single year would be very likely to drive up rates.[172] While the low rates experienced in the past decade are appealing and are expected to fall further, the consequences of a future launch failure would create an inverse response.

1.3.3 Developments in the Space Industry

1.3.3.1 Industrial Developments in Europe

The Arianespace commercial launch company expected to post a 10 % revenue increase in 2011 and to report a slight profit after 2 years of losses. The company's backlog stood at an unprecedented combined amount of €4.5 billion ($5.9 billion) for its heavy-lift Ariane 5 series and medium-lift Soyuz 2 ST vehicles. This was composed as follows: €2.5 billion ($3.25 billion) towards commercial launches aboard the Ariane 5, €1 billion ($1.3 billion) for government launches aboard the Ariane 5, and €1 billion ($1.3 billion) in backorders for Soyuz launches. Whereas in 2011, the Soyuz 2 ST conducted two launches, three Soyuz launches from French Guiana were expected in 2012, with two additional commercial launches conducted from its original Baikonur location. Meanwhile, Arianespace scheduled seven launches of the Ariane 5 series launch vehicle, and conducted the inaugural launch of the Vega small-satellite launcher in 2012. While Arianespace earned €897 million ($1.16 billion) in 2010, amounting to a fall of €83 million ($107.9 million) in its revenue; with the introduction of the Soyuz 2 in Kourou, Arianespace was able to avert a loss for the third year in a row, making a small net profit on its €985 million ($1.28 billion) in revenue. In 2011, following an audit of Arianespace and its contractors (a condition for injecting about €240 million ($312 million) into Arianespace over 2 years), ESA determined that additional savings related to the Ariane series would not be possible without wholesale restructuring. ESA released €217 million ($282 million) allocating half to Arianespace's 2011 accounts, with the remaining funds to be made available in 2012.[173]

Eutelsat expects its total revenue to grow by more than 7 % per year on average through 2014, mainly from selling short-term capacity to the U.S. Defence Department. But the company's government business is not expected to continue its accelerated pace. Additionally, Eutelsat's video business in the Balkans and its data-transmission business in parts of Africa will also face increased competition in both regions. In the second half of 2011, Eutelsat's multi-usage revenue increased by 29.9 %, to €74.4 million ($97 million), that amount accounted for 12.5 % of

[172] Id.

[173] De Selding, Peter. "Arianespace Expects to Post 2011 Profit After 2 Years of Losses." Space News 9 Jan. 2012: 10.

Eutelsat's total revenue for the period. The operator reported revenue for the 6 months of €602.4 million ($780 million), up 4.6 % over the same period a year earlier, while its earnings before interest, taxes, depreciation and amortization (EBITDA) were 79.4 % of revenue. About 68 % of Eutelsat's revenue came from video broadcasting, where the number of television channels carried on Eutelsat's fleet rose by 10 %, to 4,173 in 2011. Eutelsat's 7 % annual growth rate through 2014 will be accomplished at an EBITDA margin that averages 77 % of revenue.[174]

On 12 January 2012, the Spanish telecommunications infrastructure provider Abertis Telecom announced it would sell half its stake in Eutelsat for a price of about €1 billion ($1.3 billion). Abertis Telecom, Eutelsat's biggest shareholder, had a 31.4 % equity stake in Eutelsat. Credit Suisse, Morgan Stanley and Society General would manage the accelerated placement sale of 16 % of Eutelsat's share capital, valued at €29.75 ($37.88) per share. While no explanation was given for the expedited disposal of such a large stake, some suspect Abertis sought to raise its cash liquidity to avoid becoming an acquisition target for its competitors.[175]

Hispasat reported a 7 % increase in EBITDA over 2010. In 2011, the company reported revenue of €187.5 million ($246.7 million), an increase of 3.4 % over 2010. Its EBITDA increased to 82.5 % of revenue (up 2.8 points from 79.7 % a year earlier), and its consolidated operating profit was €70.6 million ($92.9 million), an increase of 2.2 % from 2010. As observed by the Hispasat CEO, "the results obtained in 2011 confirm the validity and strength of the business model established by HISPASAT, which is capable of increasing operating revenue even in such a complex context like the one we experienced last year, marked by the economic crisis." Of the total earned revenue in 2011, €182.4 million ($240 million) was derived from revenue from leasing space capacity, with 50 % generated from clients located in Europe and 49.1 % from clients in the Americas, where HISPASAT has consolidated a strong presence in the region, holding a significant market share. The remaining 0.9 % came from other regions. Hispasat's growth is mainly due to the commercialization of capacity from the Amazonas 2 and Hispasat 1E satellites.[176]

Telenor Satellite Broadcasting of Norway reported a 7.6 % decrease in revenue for the year ending 2011; much of the drop was said to be caused by the increased strength of the Norwegian krone. Another cause for this depressed revenue was Oslo-based Telenor's renegotiation of its contracts with its biggest customer, Canal Digital. Their agreement consolidated multiple fixed price contracts into a single contract that featured volume discounts. Telenor reported revenue of 998 million kroner ($165.7 million) for the 12-month period, down 7.6 % from the same period

[174] De Selding, Peter. "Eutelsat Forecasts Slower Pentagon Sales Growth." Space News 20 Feb. 2012: 12.

[175] "Abertis Telecom To Sell Half of its Eutelsat Stake." Space News 16 Jan. 2012: 3.

[176] "HISPASAT reaches a new maximum for revenue in 2011 of 187.5 million euros." Noticias hispasat acercando culturas 3 May 2012 http://www.hispasat.com/media//395-Resultados%202011_EN.pdf.

in 2010. Its EBITDA was 66 % of revenue, down from 70 % in 2010. In addition to serving the Nordic market, Telenor is seeking to broaden its customer base by operating in Central and Eastern Europe as a cheaper alternative to Eutelsat and SES, while also marketing to maritime consumers by adding a Ka-band payload to its Thor 7 satellite.[177]

Despite filing for bankruptcy protection after breaching several of its loan covenants, RapidEye AG remained confident in the continued support of its three main financiers, which have already expressed willingness to renegotiate the terms of the current loans. A leader in privatized satellite imagery, RapidEye has selected this procedure to place itself on a firmer financial footing and pursue its long-term business goals. With few creditors financing the company, the debt reorganisation procedure is less complex than with multiple creditors. RapidEye's main creditors KfW and Commerzbank from Germany, and Canada's Export Development Corp, provided about half of the €160 million ($232 million) that RapidEye raised to start its business. An additional €15 million ($21.6 million) came to RapidEye in the form of a grant from the Germany Space Agency (DLR); repayable over 5 years in the form of imagery. A second more substantial grant, totalling €137 million ($181.5 million), came from the Brandenburg region of Germany in return for RapidEye creating 130 jobs there for at least 5 years.[178]

Europe's Astrium space hardware and services provider reported flat revenue and lower pre-tax profit for 2011, along with lower services revenue which offset increases in satellite manufacturing and launch-vehicle sales. Astrium's revenue for 2011 totalled €4.96 billion ($6.4 billion), decreasing slightly from 2010. Pre-tax profit was 5.4 % of revenue, down from 5.7 % in 2010. The reduction in pre-tax profit was partly related to its acquisition of mobile satellite services provider Vizada; a €745 million ($965 million) transaction which closed in December 2011. Astrium has three principal divisions that contribute to revenue: Astrium Services, which sells Earth observation imagery and satellite communication services to government customers; Astrium Satellites, a major satellite builder; and Astrium Space Transportation, Europe's biggest contractor for the international space station, and the prime contractor for Europe's Ariane 5 heavy-lift rocket and French ballistic missiles. Astrium Services revenue totalled €861.8 million ($1.13 billion), down 12.7 % from 2010; however, revenue was expected to increase by 60 % with the acquisition of Vizada. Astrium Satellites revenue totalled €2.15 billion ($2.8 billion), up 12.2 %, following the delivery of 13 satellites, including Pleiades 1A and two Galileo navigation spacecraft. And Astrium Space Transportation reported revenue of €2.18 billion ($2.86 billion), up by about 4 % from 2010, after delivering the second Automated Transfer Vehicle and five Ariane 5 rockets

---

[177] "Telenor: Exchange Rates Contribute to Revenue Drop." Space News 13 Feb. 2012: 8.

[178] De Selding, Peter. "Earth Imagery Firm RapidEye Seeking Bankrupcy Protection." Space News 6 June 2011: 6.

during the year. Additionally, these divisions booked a combined total of €3.5 billion ($4.59 billion) in new orders in 2011.[179]

The Astrium space hardware and services division planned to eliminate 1,000 Astrium positions and 1,000 subcontracting positions from its pool of 17,000 employees. Described as a part of Astrium's streamlining effort, the 2015 goal is to create a company that is leaner in middle management and bureaucracy, eliminating jobs that do not directly contribute to Astrium's product and service portfolio. The company will reinvest the €400 million ($560 million) in annual savings resulting from employment streamlining into self-financed research and offering better prices to its customers. Astrium's Agile programme, an internal transformation programme designed to prepare the company to adapt to the competitive market, has been operating in an environment where, for instance, the currency exchange rate between the Euro and the Dollar has made it very hard for Europe's Ariane 5 rocket to make a profit, despite 44 consecutive launches without failure. With the Euro strong, streamlining is Astrium's best option if it is to make a profit and win business in the coming years. Astrium's workforce is dispersed among several European states with the following distribution: 43 % of Astrium's workforce is in France, 27 % in Germany, 21 % in Britain, 5 % in Spain, 1 % in Netherlands, with the final 3 % dispersed over other countries.[180]

Astrium Geo-Information Services increased the competition for its two U.S. competitors, GeoEye and DigitalGlobe in early 2012 when its Pleiades 1A high-resolution optical Earth observation satellite entered service. Launched on 17 December 2011, this lone satellite has a twin, Pleiades 1B, that was expected to join it in orbit before March 2013. In the meantime, the French government was testing Pleiades 1A's performance and capabilities. Built by Astrium Satellites and Thales Alenia Space under contract to the French Space Agency (CNES), the 1A was ready to launch aboard the Soyuz 2 ST in June 2011 at the European launch base in Kourou, but the owners decided against launching the 1A on the Soyuz 2 ST's first mission, resulting in a 6-month delay before the next launch opportunity arose. GeoEye and DigitalGlobe had jumped ahead in mid-2010 after securing a joint 10-year, $7.3 billion contract with the U.S. National Geospatial-Intelligence Agency (NGA). However, at a cost of more than €760 million ($1 billion), covered 90 % by CNES, the French government soon will have access to an output of 900 images per day once both satellites are in near-polar low Earth orbit position. Of that share, the French Ministry of Defence has priority access to 50 images per day, while civil agencies have access to 40 % of the remaining output; Astrium Geo-Information Services gets the remaining 60 % of CNES's share. As the rest of the bill was covered by Belgium, Spain, Sweden, and Austria; these states have

[179] De Selding, Peter. "Drop in Services Business Keeps Astrium Revenue Flat." 8 Mar. 2012 Space News 3 May 2012 http://spacenews.com/satellite_telecom/120308-astrium-revenue-flat.html.

[180] De Selding, Peter. "Astrium Efficiency Initiative to Eliminate 2,000 Positions by 2015." Space News 27 June 2011: 12.

access to a pro rata share of the remaining 10 % of imagery and work for their respective national industries.[181]

Thales Alenia Space, owned 67 % by Thales and 33 % by Finmeccanica, has 7,500 employees in France, Italy, Spain, Belgium, Germany and the United States. It posted total revenue of €2.1 billion ($2.72 billion) in 2011.[182] Thales, the 67 % share holder of Thales Alenia Space, reported a slight decrease in revenue, earning €13.028 billion ($16.869 billion) in 2011 compared to €13.081 billion ($16.938 billion) in the previous year. Registered in both France (Thales Alenia Space SAS) and Italy (Thales Alenia Space Italia SpA),[183] Thales Alenia Space is a key supplier of satellite and orbital infrastructure solutions. Thales Alenia Space is thus a global market leader in telecommunications, navigation, space exploration and Earth observation.[184] In telecommunications, it competes in the commercial satellite market, producing its own satellites from its Spacebus platform dedicated to geostationary satellites, while supplying additional payloads for all the major contractors in the sector, and providing satellites for the low-orbit civilian constellation market. In the defence and security segment, it offers space segments and ground telecommunications systems (Syracuse, Sicral, COMSAT-BW), military observation systems (Helios, Pleiades, CSO – rest of Helios, SAR-LUPE), and also dual civilian and military systems (COSMO-SkyMed, Athena-Fidus, Yahsat). Thales Alenia Space is also a major player in Earth observation and scientific missions. For the last three decades, it has been the prime contractor for weather satellite programmes in Europe (METEOSAT) and for environmental missions in the context of GMES (Sentinel programmes) or space altimetry. And for scientific missions, it developed Proteus, a multi-mission platform (class 500–700 kg), used in the SMOS, Jason, COROT and Calipso missions. Thales Alenia Space also plays a major role in the Galileo programme, participating in the In-Orbit Validation (IOV) phase, as well as providing system support and the Mission Ground Segment for the full constellation. In manned space flight, Thales Alenia Space is a major contributor to the ISS, supplying more than 50 % of its pressurised volume (Nodes 2 and 3, MPLM, Cupola, PMM) and is significantly involved in the ATV vehicles for ESA.[185] On the transaction side, in October 2010 Thales Alenia Space sold its stake in Indra Espacio to Indra Sistemas for €39.2 million, resulting in a €13.8 million profit on disposal.[186]

OHB Technology of Germany expected its total revenue to rise to over €600 million ($780.8 million) for fiscal year 2011. According to Reuters I/B/E/S

[181] De Selding, Peter. "With Pleiades in Orbit, Astrium Sets Sights on DigitalGlobe, GeoEye." Space News 9 Jan. 2012: 7.

[182] Thales Alenia Space 17 May 2012 http://www.thalesgroup.com/Markets/Space/Related_Activities/Thales_Alenia_Space/.

[183] 2011 Registration Document. Thales. 23 May 2012 http://www.thalesgroup.com/Group/Investors/Documents/Releases_and_Publications/2012/2011_Registration_Document/.

[184] 2010 Registration Document. Thales 23 May 2012 http://www.thalesgroup.com/Workarea/DownloadAsset.aspx?id=15726&LangType=2057.

[185] See generally 2011 Registration Document, 132.

[186] 2010 Registration Document, 41.

estimates, OHB was expected to report EBITDA of €41.77 million ($54.3 million) and EBIT of €29.25 million ($38 million) for the 2011 fiscal year.[187] During the period OHB Technology purchased the Space System division of the Swedish Space Corporation (SSC); yet again branching into another market after signing an agreement with China in 2011 in order to develop a small constellation to measure atmospheric levels of methane and carbon dioxide.[188] The transaction, concluded for 1 Swedish krona, was an asset deal whereby OHB agreed to assume the risk associated with the Swedish division's future performance. SSC is a for-profit government enterprise that will act as a subcontractor for its previously held Space System division on the OHB-led Small-Geo satellite platform. In recent years, the Space Systems division has returned revenue of about €10 million ($14 million) per year, while employing around 53 employees that will be designated under the organisational title of OHB Sweden. As this transaction occurred with another ESA Member State, by virtue of ESA's geographic return rule, the German-owned OHB Sweden will be entitled to benefit from ESA opportunities arising from Swedish space spending.[189]

RUAG Space, the largest independent supplier of space technology in Europe, developing subsystems and equipment for satellites and launch vehicles, reported a 2.8 % decrease in net sales for 2011, earning CHF 275 million ($292.6 million) compared to CHF 283 million ($300.8 million) in 2010. Nevertheless, RUAG Space increased its EBIT by 46 %, from CHF 9 million ($9.6 million) in 2010 to CHF 13 million ($13.8 million) in 2011; and its EBITDA in 2011 amounting to CHF 33 million ($35.1 million), an increase of 10 % from CHF 30 million ($31.9 million) in 2010.[190] Based in Switzerland, Sweden, and Austria, the company employs 1,113 personnel.[191] Structures and separation systems for launch vehicles again had the highest sales volume, with RUAG Space's payload fairings being used in eight launches of Ariane 5 and Atlas V launch vehicles.[192] The company is also involved in the "Pleiades" Earth observation satellite programme; the first satellite having been launched in December 2011.[193]

[187] "Financials: Ohb AG (OHBG.DE)." Reuters 3 May 2012 http://www.reuters.com/finance/stocks/financialHighlights?rpc=66&symbol=OHBG.D.

[188] De Selding, Peter. "OHB Purchases SSC's Space System Division." Space News 27 June 2011: 11.

[189] De Selding, Peter. "Astrium Efficiency Initiative to Eliminate 2,000 Positions by 2015." Space News 27 June 2011: 12.

[190] RUAG Annual Report 2011. RUAG 9 May 2012: 98 http://www.ruag.com/de/Konzern/Media/Geschaeftsberichte/2011/2011_e/RUAG_GFB_2011_E.pdf.

[191] Id. at 26.

[192] Id. at 24.

[193] Id.

1.3.3.2 Industrial Developments in the United States

Following a lack of commitment from the U.S. and other prospective government and commercial customers, Intelsat and MDA Corp. rescinded their agreement to collaborate on a satellite in-orbit refuelling system. Intelsat, the largest commercial satellite fleet operator, had agreed to invest $280 million in MDA's first mission for its proposed Space Infrastructure Service (SIS) business. Yet, even with the credibility gained from Intelsat backing, it was insufficient to coax in additional customers. Bad timing seemed partially to blame – while U.S. Air Force officials indicated interest in the idea, their current satellite fleets were undergoing replacements that had already been financed, leaving little funding for a program such as in-orbit fuelling. MDA is not wholly abandoning the SIS idea, but will also bid for an upcoming U.S. government contract calling for proposals on satellite-repair missions.[194]

With controversy swirling around LightSquared, and the potential interference its wholesale broadband service may have on highly sensitive GPS receivers, LightSquared is seeking to shift the burden of resolving the matter to others. Instead of switching from operating on the L-band to another frequency, LightSquared has urged stakeholders in the GPS industry to build better receivers that do not receive signals beyond their intended function. This debate has made officials hope that technical standards can be established that will allow companies developing broadband networks to tailor their systems so as to avoid disrupting GPS applications. Thus the U.S. Deputy Transportation Secretary will work with the Commerce Department's National Telecommunications and Information Administration (NTIA) to develop GPS spectrum standards in consultation with stakeholders in industry. However, those standards would be crafted so as not to affect emerging GPS applications that are vital for the economy, public safety and national security.[195]

The Stratolaunch project, seeking to develop the largest air-launch system in the world, has confused the industry with regard to its prospects and purpose. Unveiled on 13 December 2011 by billionaire Microsoft co-founder Paul Allen, the company's aim is to bring airport-like operations to the launch of commercial and government payloads and eventually conduct human missions. The carrier aircraft, operating from a large airport or spaceport, will be able to boost up to 6,100 kg of payload to low Earth orbit by the flying launch of a booster weighing up to 222,000 kg. However, with the large number of start-up commercial launch providers, some question whether the venture will be able to capture sufficient business to turn a profit, while recouping the costs associated with the air-carrier and re-useable mother-ship payloads. While the company would not go into detail

[194] De Selding, Peter B. "Lack of Interest Dooms MDA, Intelsat In-Orbit Servicing Deal." Space News 23 Jan. 2012: 7.

[195] Ledbetter III, Titus. "U.S. Officials Mull Standards in Wake of LightSquared Controversy." Space News 13 Feb. 2012: 6.

regarding its target customers, the prospective payloads were said to include small commercial satellites currently hitching rides to orbit as secondary payloads.[196]

GeoEye signed a multimillion-dollar deal with ScanEx, a Russian company, to complete a national map of Russian land properties. In the first stage of the agreement, GeoEye will provide ScanEx with previously acquired imagery from its GeoEye-1 high-resolution optical Earth observation satellite. In the next stage, starting in 2012, GeoEye will provide ScanEx with new imagery to supplement the creation of the nationwide map. ScanEx has participated in a series of similar contracts signed with GeoEye and with Europe's Astrium Services, working under contract for the Russian Federal Service for State Registration, Cadastre and Cartography to provide maps of all Russian localities. The company has already mapped 49 million land parcels and placed them in a Web-based system for use by local government agencies and the public; averaging 12,000 visits per day.[197]

Space Exploration Technologies (SpaceX) will launch commercial telecommunications satellites, AsiaSat 6 and AsiaSat 8, in 2014. AsiaSat of Hong Kong ordered the development of the two satellites from Space Systems/Loral of Palo Alto, California, to increase the capacity of the current AsiaSat fleet in orbit. Meanwhile, SpaceX is currently retooling its Falcon 9 rocket for missions to the geostationary orbit used by most telecommunications satellites. SpaceX now has two customers in Asia following its additional 2011 contract with Thaicom of Thailand. Interestingly, Thaicom is also paying AsiaSat $171 million over 15 years for half of the capacity of AsiaSat 6. That satellite, located at 120° east longitude and carrying 28 C-band transponders, permits Thaicom to retain its rights to the 120° slot that otherwise would have expired under International Telecommunication Union rules. Its share will be rebranded as Thaicom 7 for marketing purposes. Commercial launches now represent over 60 % of SpaceX's upcoming missions. The company has contracted with SES of Luxembourg for the launch of a single commercial telecommunications satellite in 2013 pending upgrade tests of the Falcon 9. And SpaceX will also launch the Iridium NEXT constellation of LEO communications satellites starting in 2014.[198]

SpaceX is developing a new launch pad at Vandenberg U.S. Air Force Base, in California. It plans to use that facility to launch the standard Falcon 9 rocket and its upcoming heavy-lift variant. SpaceX currently launches its Falcon 9 only from Canaveral Air Force Station, whereas the new pad will be built to also accommodate the developmental Falcon 9 Heavy launch vehicle, to debut in 2013 at the earliest.[199] The company will spend between $20 million and $30 million to renovate the site, unused since 2005; the Titan 4 was the last rocket launched from there. SpaceX will also update its launch facility in Cape Canaveral, allowing the heavy-lift rocket to launch from both coasts. Using the Falcon 9 Heavy

[196] Leone, Dan. "Stratolaunch Charts Course for a Tough Market." Space News 16 Jan. 2012: 6.

[197] "GeoEye Wins Contract for Mapping Russian Properties." Space News 9 Jan. 2012: 8.

[198] De Selding, Peter B. "SpaceX To Launch Pair of Satellites for AsiaSat." 8 Feb. 2012. Space News 3 May 2012 http://www.spacenews.com/contracts/120208-spacex-launch-asiasat-sats.html.

[199] "SpaceX Breaks Ground on West Coast Launch Pad." Space News 25 July 2011: 8.

Launcher will cost between $80 million and $125 million; [200] whereas launching low-end versions of the Delta 4 generally costs around $200 million.[201]

SpaceX is also expanding its Florida base in other respects, building additional hangars to prepare its Falcon 9 rockets and customer payloads for launch. Of the more than 40 flights on its manifest, valued at $3.5 billion, from the U.S. government, commercial and international customers, 40 % is for NASA, flying cargo to the ISS beginning in 2012. With an expected flight rate of 10–12 launches per year, the current facility (Space Launch Complex 40) will receive a 16,000 m^2 addition, including an unused Delta 2 processing building. SpaceX will receive $7.3 million from Space Florida, a state-funded agency, toward the upgrades. These upgrades include, *inter alia*, a clean room, a hazardous hypergolic fuelling facility and enough volumetric space to encapsulate a payload in a fairing in a vertical position.[202]

SpaceX had its first Falcon 9 rocket launch to the International Space Station (ISS) on 22 May 2012.[203] The date allowed NASA to check SpaceX's flight software, while also avoiding any conflict with the deployment of three new crew members, launched from Baikonur, Kazakhstan on 14 May 2012. A successful demonstration would mark the first commercial flight to ISS. Prior to latching and berthing onto the station, the launcher's Dragon capsule must successfully conduct a series of manoeuvres; if the capsule is unable to hook up with the station, another demonstration test flight would be scheduled. Once successful, the supplies would be unloaded and afterwards the capsule would detach, re-enter Earth's atmosphere, and land in the Pacific. If successful, SpaceX may then start conducting commercial resupply missions to the ISS under the terms of a $1.6 billion contract with NASA.[204]

Virgin Galactic planned to conduct its first rocket-powered test flight of SpaceShipTwo (SST) by the end of 2012. The spacecraft had already performed 16 unpowered flight tests, involving drop tests where SST would glide back to a runway after being dropped in midair from its mother ship, WhiteKnightTwo, and performing a set of manoeuvres along with ballast checks. Designed to be released from its carrier at an altitude of 15 km, its SSTs rocket engine will ignite to propel the craft at an apex altitude of about 100 km. Should this flight proceed as expected, Virgin Galactic may be able to start flying customers sometime between 2013 and 2014. Nearly 500 passengers have already signed up, each willing to spend

---

[200] Associated Press. "SpaceX Breaks Ground on California Launch Pad." 13 July 2011. ABC News 3 May 2012 http://abcnews.go.com/Technology/wireStory?id=14059673.

[201] De Selding, Peter B. "Boeing, ULA Wrangle with Air Force Over Delta 4 Launch Contract Prices." 29 Apr. 2011. Space News 9 May 2012 http://www.spacenews.com/military/110429-boeing-ula-wrangle-af-delta-prices.html.

[202] Klotz, Irene. "SpaceX Expanding Florida Facilities to Meet Launch Demand." Space News 28 Nov. 2011: 5.

[203] Space News Staff. "SpaceX Delivers Falcon 9 to Orbit." 22 May 2012. Space News 23 May 2012 http://spacenews.com/launch/120522-spacex-falcon-delivers-dragon-orbit.html.

[204] Boyle, Alan. "SpaceX station launch set for May 19." 4 May 2012. Cosmic Log on MSNBC. com 3 May 2012 http://cosmiclog.msnbc.msn.com/_news/2012/05/04/11543790-spacex-station-launch-set-for-may-19?lite.

$200,000 for the flight. Among these customers are scientists along with their experiments and space tourists. Soon, other companies might also be able to provide similar suborbital space travel products, including *inter alia* XCOR Aerospace, Blue Origin and Armadillo Aerospace.[205]

Despite experiencing weakness in the mobile satellite services market for usage of its hand-held satellite telephone, Inmarsat was expected to have record revenue growth in 2011 thanks to cash coming in from the struggling U.S. wireless-broadband start-up, LightSquared. LightSquared secured a broad portion of the L-band spectrum to provide its service by paying Inmarsat to modify its use of the same spectrum and to retrofit certain Inmarsat hardware to avoid future interference. Receiving over $268.1 million in 2011 (total funds received amount to $420.6 million), $152.8 million was to be recognized as revenue for 2011. Inmarsat's FleetBroadband maritime product has been successful beyond Inmarsat's forecasts, adopted by 23,500 ships worldwide, and its hand-held satellite telephone has met company forecasts for new subscribers with at least 10 % of the market. Yet with slowly developing usage and revenue rates, the company is investing in additional capacity, i.e. $1.2 billion in a next-generation, Ka-band service called Global Xpress, and also spending $113 million to purchase Ship Equip of Norway.[206]

1.3.3.3 Industrial Developments in Russia

Sea Launch AG is back in operation after coming out of U.S. Chap. 11 bankruptcy, gaining a Russian owner, and moving its registered headquarters from California to Switzerland. Coming out of a 30-month break, Sea Launch has returned to the market debt-free, while remaining a private international venture without privileged access to any government satellite markets. With total operating and financing costs of no more than around $50 million per year, the company will break even at three launches annually, and it aims to reach its target launch rate of five launches per year by 2014. It conducted a successful launch of Eutelsat's Atlantic Bird 7 Satellite in September 2011, followed by a subsequent land launch of the Intelsat-18 satellite on a modified version of the Sea Launch vehicle in October.[207] After having secured the backing of both Intelsat and Eutelsat (the world's first and third largest commercial satellite fleet owners), with additional

---

[205] Wall, Mike. "Virgin Galactic Aims for 1st Rocket-Powered Flight This Year." 28 Feb. 2012. SPACE.com 3 May 2012 http://www.space.com/14706-virgin-galactic-spaceshiptwo-powered-flight.html.

[206] De Selding, Peter B. "LightSquared Deal a Boon To Inmarsat's Otherwise Flat Business." Space News 7 Nov. 2011: 16.

[207] Bergin, Chris. "Zenit 3SLB launches successfully with Intelsat-18 satellite." 5 Oct. 2011. NASAspaceflight.com 2 May 2012 http://www.nasaspaceflight.com/2011/10/live-zenit-2slb-launch-intelsat-18/.

satellites scheduled to be launched in the upcoming years, Sea Launch is looking to provide launch services to SES (the second largest fleet owner) as well.[208]

1.3.3.4 Industrial Developments in Japan

NEC Corp. of Tokyo will design and build hardware for Japan's Hayabusa-2 asteroid sample-return mission, which JAXA plans to launch in 2014. The probe will be similar in design to the original Hayabusa spacecraft, weighing 600 kg when fully fuelled and it will be fitted on a larger satellite platform than its predecessor. The Hayabusa-2 will hold a more powerful sample collection system that will attempt to dig a crater in the asteroid to bring a bigger cache of samples back to Earth. Moreover, the Hayabusa-2 will be equipped with a Ka-band communications subsystem that will be faster than the original Hayabusa's X-band system, and it will carry a sophisticated camera to better capture the shape and the geography of the asteroid. The ¥16 billion ($204 million) spacecraft will be designed to visit 1999 JU3, a 920-m-diameter carbonaceous asteroid in a similar orbit to Itokawa. Carbonaceous asteroids are plentiful, rocky, and thought to contain water and organic materials, whereas asteroids like Itokawa are stony, and are thought to lack the same organic materials.[209]

Mitsubishi Electric Co. of Japan plans to double its annual satellite-related revenue to ¥152.4 billion ($1.9 billion) by 2021. The company is investing in new satellite production and test facilities as part of its strategy, already launching its first non-Japanese funded ST-2 commercial satellite on 20 May 2011; the ST-2 was funded as part of a joint venture between Singapore Telecommunications Ltd. and Chunghwa Telecom of Taiwan. The Türksat 4A and 4B telecommunications satellites will soon join the group of Japanese commercial satellites in 2013 and 2014. With its DS2000 satellite frame proven in orbit on Japanese commercial and technology-demonstration missions, Mitsubishi will be more aggressive in the commercial market, planning to spend ¥3 billion ($37 million) on the new plant and equipment to enlarge total facilities (satellite production, integration and test floor space) to 7,700 m^2. Through this pursuit, Mitsubishi hopes to realise its doubled satellite revenue goal by 2021.[210]

---

[208] De Selding, Peter B. "Sea Launch Back in Business with Successful Eutelsat Launch." Space News 3 Oct. 2011: 5.

[209] Kallender-Umezu, Paul. "NEC Tapped to Build Second Asteroid-bound Hayabusa Probe." Space News 30 Jan. 2012: 10.

[210] "Melco Expansion Aimed at Doubling Satellite Revenue." Space News 13 June 2011: 9.

1.3.3.5 Industrial Developments in China

The Chinese launcher, Long March 3B successfully placed Eutelsat's W3C Commercial Telecom Satellite into geo-orbit on October 7, 2011. Twelve years had passed since the U.S.'s ITAR ban on satellite technology exports to China. At that time, the U.S. Congress passed a law reclassifying commercial satellites as weapons systems for export purposes, making them subject to ITAR. The impetus for such action was based on the fear that China was using commercial satellite launches to perfect its missile technology. In fact, while the newly released U.S. government "1248 report" recommends allowing the government to focus its ITAR controls on technologies and capabilities that are only the most sensitive; it still recommends strengthening the U.S. ban on shipping U.S. satellite technology to China for launch on Chinese rockets.[211] Despite the export ban, Thales Alenia Space wanted access to the Chinese market, and developed an "ITAR-free" product line. Eutelsat was the first to contract for a launch, making it China's first launch for a Western satellite owner in more than 12 years. Intelsat, SES, and Telesat (the fourth largest fleet operator based in Canada) have yet to order an ITAR-free satellite, however, they have all expressed an interest in doing so. One reason for these companies' hesitation in using the Chinese vehicle is that they have U.S. government-licensed orbital slots, subjecting them to the risk of regulatory backlash that may compromise these assets.[212]

AsiaSat of Hong Kong reported an 18 % increase in revenue and a 26 % increase in operating profit for 2011. Total revenue amounted to 1.72 billion Hong Kong dollars ($221 million), while operating profit was 1.05 billion Hong Kong dollars ($135 million). SpeedCast, an AsiaSat subsidiary providing two-way broadband links to corporate and government customers, increased its revenue by 16 % during the same period and accounted for 14 % of AsiaSat's revenue. The company has agreed to sell to an undisclosed third party its 50 % stake in the money-losing Dish-HD Asia Satellite venture with EchoStar of Englewood, Colo.; a venture which never realized its promise to provide 36 enhanced standard-definition and high-definition television channels in Taiwan, and increased its losses in 2011. The sale of AsiaSat's share should close within a year after gaining regulatory approval. Besides the four satellites operated by AsiaSat, the newest being AsiaSat 7 which entered into service in February, two more satellites have already been ordered from Space Systems/Loral, i.e. AsiaSat 6 and AsiaSat 8, which should be ready for launch in the first half of 2014.[213]

---

[211] Leone, Dan. "U.S. Report Supports Sweeping Reform of Satellite Export Rules." 18 Apr. 2012. Space News 23 May 2012 http://www.spacenews.com/policy/120418-report-backs-reform-sat-export.html.

[212] De Selding, Peter B. "Chinese Rocket Launches Eutelsat's W3C Satellite." Space News 10 Oct. 2011: 4.

[213] De Selding, Peter B. "AsiaSat Reports Rising Profit, Remains Bullish on Satellite TV Market." 22 Mar. 2012. Space News 3 May 2012 http://www.spacenews.com/satellite_telecom/120322-asiasat-rising-profit.html.

1.3.3.6 Industrial Developments in India

Very high regulatory barriers in India make access to its satellite telecommunications market an unusually difficult pursuit. Nevertheless, SES of Luxemburg is investing heavily there in the hope of capturing 40 % of the Indian satellite television market within the next few years. Capacity will be increasing as part of a major capital spending program, as SES's total in-orbit transponder supply will be increased by 23 %, i.e. 293 transponders on 12 satellites will be launched between 2011 and 2014; 85 % of that capacity will be directed toward emerging markets, mostly television markets. India is by far the biggest emerging market; its six pay-TV and one free-to-air DTH satellite television providers have a combined total of 32 million subscribers, with that number growing at a rate of nearly one million per month toward 60 million. With over 300 television channels awaiting regulatory approval, the sheer demand partly explains why India has been forced to allow SES and other non-Indian satellite providers into the market. It is likely that India will remain a growing market for non-Indian providers for the long term, as the regulations will necessarily be relaxed since India will never launch enough bandwidth on its own to meet demand.[214]

Legislators in India have not come to this same conclusion, where in April 2012 the Indian government proposed a new tax that would impose a 10 % royalty fee on foreign satellite communications services that would be retroactive covering the past 36 years. The government is looking for ways to reduce its deficit without regard to the crippling effect this tax will have on broadcast and communications, when there is already a shortage of domestic satellite bandwidth. Such a move would put India in breach of its international tax treaties, where compliance with the royalty tax would amount to double taxation. The matter is now on appeal before the Supreme Court of India; however, if the tax is enforced and foreign satellite providers are unable to obtain a tax credit in their own tax homes, they will be forced to pass on the royalty withholding to their customers in India. The 130-member Cable and Satellite Broadcasting Association of Asia (Casbaa), including many of the world's biggest commercial satellite fleet operators, is attempting to persuade the Indian government that retroactivity violates basic international rules and principles of fair play.[215]

---

[214] De Selding, Peter B. "Despite Barriers, SES Invests Heavily in Indian Satellite Market." Space News 6 June 2011: 6.

[215] De Selding, Peter B. "Satellite Fleet Operators Protest Indian Tax Proposal." 5 Apr. 2012. Space News 9 May 2012 http://www.spacenews.com/policy/120405-sat-operators-protest-indian-tax.html.

1.3.3.7 World

Global orders for commercial GEO telecommunications satellites fell by about 35 % in 2011 (from 26 to 17), beginning the long-anticipated decline of spending by the biggest satellite fleet operators. The effect of this reduction in spending on affected markets remains unclear, and will depend on spending patterns of the four largest satellite fleet operators, Intelsat of Luxemburg and Washington, SES of Luxemburg, Eutelsat of Paris and Canada's Telesat. Yet the drop in satellite manufacturing orders in 2011 will likely not be felt by the global commercial launch industry until 2014 since the launcher sector lags behind the trend in the satellite manufacturing industry by 2–4 years. Moreover, as the number of launch contracts for geostationary-orbiting commercial satellites increases yearly, satellites that are not ready for launch in a previous year are pushed up to the next available launch opportunity, further swelling already saturated launch years.[216]

On 12 January 2012, the Canadian satellite component manufacturer Com Dev International reported a decrease in revenue, but higher profitability for the fiscal year ending 2011. Moreover, its new maritime surveillance subsidiary was expected to double its revenue in 2012. The company is seeking to return to its previous levels of profitability, 25 % or higher gross margin, rather than grow revenue. Twenty-six satellites were ordered in 2011, whereas 32 were ordered in 2010. Moreover, the number of transponders on the satellites that were ordered in 2011 decreased by 11 % from 2010. Despite the downturn, the company was not greatly concerned, explaining that it was the type of customer that was changing. As larger satellite fleet operators are beyond the peak of their replacement-and-expansion cycle, demand is increasing among smaller satellite manufacturers whose products can be tailored to better fit the needs of the region in which they operate.

To mitigate losses due to diminishing revenue, Com Dev shut down several of its less profitable product lines in 2010 and 2011 as it focused on better margins. By 2011, its gross profit margin was 22.9 %, up from 20.5 % in 2010. Moreover, Com Dev's subsidiary, exactEarth Ltd. (73 % ownership) was expected to be the company's star performer in 2012, as its revenue had increased exponentially since beginning operations in 2010, i.e. C$ 1.9 million ($1.8 million) in 2010, C$ 4.6 million ($4.5 million) in 2011, and was expected to double in 2012. Providing Automatic Identification System data on ships to global coastal authorities, the company had contracts with government agencies in Australia, Singapore, South Africa, Japan and Denmark in addition to the Canadian Space Agency; and had more than 35 trial programs under way with prospective customers.[217]

---

[216] De Selding, Peter B. "Satellite Orders Drop but Near-Term Launch Manifests Are Full." Space News 9 Jan. 2012: 11 and 12.

[217] De Selding, Peter B. "Com Dev to Focus on Increasing Profitability, Not Revenue, for 2012." Space News 16 Jan. 2012: 10.

Honeywell made a move to raise its profile on satellite-based asset tracking and satellite telecommunications equipment with a $491 million, all-cash purchase of Atlanta-based EMS Technologies. Honeywell paid a 33 % premium above the EMS stock trading price on 10 June 2011; and 55 % above the value of the stock on 18 April 2011, when EMS announced it was seeking a buyer. EMS reported revenue of $355.2 million in 2010, and its most recent contracts included on-board gear for the Wideband Global Satcom X- and Ka-band satellite system for the U.S. Air Force. It will also provide switches to Thales Alenia Space for the Iridium Next constellation of 81 low-orbiting mobile communications satellites.[218]

Thaicom of Thailand reported a 24 % increase in revenue from its satellite service division in 2011, with much of the growth stemming from new television broadcasters (420 channels, up from 358 in 2010), while its IPStar consumer broadband division saw much higher usage rates. After a poor outcome in 2010, the operator's Thaicom 2 and Thaicom 5 satellite revenue grew as new television broadcasters were added to its customer base. In 2011, the total revenue reported by Thaicom from its satellite services and related business amounted to 5.7 billion Thai Baht ($179 million), resulting in a 24 % increase over 2010 in Thai baht, (in dollar terms, the increase amounted to 40 %). The Thaicom 2 and Thaicom 5 satellite generated 43 % of the company's satellite revenue, while the remaining 57 % was generated by IPStar. IPStar revenue increased by 59 % in 2011, amounting to 2.6 billion Thai Baht ($81 million). However, sales of IPStar user terminals and other IPStar hardware dropped by about 8 %, to 672 million Thai Baht ($21 million). IPStar service revenue in 2011 increased particularly in India, Japan, Australia, Malaysia, Indonesia, the Philippines and Thailand.[219]

1.3.4 Industrial Overview

1.3.4.1 Launch Sector

Despite its crucial importance for the satellite industry, the launch sector is an enabler rather than a primary economic activity. The revenue it generates is far less significant than that originating from the satellite manufacturing and satellite services businesses.

The year 2011 experienced increased activity for the launch sector, with a total of 84 launches conducted by launch providers from Russia, the United States, Europe, China, India, Japan, Iran and the multinational Sea Launch AG (see Table 1.2). There were some important events, such as the five non-commercial launch failures: a Rockot launch carrying the Russian GEO IK-2 No.II payload in

[218] De Selding, Peter B. "Honeywell Buying EMS Technologies for $491 M." Space News 20 June 2011: 15.

[219] "Satellite Operator Thaicom Reports Turnaround in 2011." Space News 20 Feb. 2012: 9.

Table 1.2 Worldwide launches in 2011 per country, number of launched systems, and commercial status (Source: FAA)

Launchers	Number of launch systems	Total number of launches	Commercial launches	Non-commercial launches
Russia	8	31	10	21
China	7	19	2	17
USA	11	18	0	18
Europe	3	7	4	3
India	1	3	0	3
Japan	2	3	0	3
Iran	1	1	0	1
Multinational	2	2	2	0
Total	**35**	**84**	**18**	**66**

February; a Taurus XL launch carrying the Glory, Kysat I, Hermes and Explorer I (Prime) payloads in March; a Long March 2C carrying the Shijian II-04 in August; a Soyuz carrying the Progress M-12 M also in August; a Zenit 2 M carrying the Phobos-Grunt and Yinghuo payloads missions, failing after launch, in November; and a Soyuz-2 carrying the Meridian 5 payload in December. There was also one commercial launch failure, i.e. a Proton M carrying the Express AM4 payload in August. Four of these six launch failures resulted from Russian launches on three different launch vehicles.

When looking at specific countries, Russia was again the world leader in the number of launches, accounting for approximately 36.9 % of the total number. It was followed by China at 22.6 %, the United States at 21.4 %, Europe at 8.3 %, India at 3.6 %, Japan at 3.6 %, Iran with one launch at 1.2 %, and the Multinational Sea Launch AG at 2.4 % of the total launch figure (see Fig. 1.8).[220]

Russia launched 31 vehicles using eight different launch system configurations. The U.S. used a set of 11 different launch configurations for a total of 18 launches, China conducted 19 launches using seven configurations, and India used one launcher for its three launches. Europe used its Ariane 5 ECA and ES-ATV launchers in addition to its newly acquired Soyuz 2 launcher for its seven launches (4 Ariane 5 ECA, 1 Ariane 5 ES-ATV, and 2 Soyuz 2 ST). Japan had three launches using its two launchers, while Iran used its 1 launcher for its single launch, and the multinational provider used two launch configurations for two launches. The total of different launch system configurations is now 35 for 2011, marking an increase of three compared to 2010.

The launch market expanded from being shared between three actors in 2010, i.e. Russia, Europe, and U.S., to include China, Japan, India, Iran and the Multinational participants (see Fig. 1.9). However, it should be recalled that Europe's launcher Ariane 5 has the ability to carry two standard-size payloads.

[220] Commercial Space Transportation: 2011 Year in Review, 3.

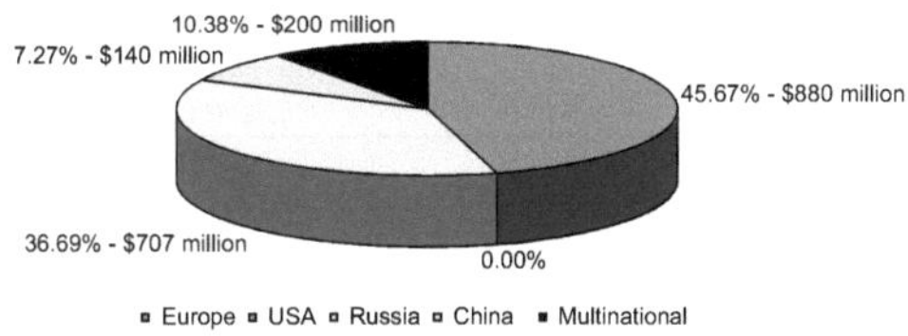

Fig. 1.8 Commercial launch revenues by country in 2011 (Source: FAA)

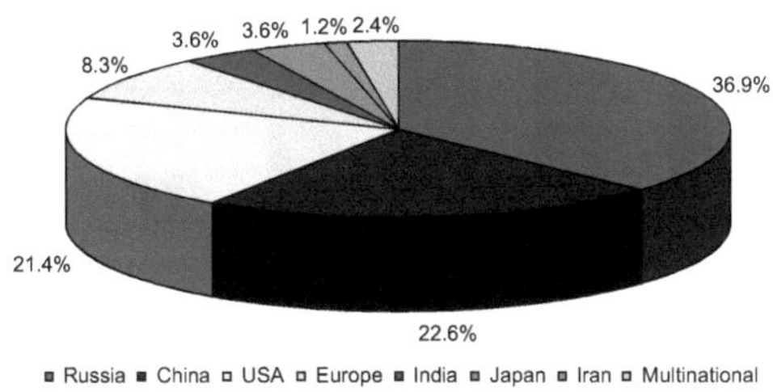

Fig. 1.9 Worldwide launches by country in 2011 (Source: FAA)

With the above payload consideration in mind, the activity of the two leaders of commercial launches (Europe and Russia) should be considered as nearly equivalent. When considering non-commercial launches, Russia's dominance is not quite apparent as it beats the U.S. by only three launches, and China by four launches. While Russia conducted 31.8 % of non-commercial launches, the U.S. followed closely behind with a 27.3 % share, along with China's 25.8 % share.

The distribution between commercial and non-commercial payloads launched seems largely to have maintained equilibrium with the previous year. The majority of non-commercial activity came from Russia, the U.S., and China. This is particularly true for China with its 17 non-commercial launches, a country that is rapidly developing its national programmes in remote sensing and navigation systems, as well as its own space station. Similarly, many of Russia's 53 payloads had non-commercial functions with 32 payloads destined for programmes such as the GLONASS system, ELISA, etc. U.S. commercial payloads were non-existent with zero launches in 2011 compared to its 28 non-commercial payloads, and also compared to the three commercial payload launches in 2010.

Regarding the global share of payloads launched in 2011 (see Fig. 1.10), Russia took first place once again, increasing its stake by 20.5 % since 2010, with 53 payloads launched; representing 39.8 % on the worldwide scale, virtually equal to its 40 % share in 2010. The U.S. position was unchanged with 28 payloads and 21.1 %, while China fortified its third place position with 21 payloads amounting to 15.8 %. Europe launched 17 payloads with a share of 12.8 %, and India followed with eight payloads, at a share of 6.0 %. Next, Japan's three payloads amounted to 2.3 %; the two Multinational payloads took a 1.5 % share; and finally, Iran's one payload resulted in a 0.8 % share. Thus, the hierarchy of

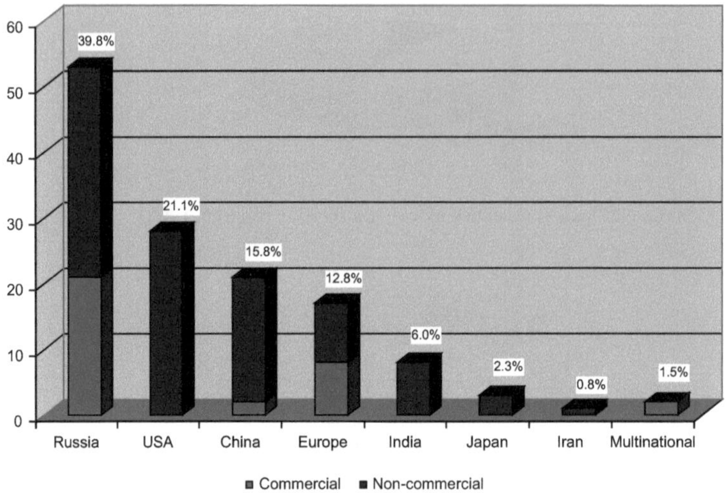

Fig. 1.10 Total payloads launched in 2011 by country, share and commercial status (Source: FAA)

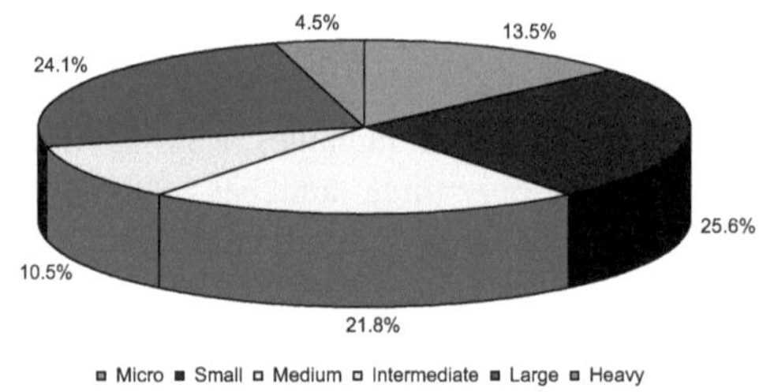

Fig. 1.11 Distribution of the payloads launched in 2011 by mass class (Source: FAA)

satellites launched between space powers stayed roughly similar to 2010, with Iran and the Multinational payloads replacing Israel and South Korea.

Concerning the distribution of payload sizes, there were some noticeable changes compared to 2010 (Fig. 1.11). "Small" was the mode with 34 payloads at 25.6 % (an increase of 7.5 percentage points relative to 2010), followed by 32 Large payloads at 24.1 % (down 3.5 points), 29 Medium at 21.8 % (up 7.1 points), 18 μ at 13.5 % (down 7.2 points), 14 Intermediate at 10.5 % (down 3.3 points), and lastly, 6 Heavy payloads at 4.5 % (down 0.7 points). Micro payloads have a mass of 91 kg or less, and are mainly science satellites, technological demonstrators or small communications satellites. Small payloads weigh between 92 and 907 kg and are very often Earth Observation satellites, similar to the Jason or the RapidEye series. Medium payloads weigh between 908 and 2,268 kg, and feature the most diverse set of satellites, including small satcoms in geostationary orbit, Earth Observation satellites, and most of the Russian military satellites from the Kosmos series.

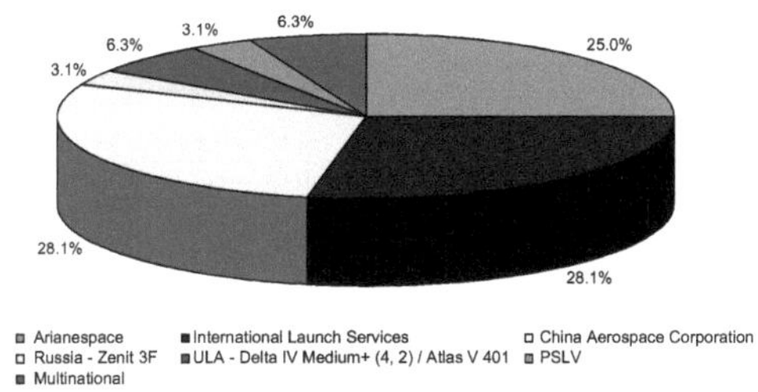

Fig. 1.12 Share of launch contracts for GEO satellites in 2011 by launch service provider

Intermediate payloads, weighing between 2,269 and 4,536 kg, encompass medium satcoms and big scientific satellites. Large payloads, between 4537 and 9,072 kg, refer to big satcoms, as well as to the Soyuz and Progress spacecrafts flying to the ISS. Finally, heavy payloads, exceeding 9,072 kg, are linked to ISS activity, such as the cargo spacecrafts, ATV, etc.[221] In 2011, while there were 13 missions to the ISS, there was a decrease in the Heavy mass class payloads destined for the station.

In 2011, the number of actors in the market for launchers for GEO satellites increased significantly with the return of the China Aerospace Corporation, Sea Launch, and the addition of India's PSLV (see Fig. 1.12). China Aerospace Corporation and the International Launch Services shared the top position, both obtaining 28.1 % market share, followed by Arianespace with a 25.0 % market share.[222]

Of Russia's commercial launches, International Launch Service launched a total of nine satellites using seven Proton M vehicles. The International Space Company Kosmostras launched eight satellites on one Dnepr into LEO and 12 Globalstar 2 satellites were deployed on two Soyuz 2 launchers. Europe launched eight GEO communications satellites on four Ariane 5 ECA launchers. China carried two commercial payloads (Eutelsat 1R and NigComSat 1R) on two Long March launchers. And Sea Launch AG, the Multinational launch provider, commercially launched Atlantic Bird 7 and Intelsat 18 from two Zenit 3SL model launchers.

1.3.4.2 Satellite Manufacturing Sector

In 2011, 133 payloads were launched (including *inter alia* an estimate of 116 satellites, and 14 crewed or cargo missions to the ISS). Russia manufactured 39.8 % of the launched payloads, while the U.S. made 21.1 %, and China produced 15.8 %. Europe accounted for 12.8 % of the payloads launched, while India produced 6.0 %, and Japan accounted for 2.3 %. Iran and Multinational replaced Israel and South Korea producing 0.8 % and 1.5 % respectively.[223]

[221] Commercial Space Transportation: 2011 Year in Review, 32.

[222] Figure 1.12.

[223] Id. at 6. Payloads are assigned to the nation that commissioned them, not according to the nationality of the manufacturer.

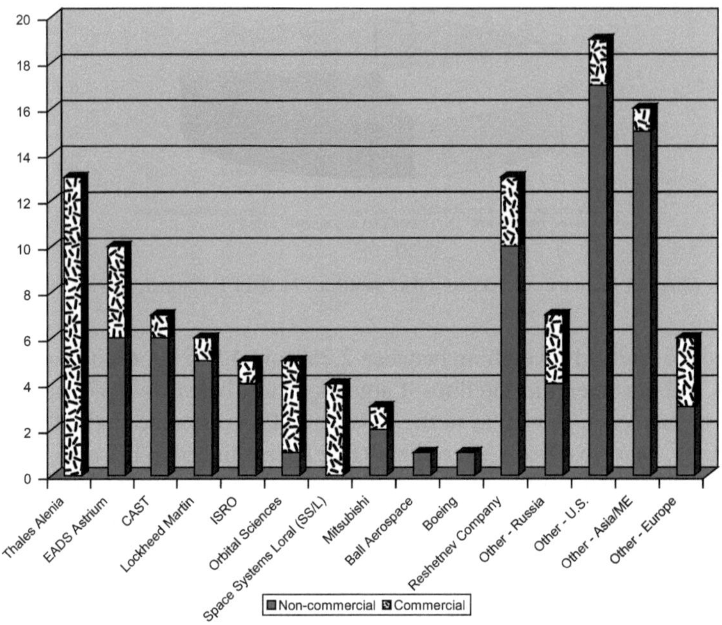

Fig. 1.13 Satellites launched in 2011 by manufacturer and commercial status (Source: Futron)

To summarize (Fig. 1.13), of the 116 satellites launched in 2011, 75 were non-commercial. The European satellite manufacturers Thales Alenia and EADS Astrium took the lead in manufacturing irrespective of whether the satellites were commercial or non-commercial with, respectively, 13 and 10 satellites launched confirming their strong share in this market. These figures result from the commercial orders these manufacturers receive, unlike satellites developed by CAST (seven satellites launched), Lockheed Martin (6), ISRO (5), Mitsubishi (3), and Reshetnev Company (13), whose satellites have mainly non-commercial origins. The majority of satellites developed by Orbital Sciences (5) and Space Systems Loral (4) were commercial in nature. Finally, 6 of the 20 Russian satellites were designated for commercial activities, along with one commercial CAST satellite.[224]

Next, 27.6 % of the 116 satellites launched in 2011 were GEO-stationary satellites (Fig. 1.14).[225] In this field, 34.4 % of the satellites came from the U.S. (4 by SS/L, 2 by Lockheed Martin, 4 by Orbital Sciences, and an additional satellite classified under Other – U.S.); while 15.6 % were from Europe (1 by Thales Alenia Space and 4 by EADS Astrium); 18.8 % from China (6 by CAST); 6.2 % from India (2 by ISRO); 3.1 % from both Japan (1 Mitsubishi) and Other – Asia/ ME; and, 18.8 % for Russia (4 Reshetnev Company, and 2 Other – Russia). In contrast, concerning the non-GEO orbiting satellites, Europe held a 28.6 % share

[224] Id. at 22.

[225] Id.

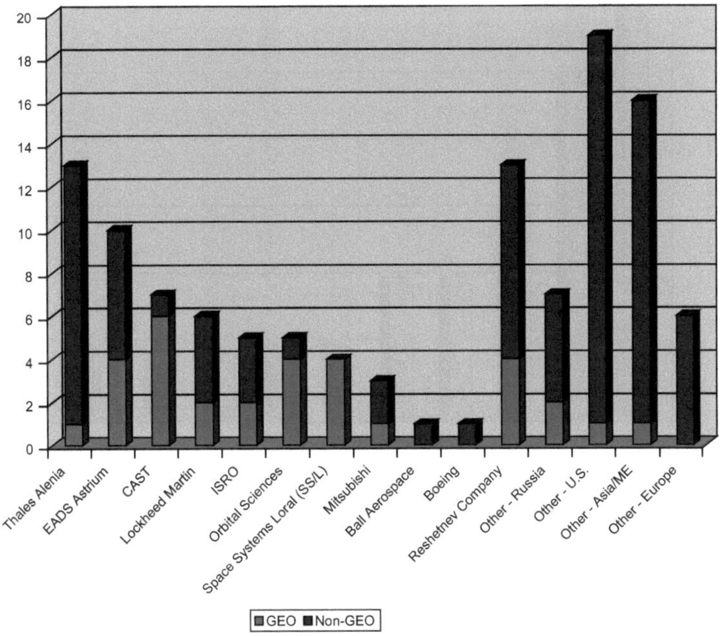

Fig. 1.14 Satellites launched in 2011 by manufacturer and orbit type (Source: Futron)

(12 Thales Alenia Space, 6 EADS Astrium, and 6 Other – Europe); Russia held 16.7 % (9 Reshetnev Company, and 5 Other – Russia); China held 1.2 % (1 CAST); India held 3.6 % (3 by ISRO); Japan held 2.4 % (2 Mitsubishi), while Other – Asia/ME held a 17.9 % share with 15 non-geo satellites. Yet, the U.S. developed the majority of non-GEO-stationary satellites in 2011 with a share of 29.8 % (4 by Lockheed Martin, 1 by Orbital Sciences, 1 by Ball Aerospace, 1 by Boeing, and 18 additional satellites classified under Other – U.S.). The strong influence of national programmes remained in 2011, considering the GLONASS constellation completion in MEO or Beidou navigation system in GEO.

Lacklustre demand for commercial GEO-stationary satellites continued in 2011, with 23 orders, even fewer than the 26 contracts awarded in 2010. Eleven manufacturers were represented in this market, an increase from the eight in 2010. The U.S. company, Boeing had no contracts, while Lockheed Martin won 1, and SS/L was awarded six GEO satellites contracts. Thales Alenia and EADS Astrium garnered respectively one and five contracts (Fig. 1.15).

The ISS-Reshetnev took one satellite order (KazSat 3), whereas its U.S. counterpart Orbital Sciences took two satellite orders (SES 8, Thaicom 6). Also, Mitsubishi took two satellite orders (Türksat 4A, 4B). ISRO is still in the market with one satellite ordered (GSAT 14). The core of the competition is thus between U.S. companies with a total of 39.1 % of the contracts awarded (nine contracts total) against 26.1 % for European ones (six contracts total). China trails

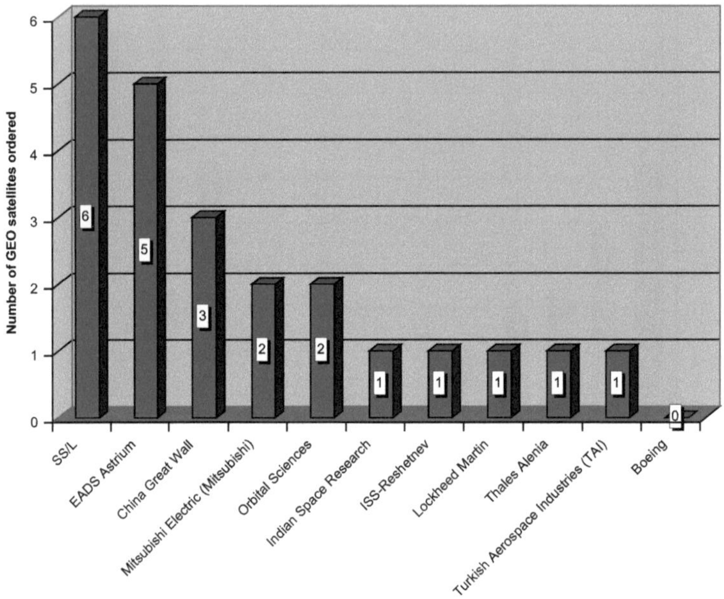

Fig. 1.15 GEO satellite orders in 2011 by manufacturer

in third place with three satellite orders (Belarus Sat 1; Chinasat 11, Chinasat 13), a share of 13.4 %.[226]

1.3.4.3 Satellite Operators Sector

Between 2011 and 2012, satellite operator business growth rates improved following the previous economic turmoil. However, all major operators were acutely aware that this development was at least partly attributable to the industry's long investment cycles. That said, global orders for commercial GEO telecommunications satellites fell by about 35 %, from 26 in 2010 to 17 in 2011, beginning the long-anticipated decline of spending by the biggest satellite fleet operators. The resulting effect of this reduction will depend on spending patterns of the four largest satellite fleet operators: i.e. Intelsat (Luxemburg, Washington), SES (Luxemburg), Eutelsat (Paris) and Telesat (Canada).

[226] "Satellite Orders Report – 2011 Year-End Summary." Futron 15 May 2012 http://www.futron.com/upload/wysiwyg/Resources/FoF/2011/FutronSM2011-EOY.pdf.

1.4 The Security Dimension

1.4.1 The Global Space Military Context

Space-related military spending has remained relatively flat since 2010, with total military spending amounting to $27.85 billion in 2011.[227] Euroconsult reported a similar outcome, estimating world government expenditures for defence space programs to total $30.04 billion in 2011, down from $32.54 billion in 2010. Whereas the previous Space Policies, Issues and Trends Report estimated total U.S. military expenditures to amount to $46 billion, there is a risk that such a figure is inflated based on military activity that was already included in larger budgets, resulting in double-counting. Moreover, while missions, often listed as civil programmes, may also serve dual-purpose military objectives, their expenditure is not included in this section. The Space Report 2012 lists United States military spending as being $26.46 billion, staying level with the 2010 budget, basing its assessment on the fact that the U.S. government failed to pass a new defence budget.[228] Interestingly, the figure only takes into account the space portion of the U.S. Department of Defense budget. Alternatively, the Euroconsult report noted a significant expenditure decrease on defence space programmes, to $22.41 billion in 2011 from $26.13 billion in 2010.[229] While there is a significant discrepancy between the two authorities, it is likely due to the availability of information at their times of publication as well as the different programmes considered. It should be noted that these are conservative estimates that do not factor in U.S. government programmes of a dual use nature, including military use of the National Oceanic and Atmospheric Administration (NOAA) weather satellite data. It should also be noted that Chap. 2 relies on solely on the Space Report 2011 values for the U.S., Russia, Japan, China, and France, while Chap. 5 will include Euroconsult 2012 defence assessments for the countries that are addressed.

The Space Report 2012 estimates the U.S. to represent about 95 % of global military space spending, whereas the remaining $1.39 billion spent in 2011 was the estimated combined military spending of the rest of the world (excluding dual-use systems).[230] However, the Euroconsult report is more conservative in this area, estimating U.S. defence expenditure to be $22.41 billion in 2011, or 75.2 % of the world total.[231] Once again, Russia and China followed the U.S. in military space spending, with 2011 defence spending estimates of $3.28 billion and $2.69 billion.[232] However, due to fluctuating exchange rates, variations in purchasing power,

[227] The Space Report 2012. Colorado Springs: The Space Foundation, 2012: 58.

[228] Id.

[229] Euroconsult Report 2012. Overview – Global Dynamics in Government Space Programs: 12.

[230] The Space Report 2012, 58.

[231] Euroconsult Report 2012, 12.

[232] Id.

and differing employment costs, a direct comparison of the budgets of these countries in fixed dollar values would not present a clear picture of their relative space defence efforts.

Europe's combined 2011 expenditure on military space activity is estimated to be $1.16 billion, staying level with Euroconsult's defence expenditure estimate in 2010 amounting to $1.15 billion.[233] However, an apparent conflict exists between the Space Report 2012 and the Euroconsult report 2012, where the former indicates that 5 % of the $27.41 billion in global military space spending ($1.39 billion) is non-U.S. spending; whereas, the latter indicates that 24.8 % of the $30.04 billion in global military space spending or ($7.63 billion) is non-U.S. spending. This considerable variation is likely due to the inclusion of dual-use systems within the census, while also remaining conservative in other areas. The best approach would be to view these two assessments as indicative of a trend relative to the sectors they assess. Moreover, as Europe's expenditure in the Euroconsult Report 2012 would represent the majority of the Space Report 2012s total $1.39 billion spent on military space activity by the rest of the space-faring world, the amount Europe expends in space defence must be considered to be a conservative estimate. The development of the GMES Earth observation constellation is expected to increase the role that dual use systems play in the European security space architecture, consequently increasing the military spending of Europe's member states. Moreover, the majority of new Earth observation revenue, having already experienced an increase of 6 % to $2.24 billion in 2011 from $2.11 billion in 2010, is expected to be funded by government and military organizations.[234]

Despite a significant decrease of 32.2 %, Japan once again maintained its position as the country with the fifth largest military space budget; its 2011 Ministry of Defence space budget amounted to ¥41.300 billion ($512.4 million) in 2011, from ¥60.933 billion ($668.41 million) in the previous year. The reduction in military spending was the main factor that caused the 8.7 % reduction in Japan's total 2011 space budget, amounting to ¥309.4 billion ($3.836 billion) in 2011 from ¥338.965 billion ($3.830 billion) in 2010.[235] Its budget signalled a shift in Japan's space priorities in view of the adverse geopolitical environment in the region. The Euroconsult was more liberal in its estimate, noting a 52.2 % increase in Japanese space defence spending, from $717 million in 2010, to $1.09 billion in 2011 (including dual-use systems).[236]

---

[233] Id.

[234] The Space Report 2012, 38.

[235] Id. at 55. Cf. The Space Report 2011. Colorado Springs: The Space Foundation, 2011: 52.

[236] Cf. Euroconsult Report 2012, 12.

1.4.2 Europe

EU Member States continue to dictate the level of spending on military space programmes. This is undertaken by all major European space faring nations, with an even larger number of European countries participating on the basis of bilateral or multilateral agreements and arrangements. European budgets in terms of military spending tended to stay level in 2011; France was the only state to decrease its military spending by a significant margin, $519 million in 2011, from $587 in 2010.[237] Italy experienced a sharp increase, from $100 million in 2010, to $148 million in 2011; while the other military space budgets in Europe remained flat compared to the 2010 military space budgets.[238] This indicates that states may be increasing their cooperation to reduce redundant expenditure on military security. Outsourcing services and reducing public budget costs through PPPs is established as an efficient alternative to individual government spending on Earth observation and dedicated military satcom services. Through cooperation and innovative funding schemes, European governments will be able to maintain current levels of security at less expense to tax payers.

In an effort to reduce expenditure, the United Kingdom, France, Italy, Poland and Romania will back the European Defence Agency (EDA) by pooling resources to purchase commercial satellite bandwidth on the spot market for military use. No more than €2 million ($2.6 million) would be placed in an EDA account to be used by Astrium Services to purchase C-band and Ku-band satellite capacity for the five participating nations. A number of similar start-up efforts are being conducted to test the method of pooling and sharing military resources and budgets to reduce the overall cost to taxpayers. Governments, while open to the prospect, seem somewhat reluctant to put this method into effect, as in the present case where 2 years passed after the agreement was made before the first pilot was put into effect. Governments may pay up to a €50 million ($64.7 million) premium when purchasing commercial satellite bandwidth individually. Pooling resources would result in cost savings as governments could aggregate their demands and share data. Britain, France, Italy, Spain and Germany have developed five individual military satellite communications networks, at a cost between €6 and €8 billion ($7.77 and $10.36 billion) each, although these governments could significantly benefit from the shared use of a single next-generation military satellite communication system, enabling a cost savings of more than €1 billion ($1.3 billion) each while still retaining much of their autonomy.[239]

Some European governments have taken a limited approach to pooling resources, such as Norway which has teamed up with Spain to build the HisNorSat military communications satellite; they are currently seeking a developer. The

[237] Id.

[238] Id.

[239] De Selding, Peter B. "Five European Nations Agree to Pool Resources for Satellite Bandwidth Buys." Space News 5 Dec. 2011: 5.

satellite will have both Norwegian and Spanish payloads. Norway will invest around $600 million in its development over a 5-year period; where $200 million will go toward building Norway's share of the satellite. Norway will procure 300 satellite communications terminals. Spain's key contribution will be an orbital slot at 29° east. The capital expenditure in building the large X- and Ka-band telecommunications satellite and operating it over a 15-year period will actually be far less expensive than continued leasing of commercial satellite capacity on the spot market. The satellite will carry almost all of Norway's military satellite communications traffic, excluding Arctic operations that exceed the range of the geostationary satellite. However, Norway will continue to get its UHF capacity from the 28-nation NATO alliance, which leases bandwidth from French, British and Italian military satellite telecommunications systems.[240]

On 17 December 2011, France launched its Electronic Intelligence Satellite (ELISA) demonstration project, consisting of four micro satellites positioned in low Earth orbit, spaced several kilometres apart. The system, lofted on the second Soyuz 2 ST to launch from Kourou, will pave the way for its larger scale successor programme ROEM (Elint), which will map the positions of radar and other transmitters throughout the world, while also providing their technical characteristics. The ELISA programme aims to achieve the following two objectives: it will update databases used for electronic warfare, and it will detect and monitor activities during operations. The ELISA satellites record all signals they intercept; thereafter, upon combining the measurements stored by each micro-satellite, it will be possible to locate and characterise the source of each signal.[241] Incidentally, onboard that same Soyuz launcher, France launched its Pleiades 1 satellite, as part of a future series. These optical satellites are designed to provide imagery covering small areas but with a resolution of 0.70 m. Pleiades 1 complements the SPOT optical satellites, providing imagery that has dual-use purposes.[242] While CNES is funding 90 % of the project,[243] other participants in the Pleiades system include Austrian, Belgian, Spanish and Swedish space agencies, as well as French, Spanish and Italian Defence ministries.

[240] De Selding, Peter B. "Norway, Spain Expected to Order Milcom Satellite Before New Year." Space News 5 Dec. 2011: 6.

[241] "Elisa – Mapping radar stations from Space." CNES 12 May 2012 http://www.cnes.fr/web/CNES-en/5940-elisa.php.

[242] "Pleiades – Imagery to Meet European Civil and Military Challenges." CNES 12 May 2012 http://www.cnes.fr/web/CNES-en/3236-pleiades.php.

[243] "First Pléiades Satellite to Launch 16 December." 9 Dec. 2011. CNES 12 May 2012 http://www.cnes.fr/web/CNES-en/9844-gp-first-pleiades-satellite-to-launch-16-december.php.

1.4.3 The United States

While exceeding the combined the military space budgets of the world by a vast margin, the U.S. Department of Defense space budget has remained relatively flat since 2010. The U.S. spent an estimated $26.46 billion in 2011, whereas it spent an estimated $26.66 billion in the previous year.[244] Compared to the $1.39 billion spent by the rest of the world in 2011, the U.S. makes up 95 % of global military space spending.[245] The Euroconsult report is more conservative in this area, estimating U.S. defence expenditure to be $22.41 billion in 2011, or 75.2 % of the world total.[246] However, this assessment does not factor in additional civilian dual-use programmes where operators provide service to military authorities and civil authorities based on need. Hence, this subsection will focus on U.S. Department of Defense (DoD) programmes as listed by the Space Report 2012, allowing readers to clearly distinguish purely DoD expenditures from that of other space faring nations.

The U.S. military space budget decreased by about $200 million between 2010 and 2011. This would be a significant change for any other space faring country, yet relative to the entire U.S. military space budget, it is a mere 0.7 % reduction. That trend may continue into 2013 as the Defense Weather Satellite System (DWSS) will likely be cancelled, along with plans for the procurement of a second Space-Based Space Surveillance satellite, designed to detect debris, spacecraft or distant space objects without interference related to weather, atmosphere or the time of day.[247] Additional cost savings may come from the U.S. Air Force conducting a block buy strategy that would involve the procurement of 6–10 rockets annually from the United Launch Alliance over a 3–5 year period. Moreover, the space program office at Kirtland Air Force Base in California, may be shut down, with its rapid-response and emergency deployment capabilities transferred to the Space and Missile Systems Center in Los Angeles.[248] The budget for the U.S. Missile Defense Agency (MDA) will likely decrease by 8 % between 2012 and 2013, from $8.42 billion to $7.75 billion. However, MDA's 2013 budget request would boost funding for certain programmes such as its proposed Precision Tracking Space System (PTSS), a satellite constellation capable of tracking ballistic missiles during the midcourse portion of flight; and its Standard Missile (SM)-3 Block 2B interceptor, a next-generation version of the Aegis Ballistic Missile Defense System.[249]

---

[244] Cf. The Space Report 2012, 44 & The Space Report 2011, 43.

[245] The Space Report 2012, 58.

[246] Euroconsult Report 2012, 12.

[247] Ledbetter III, Titus. "U.S. Military Space Spending To Decline 22 Percent in 2013." 13 Feb. 2012. Space News 12 May 2012 http://www.spacenews.com/military/120213-mil-space-spending-decline.html.

[248] Id.

[249] Ledbetter III, Titus. "Missile Defense Agency Seeks Big Increase in Space Spending." 14 Feb. 2012. Space News 12 May 2012 http://www.spacenews.com/military/120214-mda-seeks-increase-spending.html.

Back in 2009, the Obama Administration cancelled plans to deploy a missile defence system in Europe in favour of a Phased Adaptive Approach built around the Aegis Ballistic Missile Defense system. This system would be developed in stages, with the first phase involving the deployment of Naval Aegis ships, carrying SM-3 Block 2A interceptors, in European waters in 2012. The second phase would place a land-based variant of the Aegis system and SM-3 Block 2A interceptors in Romania by 2015. By 2020, the third stage would replace the current interceptor with the next-generation Aegis SM-3 Block 2B interceptor, capable of targeting larger intercontinental ballistic threats.[250]

The U.S. Air Force relied on Europe's Ariane 5 to launch an American broadcasting SES-2 commercial satellite carrying the Air Force's Commercially Hosted Infrared Payload (CHIRP); a staring, wide-field-of-view telescope designed to test infrared sensor technologies.[251] Operating for a period of 9.5 months, its controllers will select targets to validate CHIRP's ability to detect missile launches, while also assessing whether the U.S. military's use of commercial satellites and its requirements would conflict with that of the owners of the commercial host satellites. In 2011, the DoD had nine launches, four carried classified National Reconnaissance Office payloads while the other five carried DoD or DoD-sponsored payloads.[252]

Although this report focuses on space project developments, the actual use of space assets also merits occasional mention. The year 2011 marked the end of a decade-long man-hunt for Osama Bin Laden. Surveillance satellites were a catalyst in finding Bin Laden, assisting the U.S. Central Intelligence Agency and U.S. military by creating a detailed map from above. The U.S. National Geospatial-Intelligence Agency likely participated in the search, along with a number of other military and commercial satellites that may have provided relevant mapping data. Images taken over the 10 year period showed the development of a fortified compound from an empty lot in 2001. The map produced was so detailed that it enabled operation planners to build a mock-up of the compound to be used for rehearsal sieges. It is also likely that secure satellite communications were used in directing the final operation.[253]

---

[250] Turner, Brinton. "MDA Budget Would Ramp Up Spending on European Missile Shield." 28 Feb. 2011. Space News 12 May 2012 http://www.spacenews.com/military/110228-mda-budget-spending-european-missile-shield.html.

[251] De Selding, Peter B. "Ariane 5 Lofts Hosted Payload for U.S. Air Force." 22 Sept. 2011 Space News 12 May 2012 http://www.spacenews.com/launch/ariane-lofts-hosted-payload-for-air-force.html.

[252] Commercial Space Transportation: 2011 Year in Review, 9.

[253] Boyle, Alan. "How satellites helped get Osama." 2 May 2011. Cosmic Log – MSNBC 12 May 2012 http://cosmiclog.msnbc.msn.com/_news/2011/05/02/6571688-how-satellites-helped-get-osama?lite.

1.4.4 Russia

Russia has a long tradition of military space activity; however, its current activity must be assessed in its current day context. Maintaining a reputation in the military space field may be in Russia's interest in a geopolitical sense, yet it should be borne in mind that Russia's reported military space budget makes up only a portion of the total $1.39 billion spent in 2011 by all countries in the world (not including the U.S.), while the U.S. alone spent $26.46 billion. The Euroconsult Report 2012 is unclear as to the total amount Russia puts toward its military program, but reports that it should be at least $3.28 billion (including dual-use programmes).[254] In 2011, Russia launched one classified payload, the Cosmos 2472; whereas the U.S. DoD launched 9, China and Japan each had two classified launches, and France launched its military ELISA constellation.[255] However, in that same year, of the 92 commercial and non-commercial payloads launched, 11 were Russian military payloads (two failed to reach orbit), while the U.S. had 9, China had 6, and Europe had 5.[256] Russia also conducted the highest number of launches, both commercial and non-commercial, and nearly doubled all other states in the number of payloads it launched into space, i.e. whereas Russia launched 53 payloads, the U.S. launched the second highest number of payloads at 28, followed by China with 21.[257]

On 30 March 2012, Russia launched its last Oko-series missile early warning satellite, labelled Cosmos 2479.[258] The satellite is designed to detect missile launches using an infrared telescope that senses infrared radiation emitted by the exhaust of rocket engines; it is part of an eight-satellite constellation – the first of the series was launched in 1991.[259] Russia reportedly operates between 60 and 70 military satellites, and it plans to launch at least 100 additional satellites in the next decade.[260] These satellites are meant to boost the country's reconnaissance and ballistic missile detection capabilities; the influx also benefitting the military's navigation and imaging capability.

Russia's dual-use GLONASS GNSS constellation increased by another five satellites in 2011.[261] By 2010, Russia had invested $925 million in its navigation

---

[254] Cf. Euroconsult Report 2012, 12.

[255] Commercial Space Transportation: 2011 Year in Review, 22.

[256] Id. at 2, 7.

[257] Id. at 2–5.

[258] Clark, Stephan. "Russian early warning satellite orbited by Proton." 30 Mar 2012. Spaceflight Now 12 May 2012 http://www.spaceflightnow.com/news/n1203/30proton/.

[259] Graham, William. "Russian Proton-K completes 45 years of service with US-KMO satellite launch." 29 Mar. 2012. NASA Spaceflight.com 12 May 2012 http://www.nasaspaceflight.com/2012/03/russian-proton-k-rocket-launch-us-kmo/.

[260] "Russian Military Orders Missile Early Warning Satellites." 25 Apr. 2012. Defence Talk 12 May 2012 http://www.defencetalk.com/russian-military-orders-missile-early-warning-satellites-41939/.

[261] Commercial Space Transportation: 2011 Year in Review, 22.

system overtaking U.S. investments in navigation systems for that year.[262] The system is being updated with next-generation GLONASS-K satellites, the successor to the previous GLONASS-M, with 23 operational satellites as of November 2011, and additional spacecraft in production.[263]

Another Russian military communications satellite crashed into the Pacific Ocean in April 2012, highlighting Russia's increasing need to replace its aging military communication satellites. The Molniya-1-89, launched in 1996, was part of a fleet of Molniya-1 T military communications satellites that was discontinued in 2006; it was subsequently replaced by the Meridian satellite series.[264]

1.4.5 Japan

The Euroconsult Report lists Japan's space defence budget at $1.09 billion in 2011 (including dual-use systems), with 23 % of that budget directed toward satcom, and the other 77 % toward Earth observation.[265] As a leading space faring country with broadly developed space capabilities, Japan is channelling its efforts from its traditional multilateral "peaceful-use-only" position in space activity to the space security and defence areas. In the wake of several significant regional and international security events in recent years, Japan's space priorities have adapted to a more active role in the field of national security through the use of space technologies. A leader in multilateral diplomacy, it has increased even more its cooperation with multilateral organizations, as well as with its allies in building common space capabilities. In 2010–2011, Japan became acutely aware of the challenges stemming from its region both in geopolitical terms and in terms of natural disasters. Hence, Japan increased its focus on security; the results of which are new agreements, budget reshufflings and accelerated project completion times.

In 2008, Japan changed its basic law to allow active involvement in military space, whilst still respecting that this must be for peaceful purposes. Japan's 2009 5-year Basic Space Plan should also be mentioned in this context, as it calls for strengthening security through the utilisation of space. The plan recommended ¥2.5 trillion ($26 billion) in financing for civil and military space development activities between 2010 and 2014. In October 2011, Japan's Ministry of Defence requested an additional ¥260 billion in funding for military-purpose space programs in 2012. The 2011, the budget of the Ministry of Defence was reduced by 32.2 % amounting to

---

[262] According to Euroconsult 2010 data.

[263] Clark, Stephan. "Proton rocket replenishes Russian navigation system." 4 Nov. 2011. Spaceflight Now 12 May 2012 http://www.spaceflightnow.com/news/n1111/03proton/.

[264] Brown, Mark. "Dead Russian military satellite drops into the Pacific ocean." 10 Apr. 2012. WIRED.CO.UK 12 May 2012 http://www.wired.co.uk/news/archive/2012-04/10/molniya-satellite.

[265] Cf. Euroconsult Report 2012, 459.

¥41.3 billion, from ¥60.933 billion in 2010.[266] If the October request were to have been accepted, the Ministry of Defence's budget would have increased by 53 %. If approved, these funds would be used toward the purchase of satellite imagery systems and development of space-based infrared missile warning sensors. And a significant part of the budget would go toward developing a regional satellite communication network. [267]

In December 2010, Japan released a 10-year strategy document called the National Defense Program Guideline.[268] It calls for the strengthening of development efforts and the use of outer space in the field of information gathering, communications, disaster management and arms proliferation control. One of the main reasons why Japan decided to strengthen its position in space defence and security is constant security uncertainties from the Korean peninsula and East Asia military and space ambitions. On 12 December 2011, it successfully launched its next instalment of the much delayed Information Gathering Satellite (IGS) system, consisting of two satellites with optical sensors and radar monitoring.[269] But this reconnaissance satellite constellation is still far from delivering its intended operational capability. A second radar satellite was planned to be launched in early 2013 and the third-generation optical satellite constellation is in the works for late 2013 or early 2014.[270]

In January 2012, a governmental panel on Japan's space program strategy planned to revise Japan's space law to allow the Japan Aerospace Exploration Agency (JAXA) to become involved in the use of space for national security. If implemented JAXA would be allowed to cooperate in developing spy and early warning satellite systems.[271] Furthermore, on 30 April 2011, Japan and the U.S. announced plans to jointly develop a framework for sharing surveillance data as part of their expanded space related activities. Both also agreed to pursue an international code of conduct on outer space activities, but not in the form it was originally proposed by the EU.[272] Both partners agreed to assure interoperability

---

[266] Cf. The Space Report 2012, 55 & The Space Report 2011, 52.

[267] "Japan to Boost Military Space Spending in 2012." Space News, 10 Oct. 2011: 8.

[268] "National Defense Program Guidelines for FY 2011 and Beyond." Japan Ministry of Defense 11 May 2012 http://www.mod.go.jp/e/d_act/d_policy/pdf/guidelinesFY2011.pdf.

[269] Kallender-Umezu, Paul. "Japan Launches IGS Radar Reconnaissance Satellite". 13 Dec. 2011. Space News 8 May 2012 http://www.spacenews.com/launch/121311-japanlaunches-latest-radar-reconnaissance-satellite.html.

[270] "H-2A Launch Helps Rebuild Japan's IGS Constellation." Space News 3 Oct. 2011: 3.

[271] "JAXA to get National Security Role; NASA Password Leaked." 14 Jan. 2012. The Japan Times 9 May 2012. http://www.japantimes.co.jp/rss/nn20120114x2.html?utm_source=feedburner&utm_medium=feed&utm_campaign=Feed%3A+japantimes_news+%28The+Japan+Times+Headline+News+−+News+%26+Business%29.

[272] Weisgerber, Marcus. "U.S. Wants Changes to EU Space Code of Conduct." Space News 16 Jan. 2012: 3.

between Japan's Quasi-Zenith Satellite System and GPS, in addition to expanded cooperation in the use of space-based systems for environmental monitoring.[273] And Japan, as one of the five partners of the ISS, agreed to continue operations beyond 2015.

1.4.6 China

China's space defence budget is estimated to have increased by 19.4 % from $1.11 billion in 2010 to $1.32 billion in 2012.[274] Throughout 2011 and 2012, China was actively and aggressively strengthening its position as one of the leading space faring nations. This was significant not only in terms of technological and space capacity development, but also in the general increase in launch activities. At the end of 2011, China broke its own record in the number of space launches conducted in 1 year; also beating the United States in the number of launches for 2011. China performed 19 launches during 2011, with the United States following close behind with 18 launches.

The manner in which China develops its technical capabilities and the depth of its military interest in the space program merits analysis. It seem to be clear that Chinese space efforts are intimately connected to the Chinese army, principally because many space activities are under the direct control of the People's Liberation Army (PLA); moreover, all Chinese space operation facilities are all manned and operated by the PLA. In the case of manned space activities, all development and plans are directly under the control of the Chinese military and political bureau.

Analysis of China's space defence prospective is very difficult due to a dearth of unclassified sources. What is apparent is that nearly all Chinese reports about its human spaceflight program and lunar exploration mission involve the PLA. International cooperation in space activities with China will hence mostly involve cooperation with its military.[275] However, this does not necessarily impede China's aim for international cooperation; on 31 October 2011, the Shenzhou-8 capsule carried 17 biological and medical experiments developed by German and Chinese researchers.[276]

According to the Chinese government's 11th Five Year Plan, space activities are considered as one of China's major military advances. It suggests that development

---

[273] Ledbetter III, Titus. "U.S., Japan To Explore Space Surveillance Data Sharing." 1 May 2012. Space News 9 May 2012 http://www.spacenews.com/military/120501-us-japan-space-data-sharing.html.

[274] Euroconsult Report 2012, 17.

[275] Cheng, Dean. "Five Myths about China's Space Program." 29 Sept. 2011. The Heritage Foundation 10 May 2012 http://www.heritage.org/research/reports/2011/09/five-myths-about-chinas-space-program.

[276] De Selding, Peter B. "German Experiments Onboard Chinese Shenzhou-8 Capsule." Space News 7 Nov. 2011: 12.

of certain technological areas, including space related capabilities, will play an important role in the army's modernisation efforts.[277]

At the end of 2011, China issued a statement on its 5-year space program, which includes several fields that relate to security: the development of three new launch vehicles with one launcher serving as a rapid-response launch system; its contribution to space debris mitigation (usually connected to its ASAT test in 2007); and development of its fourth space launch facility, located in the southern part of the country, intended mainly for manned spaceflights.[278]

The possible dual-uses of satellites and any related space activities give China a strong opportunity to develop its civilian and military space capabilities. This is in reference to China's array of intelligence satellites for space reconnaissance and surveillance, as well as technologies used for Earth observation. An example of Chinese development of its remote sensing capabilities was the launch on 8 November 2011 of the Yaogan 12 remote sensing satellite, which will provide information and scientific experiments for better land use.[279] China is also known to have the ability to jam satellite communications and GPS receivers, and is developing directed energy anti-satellite capabilities.[280]

1.4.7 India

India has only recently begun to develop its own space military programme; whereas the majority of the activities of the Indian Space Research Organization (ISRO) still focus on civil applications. This tranquillity might continue to change in the coming years with the ebb and flow of the geostrategic positions in this region. Pakistan and China still provoke increasing concern for India for geopolitical reasons related to their proximity, in addition to China's anti-satellite (ASAT) test in 2007. Following China's ASAT test, defence scientists in India began focusing on "space security" to protect India's $12 billion (Rs 60,000 crore) space infrastructure from electronic or physical destruction.

On 19 April 2012, India launched a 5,000 km range Agni-5 missile, which reached an apex altitude of 600 km in space while following its parabolic trajectory. This was India's first Inter-Continental Ballistic Missile (ICBM). Indian Scientists claim that Agni-5 can travel beyond the entire region of China, Eastern Europe, North Eastern and Eastern Africa, and even Australia if fired from the Nicobar Islands. India plans to introduce the missile into India's strategic forces arsenal by

---

[277] Cheng, Dean. "China's Space Program in the National Security Context." 18 Jan. 2012. The Heritage foundation 10 May 2012 http://www.heritage.org/research/reports/2012/01/us-needs-to-meet-chinas-space-challenge-of-the-next-5-years.

[278] De Selding, Peter B. "China Outlines Space Priorities: Debris Mitigation, New Rocket." Space News 9 Jan. 2012: 14.

[279] "China Launches Imaging Satellite." Space News 14 Nov. 2011: 3.

[280] Houpt, Danny. "A New Policy Typology to Better Understand the Goals of China's Space Program." 31 Oct. 2011. The Space Review 10 May 2012 http://thespacereview.com/article/1958/1.

2014–2015, to function more in a deterrent capacity against China. Until recently, only the permanent members of the UN Security Council – China, Russia, France, the United States, and the United Kingdom had access to such long-range ballistic missiles.[281]

Dr. VK Saraswat, chief of the Defence Research and Development Organisation (DRDO) and scientific adviser to the Defence Minister said that India was proud to be part of the elite group of nations as a missile power. The aim of India was to build the twenty-first century missile. He also highlighted India's role as a peaceful country, against militarisation of space, and not in the missile race. But, added that the launch of Agni-5 was a "game-changer" because it created a new dimension to India's strategic warfare capacity.[282] Following its successful Agni-5 ICBM test, India intended to develop its own anti-satellite weapons.[283]

Currently, India is working on mini-satellites for battlefield use to protect India's main satellites. Dr. Saraswat emphasized the mini-satellites would provide communication and navigation to armed forces for a limited time.[284] Furthermore, the first test of a PDV interceptor was to be in late 2012. This slender two-stage missile can destroy incoming ballistic missiles at an altitude of 150 km. The Ballistic Missile Defence (BMD) project that aims to protect the country from hostile ballistic missiles has developed a long-range radar system that is able to detect these missiles and launch its own counter-projectile. All these elements have been developed under the BMD programme.[285] According to some sources, India's military equipment and systems will be valued at over $100 billion by 2015.[286]

Finally, following the Mumbai terror attacks and the successful use of Israel's spy satellites, India has becoming increasingly interested in additional cooperation with Israel Aerospace Industries (IAI) in the space field.[287] India has ten satellites including ISRO's new Radar Imaging Satellite (RISAT).[288]

---

[281] "Agni-V Test-Fired Successfully, Expands India's Missile Reach from China to Eastern Europe." 19 Apr. 2012. IBN Live 11 May 2012 http://ibnlive.in.com/news/china-to-europe-agniv-expands-indias-reach/249977-3.html.

[282] Bhatt, Sheela. "Agni-5 Has Put India in the Elite Club." 20 Apr. 2012. Rediff News 10 May 2012 http://www.rediff.com/news/report/agni-5-has-put-india-in-the-elite-club/20120420.htm.

[283] "India Developing Anti-Satellite Weapons." 21 Apr. 2012. United Press International 10 May 2012 http://www.upi.com/Top_News/Special/2012/04/23/India-developing-anti-satellite-weapons/UPI-87941335189998/#ixzz1uO2euIGT.

[284] "After Agni-V Launch, DRDO's New Target is Anti-Satellite Weapons." 21 Apr. 2012. The Economic Times 10 May 2012 http://economictimes.indiatimes.com/news/politics/nation/after-agni-v-launch-drdos-new-target-is-anti-satellite-weapons/articleshow/12772355.cms.

[285] "India Takes on China." 01 May 2012. Military & Aerospace Electronics 09 May 2012 http://www.militaryaerospace.com/news/2012/05/01/india-takes-on-china.html.

[286] Egozi, Arie. "India Successfully Launches New Long-Range Ballistic Missile." 19 Apr. 2012. Israel Defense 10 May 2012 http://www.israeldefense.com/?CategoryID=472&ArticleID=1186.

[287] Egozi, Arie. "India Expansion of Israeli Support to India in the Space Field Is on the Agenda." 14 Feb. 2012. Israel Defense 09 May 2012 http://www.israeldefense.com/?CategoryID=472&ArticleID=927.

[288] India Takes on China.

1.4.8 Iran

In 2009, Iran became the 11th country with space launch capabilities. Its inaugural launch put the Omid satellite (meaning "Hope") into orbit using the Iranian-developed launch vehicle, Safir-2.[289] Since that launch, the state has steadily invested an estimated $100 million per year in its civil space budget; increasing that amount to $120 million in 2011.[290] Iran has begun acting on its desires to be a Middle-East space power by 2020.[291] By February 2012, Iran had launched a small Earth-observing satellite into orbit, its third successful launch out of four attempts. This new Iranian satellite has a mass of 50 kg and was built by students at the Sharif University of Technology. Its applications can be used for meteorology, management of natural disasters and measuring the temperature and humidity of air.[292] Contrary to international suspicions, the Islamic Republic of Iran maintains that it has no military ambitions for its space program.[293]

However, many Western countries, especially the United States and Israel remain concerned that Iran may be seeking to strengthen its military power.[294] Some experts expect Iran to use dual-use technology for military build-up purposes, while remaining under the guise of non-military purposes.[295] Iran states it wants to put its own satellites into orbit for civil protection purposes, monitoring natural disasters in the earthquake-prone nation, in addition to improving its telecommunication infrastructure. Iranian officials were quick to point out the United States use of satellites to monitor Afghanistan and Iraq, and said they needed similar abilities for Iran's security.[296] Another development affecting Iran's regional security is its potential cooperation with China in space activity. Ahmad Motamedi,

---

[289] "Iran sends first home-built satellite into orbit." 03 Feb. 2009. AFP 10 May 2012 http://www.google.com/hostednews/afp/article/ALeqM5h6jwhaLtMvmnBIjBipoPXdLDlgpw.

[290] Cf. Euroconsult Report 2012, 8.

[291] Derakhshi, Reza. "Iran unveils missiles and satellites as warning to foes" 07 Feb. 2011. Reuters 10 May 2012 http://www.reuters.com/article/2011/02/07/us-iran-military-missiles-idUSTRE7162F520110207.

[292] "Satellite 'Promise of Science, Industry' put on orbit successfully: Defense Min." 03 Feb. 2012. IRNA10 May 2012 http://irna.ir/News/Politic/Satellite-%E2%80%98Promise-of-Science,-Industry%E2%80%99-put-on-orbit-successfully,-Defense-Min/30795827.

[293] Malik, Tariq. "Iran Launches Small Earth-Watching Satellite Into Orbit: Report." 03 Feb. 2012. SPACE.com 10 May 2012 http://www.space.com/14464-iran-launches-small-satellite-orbit.html.

[294] Brinton, Turner. "Iran's Satellite Launch a Signal of Missile Progress, Analysts Say." 12 Feb. 2012. SPACE.com 10 May 2012 http://www.space.com/5624-iran-satellite-launch-signal-missile-progress-analysts.html.

[295] Hsu, Jeremy. "Iran's Space Program: Lots of Talk, but a Chance to Shine." 09 Nov. 2011. SPACE.com 10 May 2012 http://www.space.com/9499-iran-space-program-lots-talk-chance-shine.html.

[296] "Iran Launches Rocket Capable of Carrying Satellite." 17 Aug. 2008. The Telegraph 10 May 2012 http://www.telegraph.co.uk/news/worldnews/middleeast/iran/2575063/Iran-launches-rocket-capable-of-carrying-satellite.html.

Iran's Minister of Communications and Information Technology, referred to China as an attractive partner partly because of its track record of successfully launching 40 launch vehicles since 1996.[297]

Unlike other Islamic countries that operate civilian-purpose satellites, the Iranian Defence Ministry plays a key role with potential contributions from the Islamic Revolution Guards Corps (IRGC). This military element also manages the Shahab ballistic missile program, which is capable of being modified into a space launch vehicle. The enhancement of the Shahab, with satellite-guided navigation, is a big concern for the U.S. and Israel, because this would allow Iran to strike objects with increased precision.[298] Western countries are concerned that long-range ballistic technology used to propel Iranian satellites into orbit might 1 day be used to launch atomic warheads. On the one hand, Iran asserts its nuclear work is purely for peaceful purposes, adding that it hopes to launch a human into space by 2020 and land an astronaut on the moon by 2025.[299] Yet, Iran also seeks to challenge Israel's regional monopoly in outer space. Israel is the only regional actor in the Middle East that can launch satellites onboard Israel's own Shavit launch vehicle.[300]

Analysts claim Iran's space goals are probably both scientific and militaristic in purpose, with the increase of national pride – important to legitimize the current regime's policies and actions.[301] Iran planned to launch its first remote-sensing satellites, Tolu (Sunrise) and Fajr (Dawn), in 2012.[302]

1.4.9 North Korea

Whereas in 2011, North Korea did not conduct any launches into space, the country returned to the space scene once again in April 2012, test launching a long-range ballistic rocket. However, the launch was unsuccessful, with the missile

[297] Kass, Lee. "Iran's Space Program: The Next Genie in a Bottle?" Sept. 2006. The Middle East Review of International Affairs 10 May 2012 http://meria.idc.ac.il/journal/2006/issue3/jv10no3a2.html.

[298] Kahn, Gabe. "The Iranian Space Monkey Cometh." 17 June 2011. Israel National News 10 May 2012 http://www.israelnationalnews.com/News/News.aspx/144990#.T6o56egx_zA.

[299] Iran's Space Program: The Next Genie in a Bottle?...; "Iran Plans to Send Monkey Into Space." 28 June 2011. Al Jazeera 10 May 2012 http://www.aljazeera.com/news/middleeast/2011/06/201162812040876380.html; "Iran Space Monkey Launch Failed." 13 Oct. 2011. National Turk 10 May 2012 http://www.nationalturk.com/en/iran-space-monkey-launch-failed-14383.

[300] Iran's Space Program: The Next Genie in a Bottle?

[301] Moskowitz, Clara. "Iran Says It Launched New Rocket and Capsule Into Orbit." 17 Mar. 2011. SPACE.com 10 May 2012 http://www.space.com/11153-iran-launches-rocket-space-capsule.html.

[302] "Iran Satellites to be Placed in Orbit Next Year." 20 Feb. 2012. Jamejam Online10 May 2012 http://jamejamonline.ir/en/newstext.aspx?newsnum=100804691135.

disintegrating prior to escaping Earth's atmosphere; U.S. and Japanese sources report that the spacecraft fell into the sea. Prior to the launch, North Korea's Foreign Ministry ignored the UN Security Council's condemnation of the launch of a long-range rocket and reasserted the nation's right to develop its own autonomous civilian space program. Paek Chang Ho, chief of the North Korean command centre, said the launch was for peaceful purposes.[303] He said the Kwangmyongsong-3 satellite, onboard the failed launcher, was designed to send back images and data that will be used for meteorological and Earth observation purposes.[304]

The failed Pyongyang rocket had the size and thrust profile to loft a payload into space. Its 32 m, three-stage Unha-3 booster, was powered by liquid-fuelled engines – North Korea's preferred motor for manned launches and major satellites. Most U.S. long-range missiles use solid-fuel propellant.

Western experts assessed that the launch's apparent failure "shows the weakness of the North Korea missile program" and suggested that the threat from North Korea had been "exaggerated." The United Nations and diplomats had warned that Pyongyang would face further isolation if it went ahead. For example, the White House press secretary described the failed launch as a threat to regional security, a violation of international law and a breach of its own recent commitments.[305] Shortly after the launch, South Korea convened an emergency security meeting and said that Seoul would continue to closely monitor its neighbour's actions.[306] For now, the international community may have to accept that North Korea will soon have the technical capability to reach orbit and must consider the consequences of such a development.

---

[303] "North Korea Threatens Retaliation for U.S. Aid Clawback." 17 Apr. 2012. CBCNEWS 10 May 2012 http://www.cbc.ca/news/world/story/2012/04/17/north-korea-rocket-reaction.html.

[304] "Japan on Full Alert Ahead of North Korean Launch." 12 Apr. 2012. AdelaideNow 10 May 2012 http://www.adelaidenow.com.au/news/world/japan-on-full-alert-ahead-of-north-korean-launch/story-e6frea8l-1226324759731.

[305] Schwarz, Tim. "North Korea Rocket Breaks Up in Flight." 17 Apr. 2012. CNN 10 May 2012 http://edition.cnn.com/2012/04/12/world/asia/north-korea-launch/index.html.

[306] McCurry, Justin. "North Korea's Failed Rocket Launch Triggers Indifference in Seoul" 13 Apr. 2012. The Guardian 10 May 2012 http://www.guardian.co.uk/world/2012/apr/13/north-korea-failed-rocket-launch-reaction; Young-jin, Kim. "DUP Head Urges NK Against Nuke Test." 05 July 2012. The Korea Times 10 May 2012 http://www.koreatimes.co.kr/www/news/nation/2012/05/116_110488.html.

Chapter 2
Developments in Space Policies, Programmes and Technologies Throughout the World and in Europe

Cenan Al-Ekabi

2.1 Space Policies and Programmes

All major space policy developments worldwide were presented in the previous section of Chap. 1, in an attempt to clarify the principal space faring nations' strategies in 2011 and 2012. In the section below, there will be a brief discussion on developments in technology related areas, including policies and access to space technologies. The aim of this section is to clarify how the above presented strategies interact with and influence specific space programmes, and related research and development projects.

2.2 Space Transportation

2.2.1 Europe

Europe's launcher development programmes are funded primarily by ESA. This funding has progressively decreased in the past decade, and as the Ministerial Conference in November 2012 approached, there was considerable uncertainty about the way forward for European launcher technology.

Germany wanted to commit funding toward upgrading Europe's Ariane 5 rocket, instead of developing a next-generation launcher. It argued that key decisions must first be made on Galileo and GMES prior to addressing launcher development, as Galileo lacks over €1 billion necessary to develop and launch the 30-satellite constellation, and financing for GMES was completely removed from the European Commission's 7-year budget proposal. Meanwhile, Germany and France agreed

C. Al-Ekabi (✉)
ESPI (European Space Policy Institute), Vienna, Austria
e-mail: Cenan.Al-Ekabi@espi.or.at

C. Al-Ekabi et al. (eds.), *Yearbook on Space Policy 2011/2012*, Yearbook on Space Policy, DOI 10.1007/978-3-7091-1649-4_2, © Springer-Verlag Wien 2014

tentatively to update the Ariane 5 with a re-startable upper stage that uses the Vinci engine, now in development. With a total upgrade cost of €1.5 billion, this launcher would be capable of launching 10,500 kg into geostationary transfer orbit, an increase of 1,000–1,200 kg from the present Ariane 5 model.[1] France released a space strategy report that reiterated its position on investing in space capacity, but was unclear on whether to enhance the current Ariane 5 launcher or to develop a successor to the Ariane 5 launch vehicle. Whatever the decision, the report argued for an overhaul of the Ariane 5 launcher.[2] The launcher's development was to be decided at the 2012 Ministerial Conference, with the choice between an Ariane 5 Mid-life Evolution (ME) and an Ariane 6 depending to a large extent on the demands of government and commercial customers.

While commercial operations of Arianespace led to significant losses in 2010 and 2009, and a need for aid to offset the adverse effect of currency movements,[3] Arianespace expected to post a 10 % revenue increase in 2011 and to report a slight profit. The company's backlog stood at an unprecedented combined amount of €4.5 billion ($5.9 billion) for its heavy-lift Ariane 5 series and medium-lift Soyuz 2 ST vehicles. Whereas in 2011, the Soyuz 2 ST conducted two launches, 2012 was to have three Soyuz launches from French Guiana, with two additional commercial launches conducted from its original Baikonur location. Arianespace had also scheduled seven launches of the Ariane 5 series launch vehicle, and conducted the inaugural launch of the Vega small-satellite launcher in 2012. Last year, following an audit of Arianespace and its contractors (a condition for injecting about €240 million ($312 million) into Arianespace over 2 years), ESA determined that additional savings related to the Ariane series would not be possible without wholesale restructuring. Hence, ESA released €217 million ($282 million) allocating half to Arianespace's 2011 accounts, with the remaining funds scheduled to be made available in 2012.[4]

2.2.2 *United States*

The implementation of the new NASA direction continued to preoccupy the Agency in 2011 and 2012. One of the most controversial aspects of its progress was the Agency's efforts to manage the further development of the cancelled

---

[1] De Selding, Peter B. "Germany Reaffirms Commitment to $2B Ariane 5 Upgrade." 28 Oct. 2011 Space News 8 May 2012 http://www.spacenews.com/civil/111028-germany-reaffirms-commitment-ariane.html.

[2] De Selding, Peter B. "French Strategy Paper Says ESA Should Fall Under EU Authority." Space News 2 Apr. 2012: 4.

[3] De Selding, Peter B."Arianespace Needs aid to avoid loss in 2010." Space News 4 Jan. 2010 http://www.spacenews.com/civil/110104-arianespace-needs-aid.html.

[4] De Selding, Peter B. "Arianespace Expects to Post 2011 Profit After 2 Years of Losses." Space News 9 Jan. 2012: 10.

Constellation programme's technological spin offs, the centrepiece of which was the implementation of its 2010 Authorisation Act that mandated the development of a new spacecraft and heavy-lift vehicle. The new Multi-Purpose Crew Vehicle (MPCV) would be based on the groundwork conducted on the Orion capsule, while the launcher would draw upon components designed for the Ares rocket. Throughout the reporting period, NASA's decisions regarding the implementation of this project came under intense legislative scrutiny. For example, Congress members from states where the Constellation was to be manufactured pressed NASA to make use of the programme's existing sub-contractors to develop the new spacecraft and launcher. NASA officials, on the other hand, demonstrated a clear preference to re-open the competition in an effort to cut down development costs for the Constellation spin offs, and avoid an indirect "resurrection" of the cancelled programme. Nevertheless, the cost cutting potential of using already developed components for the new spacecraft and launcher remained an important factor.[5]

In the course of the year, the large divergence of views between the executive and legislative branches of government regarding the future of the U.S. access to space programme became apparent, adversely affecting NASA appropriations. Indeed, in its 2012 NASA spending bill, Congress effectively proposed the nullification of the new NASA direction by multiplying funding for the development of the Constellation spin offs to the point of resurrecting it. For example, it recommended a tenfold increase (to $1.95 billion) to the budget for the development of the Space Launch System (SLS), NASA's future launcher mandated by its 2010 Authorisation Act. These funds were to be used for related research and development alone, meaning that total spending on the project out of NASA's budget would have been even higher. At the same time, it slashed appropriations to develop innovative access to space technologies and fund private spaceflight companies by two thirds, limiting them to $375 million.[6] According to NASA sources, if the spending bill were to be fully implemented, the development of the MPCV and SLS tandem would cost between $41.6 and $63 billion through 2025, depending on the programme's progress and the evolution of NASA's budget. In addition to this, it was foreseen that SLS would not fly before 2018, which would be too late to meaningfully support ISS operations that were scheduled to be terminated in 2020. At the same time, development of the even heavier SLS variant

[5] Leone, Dan. "Senators Call for Competitive Heavy-Lift Propulsion Procurement." 2 June 2011. Space News 25 Jan. 2012 http://www.spacenews.com/civil/110602-senators-competitive-heavy-lift-procurement.html.

[6] Leone, Dan. "Committee Cuts NASA Budget, Adds Cash for SLS, Crew Capsule." 11 July 2011. Space News 25 Jan. 2012 http://www.spacenews.com/policy/110711-committee-cuts-nasa-budget.html; Leone, Dan. "Support for Commercial Crew Transports Also Singled Out for Less Funding in 2012." 18 July 2011. Space News 25 Jan. 2012 http://www.spacenews.com/policy/110718-commercial-crew-transports-less-funding.html.

foreseen in the bill would not be completed before 2030. However, several Congress members doubted the accuracy of these estimates, qualified as an attempt to "sabotage" SLS.[7]

In an indirect reply to these accusations, on 15 September 2011 NASA announced its SLS development plans. The new rocket would have two stages, the first one making extensive use of Space Shuttle components (including its external fuel tank and spacecraft engines), and the second one based on the prototype J-2X engine developed for Ares. According to NASA, the entire development project for SLS, MPCV and their required ground infrastructure was not expected to cost more than $18 billion through 2018. Based on tested and reliable technologies, the development process was not expected to create delays or budget overruns, except in the case of the SLS first stage that would be based on an entirely new concept.[8] Congress's encouragement to rely as much as possible on existing Shuttle and Ares components and their subcontractors in order to minimise technological and budgetary risks was fully reflected in the NASA SLS procurement plan. Nevertheless, experts and some members of Congress questioned whether the adopted approach would in fact be less expensive than opening the contract to competition. In addition to this, the approach could be found in violation of U.S. federal procurement laws that laid down open market procurement rules for all public purchases.[9]

While NASA was struggling to meet Congressional demands on the exploitation of existing Space Shuttle and Constellation components for its future space transportation programme, it also moved toward exploring new propulsion technologies, as required by its new direction. Nuclear propulsion technology was one of the options under study, with an initial budget of $7.5 million allocated to it. Such technologies would prove especially useful to planetary science and exploration missions. For the realisation of further conceptual and development work in this field, NASA asked that plutonium-238 production be restarted by the Department of Energy. However, the programme faced difficulties receiving Congressional approval.[10]

NASA efforts to maximise the gains from technology spin offs during 2011 and 2012 were not limited to the field of space transportation, but also included the exploitation of the ISS. In the interest of maximising the use of in-orbit experiments

[7] Leone, Dan. "Obama Administration Accused of Sabotaging Space Launch System." 9 Sept. 2011. Space News 25 Jan. 2012 http://www.spacenews.com/policy/110909-obama-admin-accused-sabotaging-sls.html.

[8] Leone, Dan. "NASA Commits To Building Mandated Heavy-lift Rocket." 19 Sept. 2011. Space News 25 Jan. 2012 http://www.spacenews.com/civil/110919-nasa-commits-heavy-lift.html.

[9] Leone, Dan. "NASA's SLS Procurement Plans Prompt Call for GAO Investigation." 3 Oct. 2011. Space News 6 Feb. 2012 http://www.spacenews.com/civil/111003-sls-procurement-gao-investiga tion.html.

[10] Werner, Debra. "NASA Presses on with Pu-238 Restart Despite Congressional Resistance." 26 Sept. 2011. Space News 25 Jan. 2012 http://www.spacenews.com/civil/110926-nasa-presses-pu238-restart.html.

onboard the ISS, NASA increased its outreach activities to the scientific community at large, multiplying opportunities for researchers. In order to do so, it set up a non-governmental organisation to run the ISS National Lab established since 2005. Space related research areas included the creation of an analogue spacecraft for deep-space missions, as well as an engineering lab to test new materials. Other potential research objectives included work on robotic tools and systems, as well as on advanced propulsion, habitation and radiation protection technologies. In addition to this, particular emphasis was placed on broader technological spinoffs that would be able to produce benefits for society at large. Such experiments included the development of satellite application technologies through the onboard testing of cubesats, as well as pharmaceutical research.[11]

2.2.3 Russia

A key plank of Russian space policies during 2011 and 2012 was to improve the country's self-sufficiency across the board, on technological and operational levels. This was particularly true regarding the development of access to space systems, where Russian authorities expressed their desire to maintain their country's forefront position by developing an entirely new space transportation system that would include a new crew/cargo transport vehicle and a medium-class launcher to be used at the new spaceport under construction in the Russian Far East. Although the launcher was expected to be a variation of the Zenith or Angara rockets, the spacecraft would be an entirely new design, in the area of 12–14 t in weight. For the latter, a very tight development schedule was foreseen, with the first automated flight taking place in 2015 and the first manned flight in 2018, possibly with a crew of six.[12]

Another characteristic of Russian space policies during the review period was its increased involvement in international cooperation, especially regarding the present and future of ISS operations. Indeed, in June 2011 the Russian space agency announced its plans to improve ISS exploitation and multiply the scientific experiments conducted onboard. In order to do so, it engaged in improving the scientific facilities of the Russian ISS module, notably with the addition of two new science power platforms and an improved docking module. This modernisation would also end the Russian segment's reliance on the U.S. segment for supplying the power required to conduct experiments on the ISS.[13]

---

[11] Morring, Jr., Frank. "Exploration, Earth Spinoffs Planned For NASA At ISS." Aviation Week & Space Technology 20 June 2011: 130.

[12] Wall, Robert. "Russian Industry Eyes New Space Concepts."Aviation Week & Space Technology 22 Aug. 2011: 31.

[13] Pyadushkin, Maxim. "Russia To Increase Research On ISS."Aviation Week & Space Technology 20 June 2011: 135.

2.2.4 Japan

During the reporting period JAXA unveiled its long-term human space exploration strategy. Its objective is to develop a vehicle capable of human spaceflight based on the HTV cargo spacecraft created to support ISS operations. However, neither a budget was allocated to the programme, nor has it received the green light from Japan's Space Activities Committee. Since the original HTV spacecraft was operationally conceived as a part of the ISS programme, development of its human spaceflight capable variant could face a tight programmatic timeline, in view of the possible retirement of the ISS in 2020.[14]

2.2.5 China

It is noteworthy that China explicitly positions itself between industrialized and developing countries. This approach implies a twofold strategy seeking to exploit the country's comparative technological advantages vis-à-vis the latter and remedying its weaknesses compared to the former. It therefore implies that the Chinese officials are keenly aware of their country's unique position on the global space activities' scene, and they are hopeful to fully exploit it. Presumably, this would involve providing affordable space applications to their developing country partners, and engaging in high profile missions with more established space faring nations. In this context, the areas of scientific research, satellite applications, human spaceflight, technological cooperation and satellite services' commercialisation are identified as principal future cooperation areas. When considering this list, it is indeed probable that Chinese officials aspire to establish their country as a satellite applications' provider to emerging economies, and a peer partner to developed space powers for innovative space science, technology and spaceflight missions.

On 11 July 2011 China successfully launched its second data relay satellite Tianlian 1–02. The spacecraft joined the Tianlian 1–01 data-relay satellite, launched in 2008, that supports China's manned flights, while its launch was also considered related to the deployment of its first space station. Thus, China became only the third nation after the United States and Russia to build an operational data-relay service.[15]

In 2011 and 2012 China continued its intensive launching campaign. On 27–29 July 2011 for example, it orbited two satellites in two consecutive days, with the orbiting of a Beidou/Compass navigation satellite, and the experimental

[14] Perrett, Bradley. "Japan Charts Path for Manned Space Missions." Aviation Week & Space Technology 8 Aug 2011: 48; see also "Japan's H2A Rocket Lofts Quasi-Zenith Sat". Space News 20 Sept. 2011: 8.

[15] "2nd Chinese Data-Relay Satellite Reaches Orbit." Space News 18 July 2011: 9.

Shijian SJ-11-02 spacecraft, both built by China Space Co. Ltd.[16] This frantic pace continued throughout the reporting period, undeterred by temporary setbacks, such as the failure of the SJ-11-04 experimental satellite, due to a malfunction of the Long March 2C rocket that carried it.[17]

2.2.6 India

India's space launch programme primarily relies on the use of its Polar Satellite Launch Vehicle (PSLV), capable of carrying 3,700 kg payloads to Low Earth Orbit and 800 kg to Geosynchronous Transfer Orbits (GTO).[18] In 2011, ISRO performed three PSLV launches, deploying the GSAT-12 communications satellite into GEO in July 2011; the two other launches deployed seven scientific, remote sensing, and communications satellites in LEO and SSO in April and October 2011.[19] In 2012, two additional PSLV launches were performed, placing the RISAT 1 remote sensing satellite and France's SPOT 6 remote sensing satellite into SSO in April and September, respectively.[20]

ISRO has also developed its Geosynchronous Satellite Launch Vehicle (GSLV)-Mark I & II over the past decade with mixed results. Capable of launching up to 2,500 kg into GTO, the launch vehicle uses two stage liquid fuel engines and a third stage with a solid fuel one. These launch vehicles are meant to orbit India's new and heavier communication satellites of the GSAT series. The GSLV has exhibited a trend of launch failures from 2006, and has not been used following launch failures in 2010.[21] In the meantime, ISRO has been developing its new GSLV Mk III rocket, meant to launch heavier communication satellites, reaching up to 5,000 kg in weight, into GTO. Rather than upgrading the current GSLV series, this new launcher would merely share a number of components. Its completion will enable full autonomy in launching heavier communications satellites of the INSAT-4 class. This launcher will be the first of its series to field a second stage equipped with a restartable liquid fuel engine that would greatly improve the system's operational flexibility and commercial attractiveness.[22]

---

[16] "China Launches Two Satellites in Two Days." Space News 1 Aug. 2011: 8.

[17] "Long March Failure Mars China's Launch Tempo." Space News 22 Aug. 2011: 8.

[18] Federal Aviation Administration. Commercial Space transportation: 2011 Year in Review. Washington DC: FAA, Jan. 2012: 15.

[19] Id. at 14.

[20] Federal Aviation Administration. Commercial Space transportation: 2012 Year in Review. Washington DC: FAA, Jan. 2013: 18.

[21] "GSLV." ISRO 19 Apr. 2013 http://www.isro.org/launchvehicles/GSLV/gslv.aspx.

[22] "GSLV MARK III." ISRO 19 Apr. 2013 http://www.isro.org/Launchvehicles/GSLVMARKIII/mark3.aspx.

2.2.7 *Emerging Actors*

South Korea did not launched its Korea Space Launch Vehicle (KSLV) in 2011 or 2012, following the vehicle's second launch failure on 10 June 2010. The KSLV consists of a modified Angara first stage manufactured by Khrunichev, and a South Korean solid-fuelled upper stage, and is launched from the Naro Space Center.[23]

In 2012 North Korea conducted two launches of its Unha rocket with a 50 % success rate. The Unha debuted in 2009, and by 2012, had conducted three launch attempts in total. Its latest launch in December 2012 appeared to have successfully placed North Korea's first payload into Low Earth Orbit.[24]

Iran conducted a successful launch of its Safir 2 rocket on 15 June 2011.[25] In the following year, the Persian space power would made three more launch attempts, with a 33 % success rate. With a total of six launches of its Safir 2 rocket since its debut in 2009; by the end of 2012, the total success rate of the launcher stood at 50 %.[26]

2.3 Space Science and Exploration

In this section, space science covers new developments in the origin, evolution and future of the Universe, its galaxies, our Solar System, and other celestial bodies e.g. stars, exoplanets, comets, and asteroids. Space exploration reflects both human and robotic spaceflight missions in process or still developing. While traditional governmental space agencies dominate in both these fields, expanded progress in the latter category can be seen with the further development of commercial exploration, and with the new space powers demonstrating the technology needed to carry out such missions.

2.3.1 *Human Spaceflight Activities*

Human spaceflight was focused in Low Earth Orbit (LEO), with the International Space Station (ISS) at centre stage, following its formal extension to at least 2020. NASA retired its US Space Shuttle after its final STS-135 mission, launched on 8 July 2011. Roscosmos is the sole launch provider relied on to transport crew regularly to the ISS and, using Progress and Soyuz, it provided ISS cargo resupply

[23] Federal Aviation Administration. Commercial Space transportation: 2010 Year in Review. Washington DC: FAA, Jan. 2011: 15.

[24] Commercial Space transportation: 2012 Year in Review, 18.

[25] Commercial Space transportation: 2011 Year in Review, 14.

[26] Commercial Space transportation: 2012 Year in Review, 19.

services, along with Europe's Automated Transfer Vehicle (ATV) and Japan's H-II Transfer Vehicle (HTV) adding auxiliary support.

ESA's newest batch of astronauts, Samantha Cristoforetti, Alexander Gerst, Andreas Mogensen, Luca Parmitano, Timothy Peake, and Thomas Pesquet, graduated from the European Astronaut Centre (EAC) on 22 November 2010; with Luca Parmitano as the first assigned to ISS Expeditions 36/37 in 2013.[27] Alexander Gerst will fly to the space station as a flight engineer for Expeditions 40/41 in 2014.[28] However, the newly selected astronauts continued their survival training in places such as isolated complex cave systems in the Mediterranean.[29] Meanwhile, ESA astronaut Paolo Nespoli completed a 6-month mission on the ISS, between December 2010 and May 2011, carrying out an intensive programme of experiments, ranging from radiation monitoring to taking measurements that could improve oil recovery in petroleum reservoirs.[30]

ESA astronaut André Kuipers began a 6-month mission on the ISS on 23 December 2011.[31] The first Dutch astronaut to return to space, André Kuipers participated in about 50 experiments covering a wide range of disciplines, while also involved in the rendezvous and docking of ESA's third ATV Edoardo Amaldi, and in berthing SpaceX's Dragon to the ISS.[32] André Kuipers returned to Earth on 1 July 2012.

While the life-cycle of the ISS has been extended by another 5 years, ESA has cancelled the development of Automated Transfer Vehicles (ATVs) from Thales Alenia Space after 2015. Instead, the space agency has directed Thales to look for ways to supply a service module for the Multi-Purpose Crew Vehicle (MPCV) that Lockheed Martin started building under the old Constellation program.[33] Three ATVs had resupplied the ISS by 2012, with the ATV-1 Jules Verne launched in 2008, the ATV-2 Johannes Kepler launched in February 2011, and the ATV-3 Edoardo Amaldi launched on 23 March 2012. With a mass of over 20 t, the ATV-3

---

[27] "ESA – Human Spaceflight and Exploration – Astronauts – Graduation of Europe's new astronauts." European Space Agency 25 Aug. 2011 http://www.esa.int/esaHS/SEMRFLIRPGG_astronauts_0.html.

[28] "ESA Astronaut Alexander Gerst To Fly To Space Station in 2014." 18 Sept. 2011. ESA 18 Apr. 2013 http://www.esa.int/Our_Activities/Human_Spaceflight/ESA_astronaut_Alexander_Gerst_to_fly_to_Space_Station_in_2014.

[29] "Mission Accomplished: Cave Crew Returns to Earth." 19 Oct. 2011. ESA 18 Apr. 2013 http://www.esa.int/Our_Activities/Human_Spaceflight/Mission_accomplished_cave_crew_returns_to_Earth.

[30] "About the MagISStra Mission." 18 Mar. 2013. ESA 20 Apr. 2013 http://www.esa.int/Our_Activities/Human_Spaceflight/MagISStra/About_the_MagISStra_mission.

[31] "ESA Astronaut André Kuipers Arrives at the Space Station." 23 Dec. 2011. ESA 18 Apr. 2013 http://www.esa.int/Our_Activities/Human_Spaceflight/PromISSe/ESA_astronaut_Andre_Kuipers_arrives_at_the_Space_Station.

[32] "André Kuipers." 30 July 2012. ESA 18 Apr. 2013 http://www.esa.int/Our_Activities/Human_Spaceflight/Delta_Mission/Andre_Kuipers.

[33] Morring, Jr., Frank. "Spacefaring Nations Regroup For Push Beyond LEO." Aviation Week & Space Technology 10 Oct. 2011: 46.

is the heaviest payload the Ariane 5 rocket has ever launched into space.[34] ATV-4 Albert Einstein will launch in 2013, with the final ATV-5 scheduled in 2014.[35]

ESA's Inter-Directorate Exploration Scenarios Working Group which awarded Exploration Scenario Studies contracts to various European companies with the end goal of developing a Strategic Plan for Human Spaceflight and Exploration by mid-2012, has made significant progress in identifying and defining optional building blocks elements and their integration into seven optional roadmaps.[36]

In this context, national space agencies within Europe progressed in their exploration and space science efforts in addition to participating in ESA activities. DLR participated in a number of space activities in the 2011 period; its highlight involved the ATV-2 Johannes Kepler's automated docking to the ISS, with EADS of Bremen in the industrial lead in this ESA Project. And in September 2011, the DLR met with a JAXA delegation to discuss German participation in Japan's Hayabusa-2 mission; the DLR and JAXA agreed on the inclusion of the German small surface lander science package MASCOT in the mission. MASCOT has been developed with CNES science instrumentation.

In 2011, CNES began work on the 'Contract between the State and CNES for the period 2011–2015', signed in October 2010, which tasks CNES to 'make proposals to promote an international exploration programme of the Solar System in renewed governance,' with the aim of an increased role for the European Union in exploration matters. CNES participated in the writing of the ISECG Global Exploration Roadmap, released in September 2011; and was also active in the preparation of the high-level international conference on exploration which took place in November 2011, in Lucca, Italy. France has also participated in the development of ExoMars programme in addition to the exploitation and utilization of the ISS.[37]

NASA launched its final three space shuttle missions in 2011, with Discovery launching on 24 February, Endeavour on 16 May, and Atlantis on 8 July. Discovery was launched on the STS-133/ISS ULF 5 ISS assembly mission; the main payloads were the Permanent Multipurpose Module (PMM) and Express Logistics Carrier 4 (ELC4).[38] Endeavour was launched on the STS-134/ISS ULF 6 ISS assembly mission; its payloads included the Alpha-Magnetic Spectrometer (AMS-02) and

[34] Botta, Oliver. "Factsheet – ATV-3 Edoardo Amaldi begins its journey to the ISS." 19 Mar. 2012. Swiss Space Office 6 May 2012 http://www.sbf.admin.ch/htm/dokumentation/publikationen/raumfahrt/FactSheet_ATV3-e.pdf.

[35] "Ariane 5 ES." ESA Launch Vehicles. 24 Apr. 2012 http://www.esa.int/esaMI/Launchers_Access_to_Space/SEM20W67ESD_0.html.

[36] "Annual Report 2011 of the International Space Exploration Coordination Group." 16 Nov. 2012. International Space Exploration Coordination Group (ISECG) 20 Apr. 2013 http://www.globalspaceexploration.org/c/document_library/get_file?uuid=757abb46-0e23-4bfc-8c1c-dde1320faadc&groupId=10812.

[37] Id.

[38] "NASA – STS-133." NASA 25 Aug. 2011 http://www.nasa.gov/mission_pages/shuttle/shuttlemissions/sts133/main/index.html.

Express Logistics Carrier 3 (ELC3).[39] And Endeavour was launched on the STS-135/ISS ULF 7 mission, the final mission of the US Space Shuttle fleet, carrying astronauts Chris Ferguson, Doug Hurley, Rex Walheim, and Sandy Magnus. Its payloads included the Raffaello Multi-Purpose Logistics Module (RPLM); a Robotic Refuelling Mission (RRM) experiment; and a Pump Module (PM) for the station's Active Thermal Control System (ATCS) that maintains temperatures among avionics, crew members and payloads.[40]

Russia launched four Soyuz spacecraft: TMA-02M on 7 June 2011 with the Expedition 28/29 crew of Mike Fossum, Satoshi Furukawa, and Sergei Volkov[41]; TMA-22 on 14 November 2011 with the Expedition 29/30 crew of Dan Burbank, Anatoly Ivanishin, and Anton Shkaplerov[42]; TMA-03M on 21 December 2011 with the Expedition 30/31 crew of Oleg Kononenko, André Kuipers, and Don Petit[43]; and TMA-04M on 15 May 2012 with the Expedition 31/32 crew of Joseph Michael Acaba, Gennady Padalka and Sergei Revin.[44] Russia also continued its regular resupply of the ISS with four out of five successful launch attempts of the Progress cargo transfer vehicles: M-11M on 21 June 2011, M-12M on 24 August 2011 (launch failure), M-13M on 30 October 2011, M-14M on 25 January 2012, and M-15M on 20 April 2012.[45]

China launched only unmanned missions in 2011 – on 31 October 2011, following the launch of the Tiangong 1 space station, China successfully launched and later docked Shenzhou 8 to the station.[46] Shenzhou 9 will be a manned mission that launches in June 2012.

Analogue campaigns provide safe and affordable environments to test and train in conditions that are similar to those on the Moon and on Mars. In addition to aiding researchers directly in planning for future human exploration, these campaigns have also stimulated a large amount of media interest. The Mars500 520-day isolation study began on 3 June 2010 at the Institute of Biomedical Problems in Moscow. The crew was made up of two Europeans, one Chinese and three

---

[39] "NASA – STS-134." NASA 25 Aug. 2011 http://www.nasa.gov/mission_pages/shuttle/shuttlemissions/sts134/main/index.html.

[40] "NASA – STS-135." NASA 20 Apr. 2013 http://www.nasa.gov/mission_pages/shuttle/shuttlemissions/sts135/main/index.html.

[41] "Soyuz TMA-02M Crew." Russian Space Web 20 Apr. 2013 http://www.russianspaceweb.com/iss_soyuz_tma02m.html.

[42] "Soyuz TMA-22." Russian Space Web 20 Apr. 2013 http://www.russianspaceweb.com/iss_soyuz_tma22.html.

[43] "Soyuz TMA-03M." Russian Space Web 20 Apr. 2013 http://www.russianspaceweb.com/iss_soyuz_tma03m.html.

[44] "Soyuz TMA-04M." Russian Space Web 20 Apr. 2013 http://www.russianspaceweb.com/iss_soyuz_tma04m.html.

[45] "Spaceflight Now: Tracking Station: Launch Log." 20 Apr. 2013 http://spaceflightnow.com/tracking/launchlog.html.

[46] Federal Aviation Administration. Commercial Space transportation: 2011 Year in Review. Washington DC: FAA, Jan. 2012: 26.

Russians. The six crewmembers, Romain Charles, Sukhrob Kamolov, Alexey Sitev, Alexandr Smoleevskiy, Diego Urbina and Yue Wang conducted simulated Martian surface operations beginning on 14 February 2011.[47] The mission ended on 4 November 2011, with the crew having faithfully followed all phases of a real mission, and having performed more than 100 experiments, all linked to the problems of long-duration missions in deep space.[48]

The Canadian Space Agency (CSA) continued operating its Mobile Servicing System (MSS), providing robotic maintenance and resupply operations on the ISS. Dextre and Canadarm2 were utilized in the transfer of cargo from the Japanese HTV vehicle onto the station, and also performed flawlessly during a robotic refuelling demonstration in cooperation with NASA. Canadian robotics successfully replaced critical equipment on the station, and the CSA also continued the upgrade of its ground control operations for Canadarm2 to enable the handling of heavier payloads. In addition to technological upgrades, the CSA also qualified two new Canadian astronauts, Major Jeremy Hansen and Dr. David Saint-Jacques, who completed their Basic Training course requirements and are now eligible for long-duration flight assignment on the ISS.[49]

2.3.2 Lunar Exploration

The Moon stimulated a great deal of interest in this reporting period in regard to the science and exploration activities that were planned or already underway. Europe, China, Japan and India made steady progress toward a robotic and human lunar presence, however budget constraints had the potential to delay well-intentioned initiatives. The US and Russia also maintained their interest in exploring the Moon, and its potential benefit for future Mars exploration.

Looking to future interests in potential surface payloads on the Moon, ESA released a "Call for Declarations of Interest (CDI)" inviting members of the Science and Exploration community to register their interest in proposing a surface payload for the European Lunar Lander mission on 17 January 2012. This call was meant to support discussions with national agencies regarding potential contributions and to inform of preparations for an Announcement of Opportunity (AO) in early 2013. The subsequent AO would lead to the formal selection of the payload to be used as

[47] "ESA – Mars500" European Space Agency 25 Aug. 2011 http://www.esa.int/esaMI/Mars500/.

[48] "Welcome Back and Thank You, Mars500." 4 Nov. 2011. ESA 21 Apr. 2013 http://www.esa.int/Our_Activities/Human_Spaceflight/Mars500/Welcome_back_and_thank_you_Mars500.

[49] "Annual Report 2011 of the International Space Exploration Coordination Group." 16 Nov. 2012. International Space Exploration Coordination Group (ISECG) 20 Apr. 2013 http://www.globalspaceexploration.org/c/document_library/get_file?uuid=757abb46-0e23-4bfc-8c1c-dde1320faadc&groupId=10812.

part of the Lunar Lander mission.[50] The mission would land autonomously near the previously unexplored South Pole of the Moon and was described as a precursor for future human exploration. The South Pole is a region of interest due to the near-continuous illumination of the surface and potential access to water.[51]

Unfortunately, despite experiencing steady development in the previous reporting period, with Germany backing the ESA Lunar Lander as a top priority, by the end of 2012 funding issues forced the programme to be shelved. The project was put on hold in favour of launcher development, EO, ISS operations, and the joint ExoMars mission with Russia.[52]

NASA's Lunar Reconnaissance Orbiter (LRO), launched in June 2009, is scouting the Moon in preparation for future lunar exploration, including finding landing sites, locating resources such as water, ice and hydrogen, and investigating the long-term effects of the lunar environment. The mission has created the most precise and complete topographic maps of the moon yet, and has determined areas of the moon that are in perpetual darkness and in near-continuous sunlight. Helium has been detected in the Moon's atmosphere, and the LRO's mini-RF radar instrument detected small patches of ice in the permanently shadowed craters with temperatures cold enough to permit ice to accumulate. Such areas could be valuable to power hardware in support of a robotic or human mission.[53]

NASA's Acceleration, Reconnection, Turbulence and Electrodynamics of Moon's Interaction with the Sun (ARTEMIS) mission utilized two of the five spacecraft from the Time History of Events and Macroscale Interactions during Substorms (THEMIS) Earth-orbiting mission to orbit the moon to study its lunar environment. The first ARTEMIS spacecraft was inserted into lunar orbit on 27 June 2011; the second on 17 July 2011. Using minimal amounts of fuel and some complex orbit manoeuvres, involving numerous gravity assists from the Moon and the Earth, the spacecraft moved to two different Lagrange points, gravitationally semi-stable points near the Moon. The mission will collect data about the Moon's core, surface composition, and magnetic field, to understand the Moon's environment in space.[54]

NASA's Gravity Recovery And Interior Laboratory (GRAIL) mission was launched on 10 September 2011, with the primary goal of determining the total structure of the lunar interior and advancing understanding of the Moon's thermal

---

[50] "European Lunar Lander – Call for Declarations of Interest." 17 Jan. 2012. ESA 20 Apr. 2013 http://www.esa.int/Our_Activities/Human_Spaceflight/Human_Spaceflight_Research/European_Lunar_Lander_-_Call_for_Declarations_of_Interest.

[51] "ESA Portal – Next step for ESA's first Moon lander." 16 Sep. 2010. European Space Agency 19 Aug. 2011 http://www.esa.int/esaCP/SEMUV2KOXDG_index_0.html.

[52] Clark, Stephen. "ESA lunar lander shelved ahead of budget conference." 20 Nov. 2012. Spaceflight Now 21 Apr. 2013 http://spaceflightnow.com/news/n1211/20moonlander/.

[53] "NASA – Lunar Reconnaissance Orbiter (LRO)." NASA 21 Apr. 2013 http://www.nasa.gov/mission_pages/LRO/main/index.html.

[54] "Two NASA Probes Tackle New Mission: Studying the Moon." 19 Jul. 2011. NASA 21 Apr. 2013 http://www.nasa.gov/centers/goddard/news/releases/2011/11-042.html.

evolution. A secondary objective was to extend the knowledge gained from the Moon to the other planets the solar system.[55] Upcoming missions of note include NASA's Lunar Atmosphere and Dust Environment Explorer (LADEE) that will study the Moon's exosphere and dust environment, was scheduled to launch in late summer or early autumn 2013.[56] Further in the future, an International Lunar Network (ILN) is in its study phase, and is expected to launch in March 2018; the mission, involving robotic landers, orbiters, instrumentation, or other significant infrastructure contributions, will operate all upcoming lunar landing missions as nodes in a geophysical network.[57]

In respect to JAXA's lunar strategy and its "Lunar Exploration Strategy of Japan – World-Leading Robotic Lunar Exploration and Establishment of Technology Base towards Manned Space Activity" report published in July 2010, more details on JAXA's proposed roadmap became available. Japan aims to land its SELENE-2 lander and rover on the Moon's surface in 2016, followed by a SELENE-X advanced lander for South Pole missions in 2020.[58]

China's lunar exploration programme proceeded with the launch of its Chang'e-2 second lunar orbiter on 1 October 2011. Flying in a lower orbit than its predecessor, it provided higher resolution images of the Moon's surface. The spacecraft completed its primary objectives within 6 months, including a low pass of 15 km altitude to image Sinus Iridum, or the Bay of Rainbows, the proposed landing site for future Chang'e missions.[59] The Chang'e-3 lunar rover was expected to launch in 2013 as part of phase II of the lunar exploration programme; and China planned to bring 2 kg of lunar soil back to Earth sometime in 2017–2018 as part of phase III of its programme.[60] China was also considering a manned lunar landing sometime between 2025 and 2030.[61]

India's second lunar mission, Chandrayaan 2, was originally envisioned as a joint venture between India and Russia, featuring an ISRO orbiter and rover, and delivered by a Russian-supplied lander. However, following the failure of the Russian Phobos-Grunt mission in December 2011, Russia had to delay construction of the lander, and financial problems that arose afterward made Russia unable to

[55] "Missions – GRAIL – NASA Science NASA 26 Mar. 2013 http://science.nasa.gov/missions/grail/.

[56] "Missions – LADEE – NASA Science." NASA 26 Mar. 2013 http://science.nasa.gov/missions/ladee/.

[57] "Missions – ILN – NASA Science" NASA 26 Mar. 2013 http://science.nasa.gov/missions/iln/.

[58] Sato, Naoki. "JAXA Status of Exploration and Human Space Program." 14 Nov. 2011. JAXA 21 Apr. 2013: 9 http://www.nasa.gov/pdf/605307main_JAXA-Status-(Final)-A-Sato.pdf.

[59] "China announces success of Chang'e-2 lunar probe mission." 8 Nov. 2010. Xinhua News 21 Apr. 2013 http://news.xinhuanet.com/english2010/china/2010-11/08/c_13596211.htm.

[60] Laxman, Srinivas. "China's Unmanned Moon Mission To Bring Back Lunar Soil To Earth." 21 Mar. 2012. Asian Scientist 21 Apr. 2013 http://www.asianscientist.com/topnews/china-unmanned-moon-mission-to-bring-back-lunar-soil-2012/.

[61] "China considering manned lunar landing in 2025–2030." 24 May 2009. China View 21 Apr. 2013 http://news.xinhuanet.com/english/2009-05/24/content_11425131.htm.

provide the lander even if the Chandrayaan 2 were to be launched in 2015. The natural consequence of this factor will mean that the mission profile will have to change moderately. The Chandrayaan-2 mission has five primary payloads on the orbiter, two of which will be improvements on instruments that were onboard the previous Chandrayaan-1 mission; the rover too will carry two additional instruments. And Chandrayaan-2 will be launched on a GSLV using an indigenous cryogenic engine.[62]

Russia continued work on its Luna-Glob and Luna-Grunt series of missions, with the former scheduled to launch in 2014. These missions include a lunar orbiter and surface penetrators, and will contribute to knowledge about the moon's formation. The Luna-Grunt mission would comprise a lunar rover and the Earth return vehicle. The Luna-Glob probe will study the Moon's Polar Regions following NASA's LRO discovery of the presence of water ice in polar craters that are constantly in the sun's shadow. Four high speed penetrators, and a polar probe equipped with a radio beacon to facilitate future landings will be deployed on the Moon's surface.[63] While Roscosmos was unable to provide a lunar lander in time for the 2015 launch of India's Chandrayaan-2 mission, that mission was intended to be part of Roskosmos' Luna-Glob moon exploration programme, and hence, collaboration with India will likely be ongoing.[64]

The Google Lunar X PRIZE is a competition for a total of $30 million in prizes for the first privately-funded teams to safely land a rover on the moon by the end of 2015. To win, the rover must travel at least 500 m on the Moon's surface and send high-definition video, images, and data back to the Earth. To provide additional incentive for accelerated development, the prize will reduce in value after a government-funded mission explores the lunar surface. Team registration closed on 31 December 2010, and there are now 25 teams fundraising, mission planning, and building robots for this race.[65] In press releases on 15 October and 20 December 2010, NASA announced that it would purchase data and contract with some of the teams to demonstrate technology in high technical risk areas associated with low-cost lunar missions.[66]

---

[62] Ramachanran, R. "Chandrayaan-2: India to go it alone." 22 Jan. 2013. The Hindu 21 Apr. 2013 http://www.thehindu.com/news/national/chandrayaan2-india-to-go-it-alone/article4329844.ece.

[63] Pavlishev, Boris. "Lunar probe to search for water on Moon." 18 Oct. 2011. radio – The Voice of Russia 21 Apr. 2013 http://english.ruvr.ru/2011/10/18/58931510/.

[64] Ramachanran, R. "Chandrayaan-2: India to go it alone."

[65] "Google Lunar X PRIZE." Google Lunar XPRIZE 26 Mar. 2013 http://www.googlelunarxprize.org/.

[66] Braukus, Michael, Lynnette Madison, and Josh Byerly. "NASA Awards Contracts For Innovative Lunar Demonstrations Data." 15 Oct. 2010. NASA Press Releases 26 Mar. 2013 http://www.nasa.gov/home/hqnews/2010/oct/HQ_10-259_ILDD_Award.html; *see also* Harrington, J.D., and Josh Byerl. "NASA Selects Companies for Further Lunar Demonstrations Data." 20 Dec. 2010. NASA Press Releases 26 Mar. 2013 http://www.nasa.gov/home/hqnews/2010/dec/HQ_10-344_ILDD_Selections.html.

2.3.3 Mars Exploration

The focus for Mars science has for decades remained the investigation of the planet's habitability, in searching for the presence of water. The collected data continues to suggest that Mars was once partially covered by large oceans, and that life could have been possible in many locations on the planet's surface.

ESA's Mars Express orbiter, launched in June 2003, continued its mission imaging the entire surface of the planet at high resolution, including maps of the mineral composition and atmosphere, and determining the structure of the sub-surface to a depth of a few kilometres, the effect of the atmosphere on the surface, and the interaction of the atmosphere with the solar winds. The spacecraft observed a crater that was deemed to have once been filled by a lake, revealed by the presence of a delta of dark sediments on the crater's edge. Radar has shown that there may be glaciers hidden beneath the surface of Mars' Phlegra Montes mountain range; and it has also detected sediments that are reminiscent of an ocean floor within the previously identified boundaries of ancient shorelines. Moreover, a planetary alignment between Earth and Mars, both passing through a gust of the same solar wind, allowed researchers to compare the protective effects of Earth's magnetic field with Mars' lack of a magnetic field, showing that the existence of a magnetic field is vital for keeping an atmosphere in place. And gravity mapping data collected over a period of 5 years allowed researchers to determine that Martian volcanic lava grew denser over time and that the thickness of the planet's rigid outer layers varies in Mars' Tharsis volcanic region.[67]

The ESA ExoMars mission continued its development, undergoing a revision in spring 2011 following the uncertainties in funding that arose from NASA's constraints and eventual withdrawal.[68] Envisaged as an orbiter to be launched in 2016 with a rover following 2 years later, its payload selection was completed on 2 August 2010.[69] The orbiter features a 1,000-fold increase in sensitivity over previous Mars orbiters, and will study the chemical composition of the planet. The System Preliminary Design Review for the orbiter and the rover was completed by NASA in December 2010.[70] On 29 March 2011, NASA advised ESA that it

---

[67] "ESA – Mars Express." European Space Agency. 26 Mar. 2013 http://www.esa.int/esaMI/Mars_Express/index.html.

[68] "Annual Report 2011 of the International Space Exploration Coordination Group" International Space Exploration Coordination Group (ISECG) 26 Mar. 2013 http://www.globalspaceexploration.org/c/document_library/get_file?uuid=757abb46-0e23-4bfc-8c1c-dde1320faadc&groupId=10812.

[69] "NASA – NASA and ESA's First Joint Mission To Mars Selects Instruments." 2 Aug. 2010. NASA 11 Aug. 2011 html://www.nasa.gov/home/hqnews/2010/aug/HQ_10-181_Joint_Mars_Mission.html.

[70] "Annual Report 2010 of the International Space Exploration Coordination Group" International Space Exploration Coordination Group (ISECG) 26 Mar. 2013 http://www.globalspaceexploration.org/c/document_library/get_file?uuid=927d699a-2126-4205-b182-2fc8409eff88&groupId=10812.

could not fund its rover, leading ESA to issue stop-work orders for all ExoMars contracts. Contract payments were expected to resume in July.[71] By 13 February 2012, NASA announced on that it would have to withdraw entirely for budgetary reasons, with Roscosmos replacing the NASA as a main partner in the mission at the end of the reporting period.[72]

NASA's Mars Odyssey mission, launched on 7 April 2001, is the longest-operating spacecraft to be sent to Mars. Orbiting the planet since 24 Oct. 2001, some of its contributions include: confirming the mineral exposure that was selected as the landing site for NASA's Mars Exploration Rover Opportunity and helping to identify safe landing sites for NASA's Mars Phoenix lander; discovering carbon-dioxide gas jets at the south polar ice cap during the spring season; finding chloride salt deposits across the planet; and producing the best available global image map of Mars. At the end of this reporting period, the spacecraft put itself into temporary standby safe mode after unexpected characteristics of movement in one of its reaction wheels were detected. The spacecraft also served as a communications relay for the two Mars Exploration Rovers.[73] The Mars Exploration Rover (MER) Opportunity reached the Endeavour crater, examining scientific targets in the surrounding area and finding mineral veins that were deposited by water. NASA's MER Spirit, was deemed to have completed its mission on 25 March 2011, following failed attempts to communicate with the rover beyond the last transmission received on 22 March 2010.[74]

NASA's Mars Reconnaissance Orbiter (MRO) continued to provide valuable data for the purpose of determining whether or not life has existed on Mars, characterising the climate and geology, and preparing for future human exploration. Launched in 2005 and undergoing a 7 month journey to the red planet, the MRO reached its 5-year mark orbiting Mars on 10 March 2011. During a very productive year, it returned data that suggests that water still flows in some places on Mars, depicted as dark, finger-like features which appear and extend down some Martian slopes that change during the seasons. The MRO's highest resolution camera was shut down for a period of 2 weeks as a precaution due to improperly received commands for an observation by the memory module controlling one of the instrument's 14 electronic detectors; the camera began making do with 13 detectors while running diagnostics on the malfunctioning detector. A new study suggests that if Mars ever incubated life, the longest lasting habitats were most likely below the surface, in the clay minerals that formed in the shallow subsurface all over the planet. Sand dunes, once thought stationary when observed with lower resolution cameras in previous decades, were caught migrating by several meters per year in

---

[71] "ESA Expects by July To Restart Work on 2016 Mars Mission." Space News 30 May 2011: 6.

[72] De Selding, Peter. "ExoMars Wins One-month Reprieve." SpaceNews 21 May 2012: 8.

[73] "Mars Odyssey." NASA Jet Propulsion Laboratory, California Institute of Technology" 26 Mar. 2013 http://mars.jpl.nasa.gov/odyssey/.

[74] "Mars Exploration Rover Mission: Home." NASA Jet Propulsion Laboratory, California Institute of Technology 26 Mar. 2013 http://marsrover.nasa.gov/home/index.html.

dozens of locations on the planet, following years of monitoring the Martian surface. In fact, some dunes similar in scale to those on Earth, i.e. some as thick as 61 m, were observed to be moving as coherent units across the landscape.[75]

The NASA Mars Science Laboratory (MSL), nicknamed Curiosity, launched on 26 November 2011. On completing its 8-month journey, the rover reached Mars in August 2012. As the largest rover to ever land on Mars, its mission will run for at least 687 Earth days (a full Martian year), and will study Mars's habitability. The rover has eight scientific objectives, i.e. determining the nature and inventory of organic carbon compounds; conducting an inventory of the chemical building blocks of life; identifying features that may represent the effects of biological processes; investigating the chemical, isotopic, and mineralogical composition of Martian geological materials; it will interpret the processes that have formed and modified rocks and soils; assess four-billion-year timescale atmospheric evolution processes; determine the present state, distribution, and cycling of water and carbon dioxide; and it will also characterize the broad spectrum of surface radiation, including galactic cosmic radiation, solar proton events, and secondary neutrons.[76]

Next, NASA was to launch the Mars Atmosphere and Volatile EvolutioN (MAVEN) mission on 18 Nov. 2013, to reach Mars on 22 Sept. 2014. MAVEN aims to explore the planet's upper atmosphere, ionosphere and interactions with the Sun and solar wind, which will be used to determine the role that the loss to space of volatile compounds from the Mars atmosphere has played in the history of Mars' habitability.[77]

The Phobos-Grunt mission missed its window in travelling to Mars, as it did not perform its scheduled burn to begin the needed trajectory. The joint Russian-Chinese Phobos-Grunt and Yinghuo-1 mission aimed at collecting approximately 100–200 g of samples from the moon Phobos, to be deposited in a return capsule and launched back to Earth.[78] Since missing its further burn, thus only reaching an elliptical Earth orbit, it re-entered the atmosphere on 15 Jan. 2012.[79]

[75] "Mars Reconnaissance Orbiter." NASA Jet Propulsion Laboratory, California Institute of Technology 26 Mar. 2013 http://marsprogram.jpl.nasa.gov/mro/.

[76] Mars Science Laboratory (MSL). 14 May. 2012. NASA NSSDC 14 Jan. 2013 http://nssdc.gsfc.nasa.gov/nmc/spacecraftDisplay.do?id=2011-070A.

[77] "MAVEN." University of Colorado at Boulder, Laboratory for Atmospheric and Space Physics 14 Jan. 2013. http://lasp.colorado.edu/home/maven/.

[78] Phobos-Grunt. 14 May. 2012. NASA NSSDC 14 Jan. 2013 http://nssdc.gsfc.nasa.gov/nmc/spacecraftDisplay.do?id=2011-065A.

[79] Amos, Jonathan. "Phobos-Grunt: Failed probe 'falls over Pacific'." 15 Jan. 2012. BBC News 14 Jan. 2013 http://www.bbc.co.uk/news/science-environment-16491457.

2.3.4 Saturn Exploration

The Cassini-Huygens mission, a joint mission of NASA, ESA and ASI, was launched in 1997. Reaching Saturn in 2004, Cassini went on to drop the Huygens probe onto Saturn's moon, Titan. The renamed Cassini Solstice Mission was supposed to end in June 2008, however, funding was provided to allow continued operation providing new insights on Saturn and its moons; it is now slated to explore Saturn until 2017.[80] By 2011, the 14 year old mission had returned images of a storm that was 500 times larger than the storm Cassini witnessed between late 2009 and early 2010; this storm covered approximately four billion square kilometers and was wrapped around the entire planet.[81]

On Enceladus, plumes of 'dusty plasma' emanating from the icy geyser moon were believed to have been observed. In addition to recording the results that were previously only theoretical, Cassini's instruments showed that the 'heavy' and 'light' species of charged particles in normal plasma were actually reversed near the plume spraying from the moon's south polar region.[82] The complexity of the plasma was increased by the presence of ionized water vapour that attached to dust particles; changing its properties and producing a new collective behaviour. Being a rare opportunity, as dusty plasma is thought to exist in comet tails and dust rings around the Sun, Cassini flew through the dusty plasma and directly measured its characteristics in place.[83] Moreover, recent scientific results from a flythrough of the plumes also show strong evidence of the existence of large-scale saltwater reservoirs beneath Encelaus' crust.[84]

2.3.5 Venus Exploration

ESA's Venus Express mission was launched in 1995 and reached Venus in 2006. It studies Venus's atmosphere, including its dynamics and chemistry, atmosphere-surface interactions, and interactions with solar wind, to address open questions such as the workings of the complex global dynamics of the planet, its cloud system, processes that govern the chemical state of the atmosphere, and the 'green-house effect' in its global climate. While Venus Express already detected the escape of ionic hydrogen and oxygen in the ratio of two to one in the previous

---

[80] Mason, Betsy. "Cassini Gets Life Extension to Explore Saturn Until 2017." 3 Feb. 2010. WIRED 18 Dec. 2012 http://www.wired.com/wiredscience/2010/02/cassini-life-extension-2017/.

[81] SpaceNews Staff. "NASA's Cassini Spacecraft Witnesses Big Saturn Storm." 11 Jul. 2011. SpaceNews 18 Dec. 2012 http://www.spacenews.com/article/nasas-cassini-spacecraft-witnesses-big-saturn-storm.

[82] "Enceladus plume is a new kind of Plasma Laboratory." 31 May 2012. NASA Cassini 18 Dec. 2012 http://saturn.jpl.nasa.gov/news/newsreleases/newsrelease20120531/

[83] Id.

[84] Id.

reporting period; from this, it was inferred that solar ultraviolet radiation streams into the atmosphere and breaks up the water molecules into atoms. Now, Venus Express has discovered an ozone layer high in Venus's atmosphere. Previously detected only on Earth and Mars, this discovery offers a useful comparison for searching for life outside of Earth. Three oxygen atoms make up the ozone molecule, which in the Venus atmosphere is thought to be formed when sunlight breaks up carbon dioxide molecules, releasing oxygen atoms which are then swept to the dark side of the planet by atmospheric winds; they subsequently combine to form two-atom oxygen molecules, and occasionally, three-atom ozone molecules.[85]

Moreover, Venus Express discovered that the planet rotated at a slower rate than first determined by NASA's Magellan orbiter in the early 1990s. Since last being studied, surface features on Venus had been displaced by up to 20 km from where they were expected. Over a 4 year period, Magellan enabled scientists to determine the length of the day on Venus to be equal to 243.0185 Earth Days. Nearly two decades later, those surface features could only be lined up with those observed by Magellan if the length of the Venus day is on average 6.5 min longer than Magellan's measurements. These measurements help to determine whether Venus has a solid or liquid core; if it had a solid core, the planet's rotation would react less to external forces because its mass would be more concentrated towards the centre. Venus's dense atmosphere (i.e. more than 90 times the pressure of Earth's) and high-speed weather systems are the most important of those forces, and they are believed to change the planet's rotation rate by causing friction with the planet's surface. Earth experiences a similar but vastly diminished effect (largely caused by wind and tides), where the length of a day can change by roughly a millisecond, depending on wind patterns and temperatures occurring over the course of a year.[86]

The Russian Federal Space Programme is planning to further build on its Venera programme – first initiated in the early 1960s. Throughout the following decades, the Venera programme launched a series of probes, landers, orbiters, and conducted repeated impact experiments and flybys up to 1985.[87] The Venera-D mission is currently being developed by the Russian Federal Space Programme, and it is scheduled for launch in 2017. The mission will comprise of a lander, orbiter and a subsatellite. The lander will study the formation and evolution of Venus, in particular the elemental and mineralogical composition of the surface, geology, iron-containing phases and the distribution of iron oxidation states. During its descent, the lander will make meteorological measurements, record the isotopic composition of the atmosphere, measure the structure, chemistry and microphysics of clouds, and monitor electromagnetic radiation. The orbiter will be in a daily polar

---

[85] Venus Express. "ESA finds that Venus has an ozone layer too." 6 Oct. 2011. ESA 9 Jan. 2012 http://www.esa.int/Our_Activities/Space_Science/Venus_Express/ESA_finds_that_Venus_has_an_ozone_layer_too.

[86] Venus Express. "Could Venus be shifting gear?" 10 Feb. 2012. ESA 9 Jan. 2012 http://www.esa.int/Our_Activities/Space_Science/Venus_Express/Could_Venus_be_shifting_gear.

[87] Williams, David R. "Chronology of Venus Exploration." 29 Jun. 2011. NASA 10 Jan. 2013 http://nssdc.gsfc.nasa.gov/planetary/chronology_venus.html.

orbit and will study the atmosphere from the surface to an altitude of 160 km, using spectrometers from the ultraviolet to millimetre ranges. The subsatellite will allow the simultaneous measurements of plasma and magnetic fields. Europe and China have been invited to participate in the project.[88] The next Russian mission to Venus after Venera-D is called Venus Globe. It is targeted for a launch date of 2021. The major difference will be a long-life lander.[88]

2.3.6 *Mercury Exploration*

Running as a partnership between ESA and JAXA, BepiColombo will be Europe's first mission to Mercury. The mission is targeted for a launch in 2015 with arrival at Mercury in 2022 for a 1-year mission with a possible 1-year extension. To be executed under ESA leadership, BepiColumbo is currently in the implementation stage, to be eventually launched on the Ariane 5 launch vehicle. The mission is made up of two spacecraft: the Mercury Planetary Orbiter (MPO), a three-axis stabilized spacecraft provided by ESA that will study the planet's geology, composition, inner structure, and exosphere, and the Mercury Magnetospheric Orbiter (MMO), a spin-stabilised spacecraft provided by JAXA that will study the planet's magnetic field, atmosphere, magnetosphere and inner interplanetary space.[89]

MErcury Surface, Space ENvironment, GEochemistry and Ranging (MESSENGER), a NASA discovery-class mission, was launched in August 2004. On 18 March 2011, it became the first spacecraft to orbit the planet, following three flybys. By 17 March 2012, MESSENGER had completed its year-long campaign to perform the first complete reconnaissance of the geochemistry, geophysics, geological history, atmosphere, magnetosphere, and plasma environment of Mercury. On 18 March 2012, it began its extended mission to build on its discoveries.[90] The mission is designed to address six broad scientific questions: why Mercury is so dense, the planet's geological history, the nature of its magnetic field, the structure of its core, the nature of the unusual materials at the poles, and what volatiles are important on Mercury. MESSENGER was designed and built by the Johns Hopkins University Applied Physics Laboratory (APL).[91]

---

[88] "VENERA-D: ВЕНЕРА: Изучение продолжается." 10 Mar. 2011. Roscosmos 25 Aug. 2011 http://venera-d.cosmos.ru/index.php?id=692&tx_ttnews[tt_news]=1288&cHash=f9bfd2c6e7616171412b316d206d73a4.

[89] BepiColombo. ESA Science & Technology 13 Jan. 2013 http://sci.esa.int/science-e/www/area/index.cfm?fareaid=30.

[90] MESSENGER Completes Primary Mission at Mercury, Settles in for Another Year. 19 Mar. 2012. MESSENGER 13 Jan. 2013 http://messenger.jhuapl.edu/news_room/details.php?id=197.

[91] "MESSENGER: Mercury Surface, Space Environment, Geochemistry, and Ranging: Mercury Orbit Insertion." Press kit. NASA 25 Aug. 2011 http://www.nasa.gov/pdf/525164main_MercuryMOI_PK.pdf.

2.3.7 Jupiter Exploration

The Europa Jupiter System Mission (EJSM), a proposed joint NASA-ESA mission also known as Laplace, was placed under review due to the combined effect of not being ranked as a top priority in the U.S. National Research Council (NRC)'s decadal survey of astronomy and planetary science, and the overall U.S. budget outlook. The mission has been placed in the L-class mission category, to be launched in the early 2020s. The primary goals of the mission are to characterize Ganymede as a planetary object including its potential habitability and to study the Jupiter system as an archetype for gas giants.[92]

NASA's new frontiers mission to Jupiter, Juno, was launched on an Atlas V55 on 15 August 2011. The $1.1 billion spacecraft carries an assortment of instruments, including a Gravity Science Experiment, a Magnetometer (MAG), a Microwave Radiometer (MWR), a Jupiter Energetic Particle Detector Instrument (JEDI), Jovian Auroral Distributions Experiment (JADE), Waves, a Jovian Infrared Auroral Mapper (JIRAM), Ultraviolet Imaging Spectrograph (UVS), and a JunoCam. The mission's objectives are to determine how much water is in Jupiter's atmosphere, measure the atmospheric composition, temperature, cloud motion and other properties, map the magnetic and gravitational fields, and explore the magnetosphere near the poles, especially the planet's auroras.[93]

2.3.8 Solar Observation

Continued observation of the Sun's external activity has the benefit of improving our understanding of its interior, its corona, the monitoring of solar wind and its consequences on Earth and its neighbouring planets. Coronal mass ejections (CMEs) from the Sun emit surges of charged particles in directions that may cross Earth's path and can damage satellites, impede space-based services and affect the terrestrial electrical infrastructure.

ESA's PRoject for OnBoard Autonomy (PROBA)-2 microsatellite continued its solar observation activity, tracking spikes in CMEs ejecting from the Sun at the beginning of the reporting period that just skimmed Earth on 9 June 2011, and brought with it a burst of radio energy.[94] It also tracked Comet Lovejoy's 120,000 km perigee with the Sun, providing a close-up extreme ultraviolet view

---

[92] Mission Summary. ESMJ-Laplace. 16 Mar. 2012. ESA Science & Technology 14 Jan. 2013 http://sci.esa.int/science-e/www/area/index.cfm?fareaid=107.

[93] "NASA – Juno." NASA 4 Mar. 2013 http://www.nasa.gov/mission_pages/juno/main/index.html.

[94] "Small Sun-Watcher Proba-2 Offers Detailed View of Massive Solar Eruption." 9 June 2011. ESA 17 Apr. 2013 http://www.esa.int/Our_Activities/Technology/Small_Sun-watcher_Proba-2_offers_detailed_view_of_massive_solar_eruption.

of the comet's survival.[95] By the end of 2011, the mission had gathered about 400,000 images of the Sun, and made nearly 20 million in-situ ionospheric observations.[96]

CNES' solar metrology mission 'PICARD', launched on 15 June 2010, aims to improve our knowledge of how the Sun functions and of the influence of solar activity on the Earth's climate. PICARD will accomplish this by measuring absolute total and spectral solar irradiance, solar diameter and shape, and by probing the interior of the Sun using the helioseismology method. In this reporting period, among additional findings, the spacecraft observed a partial Sun eclipse and an unusual Sun-spot that extended eight times Earth's diameter.[97]

The Solar Dynamics Observatory (SDO) is the first NASA mission to operate under its Living With a Star (LWS) programme. Launched on 11 February 2010, its objectives are to determine how the Sun's magnetic field is generated and structured, and how this stored magnetic energy is released in the form of the solar wind, energetic particles and variations in the solar irradiance. The spacecraft is comprised of three scientific experiments: the Atmospheric Imaging Assembly (AIA), EUV Variability Experiment (EVE) and the Helioseismic and Magnetic Imager (HMI).[98] The SDO's global view of the Sun facilitates research that focuses on the previously unrecorded real fine structure of the star.[99]

The SOlar and Heliospheric Observatory (SOHO) continued to operate during this reporting period. As an international cooperation project between ESA and NASA, this EADS Astrium-et al.-built spacecraft was launched on 2 December 1995. The spacecraft orbits around the Sun in step with the Earth, at a distance of 1.5 million km from Earth, enabling an uninterrupted view of the star.[100] Its scientific objectives are to investigate the solar interior and explain the extreme heating of the solar corona and the mechanism by which the solar wind is produced and accelerated. Some of its key results include discovering new dynamic solar phenomena such as coronal waves and solar tornadoes, vastly improving our ability to forecast space weather, by giving up to 3 days' notice of adverse space weather, and monitoring the total solar irradiance, which is important in understanding the

---

[95] "Proba-2 Tracks Comet Lovejoy Through Sun's Fiery Corona." 22 Dec. 2011. ESA 17 Apr. 2013 http://www.esa.int/Our_Activities/Technology/Proba-2_tracks_Comet_Lovejoy_through_Sun_s_fiery_corona.

[96] "ESA's Space Weather Station Proba-2 Tracks Stormy Sun." 2 Dec. 2011. ESA 17 Apr. 2013 http://www.esa.int/Our_Activities/Technology/ESA_s_space_weather_station_Proba-2_tracks_stormy_Sun.

[97] "PICARD NEWS." CNES 17 Apr. 2013 http://smsc.cnes.fr/PICARD/GP_actualites.htm.

[98] "SDO | Solar Dynamics Observatory." NASA Goddard Space Flight Center 4 Mar. 2013 http://sdo.gsfc.nasa.gov/.

[99] SpaceNews Staff. "NASA Boasts Big Results from 5-minute Spaceflight." 28 Jan. 2013 SpaceNews 4 Mar. 2013 http://www.spacenews.com/article/nasa-boasts-big-results-from-5-minute-spaceflight.

[100] About the SOHO Mission. "SOHO Fact Sheet." SOHO – Solar and Heliospheric Observatory 4 Mar. 2013 http://sohowww.nascom.nasa.gov/about/docs/SOHO_Fact_Sheet.pdf.

impact of solar variability on the Earth's climate.[101] SOHO has helped to define what occurs during CMEs by providing simultaneous images of reactions on the sun and further out in the corona.[102] The SOHO mission has very significantly exceeded its expected lifetime of 2 years and on 19 November 2010, it was extended until 31 December 2014.[103]

In addition to SOHO, NASA's Solar TErrestrial RElations Observatory (STEREO) continued to operate. STEREO is made up of two space-based observatories, i.e. STEREO-A travelling in a smaller and faster orbit (ahead of Earth's orbit), and STEREO-B trailing behind with a larger and slower orbit; these spacecraft are now 180° apart relative to the Sun.[104] They provide new insights into CMEs, including detecting and processing data that enables the tracking of CMEs headed toward Earth. STEREO also contributes to crowd-sourced data analysis, using data analysed by the public to make a predictions of solar storms that can reach Earth, it has captured the first-ever images of the entire surface of the Sun, and has been used to discover more than 122 new eclipsing binary stars and hundreds more variable stars.[105] The Deep Space Climate ObserVatoRy (DSCOVR), originally built to conduct observations of the Earth's climate will be re-purposed as a space weather and solar storm warning satellite,[106] to be launched by SpaceX in 2014.[107]

The Hinode (Solar-B) probe, led by JAXA in collaboration with NASA, the Science and Technology Facilities Council (STFC, U.K.) and ESA, was launched in September 2006, with the mission of studying the solar magnetic field. The project explores the solar magnetic fields of the Sun to better understand the mechanisms that power the solar atmosphere and drive solar eruptions. In this period, its solar optical telescope followed a special observation programme taking images of Comet Lovejoy as it passed by the Sun.[108] The Advanced Composition Explorer (ACE), a NASA mission launched in August 1997, with its primary science objective of measuring the composition of the solar corona, wind, interplanetary

---

[101] "Solar and Heliospheric Observatory Homepage." NASA 25 Aug. 2011 http://sohowww.nascom.nasa.gov/.

[102] "Approaching 17 Years Of Observations For ESA/NASA's SOHO Spacecraft." 5 Dec. 2012. redOrbit 4 Mar. 2013 http://www.redorbit.com/news/space/1112742788/17-years-observations-esa-nasa-soho-spacecraft-120512/.

[103] SOHO. The Project Main Steps. 29 Nov. 2012. CNES 5 Mar. 2013 http://smsc.cnes.fr/SOHO/.

[104] "First Ever STEREO Images of the Entire Sun." 6 Feb. 2011. NASA 5 Mar. 2013 http://www.nasa.gov/mission_pages/stereo/news/entire-sun.html.

[105] "NASA – STEREO." NASA 5 Mar. 2013 http://www.nasa.gov/mission_pages/stereo/main/index.html.

[106] Clark, Stephen. "Spaceflight Now | Breaking News | NOAA taps DSCOVR satellite for space weather mission." 2 Feb. 2011. Spaceflight Now 25 Aug. 2011 http://spaceflightnow.com/news/n1102/21dscovr/.

[107] Leone, Dan. "SpaceX To Launch Two Military Satellites." 5 Dec. 2012. SpaceNews 5 Mar. 2013 http://www.spacenews.com/article/spacex-to-launch-two-military-satellites.

[108] "Hinode – News Archive." 16 Jan. 2012. NASA 17 Apr. 2013 http://solarb.msfc.nasa.gov/news_resources_2012.html.

particles, the local interstellar medium and galactic matter, continued to collect data to improve forecasts and warnings of solar storms, and continues to be expected to maintain its orbit until 2024.[109] The Reuven Ramaty High Energy Solar Spectroscopic Imager (RHESSI), a NASA SMall EXplorer (SMEX) mission with the objective of exploring the basic physics of particle acceleration and energy release in solar flares also continued to operate. By 22 Feb 2012, the spacecraft had undergone its third 5-week anneal procedure to rejuvenate its detectors from the effects of radiation damage, and had resumed collecting solar X-ray and gamma-ray data.[110]

2.3.9 *Outer Solar System Exploration and Observation*

Near-Earth Objects continue to be widely discussed as a target for human exploration, with renewed interest in the study of comets and asteroids. In addition to this topic, the following section aims, to cover some of the significant discoveries made about celestial bodies outside the Solar System.

Following Rosetta's successful flyby of the asteroid Lutetia on 10 July 2010, and its discovery that an object previously thought to be a comet is in fact a pair of colliding asteroids, the probe was put into hibernation while transiting in deep space on 8 June 2011. In 2014, Rosetta will release a lander that will make the first landing on a cometary nucleus, that of 67P/Churyumov-Gerasimenko.[111] The Herschel infrared telescope continued making significant discoveries in this reporting period, finding Saturn's moon Enceladus to be the only moon in the solar system to influence the chemical composition of its parent by continuously expelling jets of water into Saturn's upper atmosphere.[112] It made the first confirmed detection of oxygen molecules created in the nearby Orion star-forming complex.[113] In one finding, Herschel detected water in a comet with nearly the exact level of deuterium as water on Earth, adding support to the idea of comets as the primary source of Earth's oceans,[114] while further observations revealed evidence of a vast amount of water vapour emanating from ice on dust grains in a disc around the TW Hydrae

---

[109] "Advanced Composition Explorer (ACE) Mission Overview." California Institute of Technology 17 Apr. 2013 http://www.srl.caltech.edu/ACE/ace_mission.html.

[110] "RHESSI STATUS." 22 Feb. 2012. NASA 17 Apr. 2013 http://hesperia.gsfc.nasa.gov/rhessi2/home/news-ressources/status/.

[111] "ESA – Rosetta." European Space Agency 25 Aug. 2011 http://www.esa.int/esaMI/Rosetta/.

[112] "Enceladus rains water onto Saturn." 26 July 2011. ESA 17 Apr. 2013 http://www.esa.int/Our_Activities/Space_Science/Herschel/Enceladus_rains_water_onto_Saturn.

[113] "Astronomers Searching for Oxygen Can Breathe More Easily." 1 Aug. 2011. ESA 17 Apr. 2013 http://www.esa.int/Our_Activities/Space_Science/Herschel/Astronomers_searching_for_oxygen_can_breathe_more_easily.

[114] "Did Earth's Oceans Come From Comets?" 5 Oct. 2011. ESA 17 Apr. 2013 http://www.esa.int/Our_Activities/Space_Science/Herschel/Did_Earth_s_oceans_come_from_comets.

star.[115] Herschel's counterpart Planck, a microwave observatory, was launched in May 2009 to study the Cosmic Microwave Background (CMB) – the ancient radiative 'fingerprint' of the Big Bang. In this reporting period, Planck completed its survey of the remnant light from the Big Bang, revealing previously undiscovered islands of cold gas and a haze of microwaves throughout the galaxy.[116]

The COnvection, ROtation and planetary Transits (COROT) space telescope operated by CNES probes the inner structure of stars using stellar seismology and is used to detect extrasolar planets. This astronomy mission, launched on 27 December 2006, announced the discovery of ten new planets at the beginning of this reporting period.[117]

On 4 November 2010, the Extrasolar Planet Observations and characterisation/ deep impact eXtended Investigation (EPOXI) flew past comet Hartley 2, returning images of the comet that provide new information on the role comets may have in planetary formation.[118]

Exoplanets are planets that orbit stars other than our solar system's Sun. NASA's Kepler space telescope mission was recently extended to 2016 with the continued aim of finding Earth-sized planets in the habitable zone of other solar-like oscillating stars, where liquid water could exist on their surfaces.[119] This reporting period resulted in the discovery of Kepler-14b, a planet 8.4 times the mass of Jupiter with an orbital period of nearly 7 days.[120] The recently discovered TrES-2b was found to be the darkest known exoplanet reflecting less than 1 % of the starlight falling on it; believed to be due to extremely high temperatures and tidal lock with its star.[121] Kepler-22b was found to be the first planet in the habitable zone of a sun-like star.[122] And a new class of circumbinary planets was established when the Kepler-

[115] "Herschel Detects Abundant Water in Planet-Forming Disc." 20 Oct. 2011. ESA 17 Apr. 2013 http://www.esa.int/Our_Activities/Space_Science/Herschel/Herschel_detects_abundant_water_in_planet-forming_disc.

[116] "Planck Steps Closer to the Cosmic Blueprint." 13 Feb. 2012 ESA 17 Apr. 2013 http://www.esa.int/Our_Activities/Space_Science/Planck/Planck_steps_closer_to_the_cosmic_blueprint.

[117] "CaRoT discovers 10 new extra-solar planets." 14 June 2011. CNES 17 Apr. 2013 http://smsc.cnes.fr/COROT/GP_actualite.htm.

[118] "SOHO Watches a Comet Fading Away." 28 Jul. 2011. NASA 17 Apr. 2013 http://www.nasa.gov/mission_pages/epoxi/index.html.

[119] "NASA Approves Kepler Mission Extension." 4 Apr. 2012. NASA 16 Apr. 2013 http://kepler.nasa.gov/news/nasakeplernews/index.cfm?FuseAction=ShowNews&NewsID=199.

[120] "Kepler-14b." 25 Oct 2011. NASA 16 Apr. 2013 http://kepler.nasa.gov/Mission/discoveries/kepler14b/.

[121] "Alien World is Blacker than Coal." 11 Aug. 2011. Harvard-Smithsonian Center for Astrophysics 16 Apr. 2013 http://www.cfa.harvard.edu/news/2011/pr201121.html.

[122] "Kepler-22b, our first planet in the habitable zone of a Sun-like Star." 5 Dec. 2011. NASA 16 Apr. 2013 http://kepler.nasa.gov/news/nasakeplernews/index.cfm?FuseAction=ShowNews&NewsID=165.

35 planet was found to orbit two sun-like stars that eclipsed one another at irregular intervals.[123]

In the reporting period, NASA's NEO Wide-field Infrared Survey Explorer (WISE) mission observed Earth's first Trojan asteroid that shares its orbit at a stable distance.[124] It also discovered the coolest class of stars yet, with some of its confirmed 100 brown dwarfs bearing estimated atmospheric temperatures that are near 25 °C.[125] And data from the WISE mission has lead to the best estimate of potentially hazardous asteroids (PHAs) that represent an impact hazard with Earth; finding roughly 4,700 ± 1,500 PHAs with diameters larger than 100 m.[126]

In addition to discovering 14 of the coldest stars known in our universe,[127] and contributing to observations of dust clouds that may help identify Earth-like planets around other stars,[128] the recently-extended Spitzer infrared space telescope mission observed infrared light emanating from a "super-Earth" planet in another solar system[129] and observed an unprecedented simultaneous elliptical galaxy with another thin disk existing in its interior.[130]

The James Webb Space Telescope (JWST), the successor to the Hubble Space Telescope (HST), survived the broad US budget cuts that threatened to derail the spacecraft from its planned 2018 Ariane 5 launch date. The spacecraft will investigate the formation of the first galaxies, planetary systems, and stars. Segments of the large primary mirror already underwent cryogenic testing throughout 2011.[131] And on 9 May 2012, NASA received the JWST's first completed instrument – the highly light-sensitive European-built Mid-Infrared Instrument (MIRI).[132]

---

[123] "Kepler Discovery Establishes New Class of Planetary System." 11 Jan. 2012. NASA 16 Apr. 2013 http://kepler.nasa.gov/news/nasakeplernews/index.cfm?FuseAction=ShowNews&NewsID=180.

[124] "NASA's WISE Mission Finds First Trojan Asteroid Sharing Earth's Orbit." 27 Jul. 2011. NASA 16 Apr. 2013 http://www.nasa.gov/mission_pages/WISE/news/wise20110727.html.

[125] "NASA's WISE Mission Discovers Coolest Class of Stars." 23 Aug. 2011. NASA 16 Apr. 2013 http://www.nasa.gov/mission_pages/WISE/news/wise20110823.html.

[126] "NASA Survey Counts Potentially Hazardous Asteroids." 16 May 2012. NASA 16 Apr. 2013 http://www.nasa.gov/mission_pages/WISE/news/wise20120516.html.

[127] "The Coolest Stars Come Out of the Dark." 24 June 2010. NASA 15 Apr. 2013 http://www.spitzer.caltech.edu/news/1137-feature10-08-The-Coolest-Stars-Come-Out-of-the-Dark.

[128] "Catch a Planet by the tail." 9 July 2010. NASA 15 Apr. 2013 http://www.spitzer.caltech.edu/news/1150-feature10-10-Catch-a-Planet-By-the-Tail.

[129] "NASA's Spitzer Sees The Light of Alien 'Super Earth'." 8 May 2012 NASA 16 Apr. 2013 http://www.spitzer.caltech.edu/news/1419-ssc2012-07-NASA-s-Spitzer-Sees-The-Light-of-Alien-Super-Earth-.

[130] "NASA's Spitzer Finds Galaxy with Split Personality." 24 Apr. 2012. NASA 16 Apr. 2013 http://www.spitzer.caltech.edu/news/1412-ssc2012-06-NASA-S-Spitzer-Finds-Galaxy-with-Split-Personality.

[131] "The James Webb Space Telescope." NASA 25 Aug. 2011 http://www.jwst.nasa.gov/.

[132] SpaceNews Staff. "First JWST Instrument Handed Over to NASA." 14 May 2012. SpaceNews 15 Apr. 2013 http://www.spacenews.com/article/first-jwst-instrument-handed-over-nasa.

The 'Peter and Patricia Gruber Foundation', in association with Yale University, awarded the 2011 Gruber Cosmology Prize to four astronomers for finding evidence of dark matter. The team, led by Marc Davis of the University of California, Berkeley, with George Efstathiou (Cambridge, England), Carlos Frenk (Durham University, England), and Simon White (Max Planck Institute, Germany), created key computer simulations in the 1990s that mapped the large-scale distribution of matter in the universe. On realising that the universe was not just a uniform scattering of galaxies, but a cosmic web of galaxies grouped into filaments separated by expansive voids, Davis and his team developed computer models that conducted vast calculations to simulate billions of years of galaxy evolution; these models show that a type of cold dark matter could produce the topography observed in space. These 20+ year old simulations convinced most experts of the existence of dark matter, despite the subsequent fruitless search to find out what dark matter is. Ten years later, "dark energy" was discovered to permeate the universe. Today, cosmologists estimate that the universe is composed of 4.6 % 'ordinary' matter (i.e. stars, planets, life, etc.), 23.3 % dark matter, and 72.1 % dark energy.[133]

In late 2010, NASA's Rossi X-ray Timing Explorer (RXTE) satellite detected what astronomers believe is the first neutron star to burst the way that models predict. Researchers had developed models to theoretically predict how a neutron star bursts, but the X-ray observations from nearly 100 exploding neutron stars over a period of three decades failed to validate those predictions. Neutron star bursts were mostly seen at low mass-accretion rates, while bursts at high-mass accretion rates were rare, i.e. they had not observed more plasma exiting at a more frequent rate in these bursts. However, the RXTE satellite detected X-ray spikes from a binary star system in Terzan 5, whose data showed evidence of higher mass-accretion rates, validating the theoretical prediction. It should be noted that the rotation of a neutron star may affect its thermonuclear burning, possibly causing friction between layers of plasma and the neutron star's surface, thereby affecting the rate of nuclear burning. In this latest case, the observed neutron star had a much slower rate of rotation than previously observed neutron stars, i.e. 11 rotations per second compared with typical neutron stars rotating between 200 and 600 times per second. Newer models may need to consider rotation when predicting how a neutron star bursts.[134]

---

[133] Sanders, Robert. "Gruber Cosmology Prize honors 'dark matter' astronomers." 1 June 2011. UC Berkeley News Center 18 Mar. 2013 http://newscenter.berkeley.edu/2011/06/01/gruber-cosmology-prize-honors-dark-matter-astronomers/.

[134] SpaceNews Staff. "In a First, Neutron Star Bursts as Models Predicted." SpaceNews 5 Mar. 2012: 8.

2.3.10 International Cooperation in Space Exploration

While most states desire autonomy in space, the benefits of shared capabilities, including efficiencies of scale in component development make international cooperation the better approach in the current age of space exploration. This realization is almost universally acknowledged on a political level, despite the fact that the U.S., Europe, Russia and Japan have developed their own cargo transfer vehicles for the ISS.

The Third International Conference on Space Exploration, jointly organised by ESA, the European Commission, the Polish Presidency of the EU; and hosted by the Italian government, highlighted the importance of international cooperation in the pursuit of sustainable space exploration for the benefit of humankind.[135] From that conference, government representatives committed to begin a continued open structured high-level policy dialogue on space exploration at the intergovernmental level for the benefit of humankind.[136]

The International Space Exploration Coordination Group (ISECG) continued to discuss architecture options for future human space exploration. On 30 August 2011, senior managers representing 10 of its 14 member space agencies, met to advance the group's Global Exploration Roadmap (GER) for coordinated space exploration. The GER's first iteration will inform and help focus planning in the areas of planetary robotic exploration, advanced technology development and use of the ISS to help in preparation for future exploration.[137]

2.4 Satellite Applications

2.4.1 Space-Based Communications

In 2011 and 2012, the satellite services industry continued to function with remarkable resilience in view of the adverse global financial conditions. The industry's upward trend is credited to its inherently global nature, allowing it to profit from the quick economic recovery of emerging markets (e.g. in South East Asia and South

[135] "Europe opens the way for a political dialogue on space exploration." 10 Nov. 2011. European Commission – Enterprise and Industry 15 Apr. 2013 http://ec.europa.eu/enterprise/newsroom/cf/itemdetail.cfm?item_id=5555&lang=en.

[136] "Third International Conference on Exploration – First meeting of the High-level International Space Exploration Platform – Declaration." 10 Nov. 2011. European Commission – Enterprise and Industry 15 Apr. 2013 http://ec.europa.eu/enterprise/newsroom/cf/_getdocument.cfm?doc_id=7040.

[137] "Space Agency Senior Managers Meet to Discuss a Global Exploration Roadmap." 30 Aug. 2011. International Space Exploration Coordination Group 15 Apr. 2013 http://www.globalspaceexploration.org/web/isecg/news/2011-08-31.

America). The industry continues to expand its technology development programmes with further investments through the ordering or launch of larger spacecraft with enhanced signal power and transponder capacity. This industry demonstrates an acute ability to achieve the right mixture of investing in innovative technologies and new services while consolidating current operations, which has boosted the industry's revenue for one more consecutive year.[138]

2.4.2 Space-Based Positioning, Navigation and Timing Systems

The development of GNSS systems continued during this period, with the relevant actors increasing their efforts to complete their full satellite constellations.

In Europe, a milestone was reached in the second half of 2011, with the launch of the first two Galileo GNSS in-orbit validation (IOV) satellites onboard a Europeanized Soyuz rocket launched from the European spaceport in French Guiana on 21 October 2011. Four Galileo IOV satellites were being developed during the reporting period, with the other two IOV satellites set for launch on 12 October 2012. The system, with the European Commission as a major stakeholder, is being constructed by a consortium led by Astrium Satellites and Thales Alenia Space. The successful launch signified the opening of the system's operational deployment phase, with the scheduled launch of 14 satellites out of a total of 30 required to achieve full operational capability.[139]

Russia's Glonass GNSS constellation is being restored, with its operational capacity increasing by five satellites in 2011. Next-generation model GLONASS-K satellites will succeed the GLONASS-M model of which 23 were operational as at November 2011, with additional spacecraft in production.[140] The new spacecraft would most likely incorporate significant technical improvements, including a new more accurate timing device and a non-pressurized structure, bringing its operational performance close to U.S. and European standards. The constellation requires 24 operational satellites to provide complete global navigation coverage, with three sets of eight satellites operating on three orbital planes.[141]

China intended to field its regional satellite navigation capability by the end of 2012 and to complete the deployment of its entire 35 satellite Beidou GNSS constellation by 2020. In light of that expedited development, Japan decided to proceed with the development of the Quasi-Zenith Satellite System (QZSS), the

[138] See generally State of the Satellite Industry Report.

[139] De Selding, Peter B. "Soyuz Lofts Two Galileo Satellites In Debut from European Spaceport." Space News 24 Oct. 2011: 1.

[140] Clark, Stephan. "Proton rocket replenishes Russian navigation system." 4 Nov. 2011. Spaceflight Now 12 May 2012 http://www.spaceflightnow.com/news/n1111/03proton/.

[141] "Glonass System." Glonass.it 12 May 2012 http://www.glonass.it/eng/glonass-story.aspx.

country's future regional satellite navigation constellation. QZSS is likewise scheduled to be fully operational by 2020, but is designed to augment the regional accuracy of the GPS signal. The system scheme would provide Japanese authorities with a more accurate, secure and independent service. The future deployment of four spacecraft would allow for 24 h regional coverage, while deployment of the full seven satellite constellation would dramatically decrease Japan's dependence on GPS for regional coverage.[142]

2.5 Technology Developments

The new developments in space-related technologies merit discussion, as they indicate current trends in space technology and reveal the focus of relevant policy decisions. The following chapter provides an overview of recent activities, spanning across all major space faring powers and institutions, both established and emerging ones.

2.5.1 Propulsion

Significant advancements have occurred within Europe and the US in the field of propulsion. Newly developed rockets and other methods of propulsion are being explored with favourable results.

In Europe, the Italian-led Vega small-satellite launcher made its successful inaugural launch on 13 Feb, 2012. The Vega programme, approved by ESA in 1998, spent 9 years developing the launcher at a cost of 710 million Euros.[143] Vega is a single body launcher with three solid propulsion stages and an additional liquid propulsion upper module used for attitude and orbit control, and satellite release. The launcher is capable of placing a 1,500 kg satellite in a 700 km low Earth orbit.[144]

And ESA's Proba-2 micro-satellite demonstrated a whole new space technology when its experimental xenon gas 'resistojet' engine was re-pressurized with nitrogen gas produced from a solid material rather than itself being stored in a pressurized state. The four 'cool-gas generators' containing the solid material are on the

[142] Kallender-Umezu, Paul. "Japan Commits To Deploying Satellite Navigation System by 2020." Space News 17 Oct. 2011: 14.

[143] De Selding, Peter. "Europe's Italian-led Vega Rocket Succeeds in Debut." SpaceNews 20 Feb. 2012: 10.

[144] Launch Vehicles – Vega. 30 Oct. 2012. ESA 8 Mar. 2013 http://www.esa.int/Our_Activities/Launchers/Launch_vehicles/Vega.

Proba-2, and once triggered, each produces more than 250 times its own volume in pure nitrogen gas.[145]

In the U.S., Pratt & Whitney Rocketdyne (PWR) began test-firing its J-2X upper-stage rocket engine in mid-July. The liquid propellant rocket engine maker is building nine J-2X engines: seven for development tests and two for certification tests.[146] The engine, previously envisioned as the upper stage engine for the Ares 5 rocket to be used in the cancelled constellation programme, is the upper stage of the heavy-lift U.S. Space Launch System.

Space Exploration Technologies (SpaceX) developed a launch escape system engine to be used when flying astronauts in its Dragon capsule to the International Space Station. The engine is an advanced version of the Draco engines that the Dragon uses to manoeuvre while in orbit and on reentry. These "SuperDraco" engines are built into the Dragon's side walls, and the eight engines will produce up to 120,000 lb of axial thrust. [147] While the propellant used in its hypergolic engines spontaneously ignites when coming in contact with an oxidizer; hypergolic propellants tend to be corrosive and/or extremely toxic to handle.

In an effort to move away from harmful fuel sources, Innovative Space Propulsion Systems (ISPS) is developing rocket engines that run on an environmentally benign propellant, "NOFBX". SpaceX will the fly the thruster test bed, 'ISPS NOFBX Green Propellant Demonstration', to the ISS in mid-2013; which will be placed on the outside of the European Columbus module to undergo a series of in-space performance validation tests.[148] Developed by Firestar Technologies, NOFBX is a high-performance nitrous oxide/fuel/emulsifier blended monopropellant that is non-toxic, low cost and easy to produce. Capable of production from widely available chemicals, the fuel can be transported without excessive precautions.[149] In fact, NOFBX surpasses solid and bipropellants in many characteristics.[150] The current fuel used on most types of spacecraft is hydrazine, a highly toxic chemical that produces environmental hazards and pollution in addition to operational hazards and longer launch processing times.[151]

[145] "Proba-2 Fuel Tank Refilled from 'Solid Gas'." 24 Aug. 2011. ESA 17 Apr. 2013 http://www.esa.int/Our_Activities/Technology/Proba_Missions/Proba-2_fuel_tank_refilled_from_solid_gas.

[146] SpaceNews Staff. "J-2X Engine Test Fired at NASA Stennis Space Center." SpaceNews 1 Aug. 2011: 9.

[147] SpaceNews Staff. "SpaceX Test Fires SuperDraco Engine for Dragon Launch Escape System." SpaceNews 6 Feb. 2012: 9.

[148] SpaceNews Staff. "ISS-bound Propellant Demo Passes NASA Safety Review." SpaceNews 28 May 2012: 9.

[149] Messier, Doug. "A Non-Toxic Fuel From the Mojave Desert." 9 Aug. 2011. Parabolic Arc 8 Mar. 2013 http://www.parabolicarc.com/2011/08/09/a-non-toxic-fuel-from-the-mojave-desert/.

[150] Mungas, Greg. "NOFBX® Monopropulsion Overview." 14th Annual FAA Commercial Space Transportation Conference 9 Feb. 2011 https://www.aiaa.org/uploadedFiles/About-AIAA/Press_Room/Key_Speeches-Reports-and-Presentations/Greg_Mungas.pdf.

[151] SpaceNews Staff. "NASA Seeks Green Alternative to Highly Toxic Hydrazine." SpaceNews 13 Feb. 2012: 8.

While many telecommunications satellites have relied on electric propulsion to maintain steady orbit in the last decade, a new trend involves the use of electric thrusters to carry the satellite from transfer orbit to final geostationary position. In order to obtain a significant reduction in weight at launch, Boeing's ABS and Satmex satellites will carry between 300 and 350 kg of xenon propellant for the electric thrusters, rather than use 2,000 kg of conventional fuel, as used with comparable satellites. Its 25 cm xenon-ion propulsion systems are installed on 18 larger satellites that are already in orbit, while subsequent systems will be upgraded with more fuel capacity. However, a downside in relying on electric thrusters stems from the extended time delay in getting a satellite into its final orbital position.[152] Nevertheless, the European Space Agency has also pursued electric propulsion in recent years, with missions including ESA's Smart satellite, and its Artemis technology demonstration satellite. And Astrium Satellites has electric thrusters on six commercial telecommunications satellites, purely for orbit maintenance.[153]

2.5.2 *Information Technology*

At the beginning of the reporting period, concerns on the possible interference by LightSquared's hybrid satellite-terrestrial broadband network on L-band signals had already raised concern among the GPS Community. LightSquared is licensed to use two specific 10-MHz blocks within the L-band frequency range; however GPS signals are also transmitted using the L-Band spectrum.[154] The ground-based transmissions by the former overpower the generally weak GPS signals from space. The consequence of such interference could result in aviation users losing GPS capabilities around densely populated areas with stations spaced out 400–800 m apart, and aircraft below 3,040 m in altitude could not rely on GPS in some areas; additionally, police cars could not acquire GPS signals within 182 m of a LightSquared tower that broadcasts at 15 kW.[155] Of the various mitigation options considered, the only viable option would be for LightSquared to acquire rights to another part of the electromagnetic spectrum.[156] On 14 May 2012, LightSquared

[152] De Selding, Peter. "ABS, Satmex Banding Together for Boeing Satellite Buy 13 Mar. 2012. SpaceNews 8 Mar. 2013 http://www.spacenews.com/article/abs-satmex-banding-together-boeing-satellite-buy.

[153] De Selding, Peter. "Europeans Vow To Check Boeing Advantage in All-electric Sats." SpaceNews 14 May 2012: 4.

[154] Leone, Dan. "LightSquared Says Altered Plan Would Limit GPS Disruption." Space News 27 June 2011: 14.

[155] Brinton, Turner. "Reports: LightSquared Plan Poses Unacceptable Risk to GPS Service." Space News 13 June 2011: 1.

[156] Id. at 4.

filed for bankruptcy protection following earlier FCC license restrictions on the proposed mobile satellite-terrestrial broadband network.

ViaSat Corp.'s ViaSat-1 satellite, launched on 20 Oct. 2011, will deliver a total of 140 gigabits per second of throughput capacity by employing the largely unused Ka-band portion of the radio spectrum and by separating coverage into dozens of spot beams, i.e. 63 spot beams directed at the US.[157] In Europe, Eutelsat's Ka-Sat satellite was launched on 26 Dec. 2010, and began providing commercial broadband service on 31 May 2011.[158] This satellite, the first of a new generation of high throughput satellites under development by Astrium, has a throughput of about 70 gigabits per second. Ka-Sat distributes its bandwidth through 82 spot beams, each carrying 900 megabits per second of capacity. These beams are evenly distributed to provide services in Europe and Africa. The throughputs of these satellites are respectively 14 and 7 times that of the Spaceway 3 satellite, the previous broadband capacity record-holder.[159] Advantages from the use of high capacity satellite systems include the economic delivery of Internet media with 100 times the capacity of Ku-band and 10 times the throughput of any previous Ka-band satellite, and in-orbit costs per gigabyte at a fraction of even the newest satellites in orbit.[160]

Raytheon Space and Airborne Systems has increased its rate of data exploitation from its new space-based Advanced Responsive Tactically Effective Military Imaging Spectrometer (ARTEMIS) payload. Operating as the primary instrument on the TacSat-3 satellite, ARTEMIS's hyperspectral imaging capability enables the U.S. military to identify materials on the ground based on the way they reflect and absorb sunlight. The imaging capability is applicable to both military and humanitarian uses, e.g. while conducting a 13-month demonstration of its imaging capability, TacSat-3 was also used to aid in relief efforts following devastating earthquakes in Haiti and Chile in 2010. Operating in the short-wave infrared region of the electromagnetic spectrum, the ARTEMIS payload develops about 100 hypercube data products per month, where each hypercube is composed of 400 image slices; since its launch in May 2009, hypercube processing speed has increased tenfold, providing useful information to troops in a dramatically reduced timeframe. This improved analysis of space-based hyperspectral data is partly attributable to Raytheon's work on exploitation algorithms for aerial hyperspectral data. While the short-wave infrared region of the magnetic spectrum makes the ARTEMIS capable of observing solid materials, a future upgrade to the sensor to

[157] De Selding, Peter. "Long-awaited ViaSat-1, Touted As DSL Competitor, is in Orbit." SpaceNews 24 Oct. 2011: 4.

[158] De Selding, Peter B. "Ka-Sat Enters Services as European Broadband Market Heats Up." 6 June 2011: 16.

[159] De Selding, Peter. "Long-awaited ViaSat-1, Touted As DSL Competitor, is in Orbit." SpaceNews 24 Oct. 2011: 4.

[160] "High-Capacity Satellite System and ViaSat-1." ViaSat 9 Mar. 2013 http://www.viasat.com/broadband-satellite-networks/high-capacity-satellite-system.

also collect long-wave infrared imagery would enable it to detect and characterize gas plumes and atmospheric content as well.[161]

2.5.3 Spacecraft Operations and Design

In Europe, the European Space Agency's (ESA) Rosetta comet chasing spacecraft had to be put into hibernation for a period lasting 2.5 years on 8 June 2011. Now switched off and placed into an unprecedented 90 s rotation, the satellite's two 14-m-long solar arrays produce about 500 W of power to keep the thermal-control system and the on-board computer functioning as the spacecraft journeys to a point some 790 million km from the sun. The spacecraft was launched in March 2004 to rendezvous with the 67/P Churyumov-Gerasimenko comet in mid-2014. While en route, Rosetta developed attitude-control and propulsion system problems: two of its four reaction wheels started showing signs of degradation; and there was a leak in a helium-pressurization system which enables the propellant reservoir to direct fuel to Rosetta's on-board thruster engines. In response to the former issue, the Rosetta operations team will spend the hibernation period developing software to operate the satellite with three, and even two reaction wheels if necessary. Currently, the Rosetta spacecraft needs only three wheels to function, so when awakened in January 2014, one of the degraded wheels can be used as a spare. With the latter propulsion problem, ESA's first plan involved re-pressurizing Rosetta for future operations, allowing for maximum fuel efficiency; however that route had the potential to aggravate the current leak. The alternate approach adopted by ESA's European Space Operations Center (ESOC) was to allow Rosetta to use more fuel than originally planned and fly a less-efficient route to the comet. Nevertheless, with this approach, Rosetta would still have enough fuel to complete its comet rendezvous by mid-2014.[162]

ESA's Planck space observatory reached the end of its operational ability, completing five full-sky surveys of the cosmic microwave background (CMB) since its launch in May 2009. Viewed as a strong success, the mission had called for a minimum of two such surveys to collect data that might allow scientists to get a better understanding of the Big Bang and the very early universe. Planck uses two key sensors, the High Frequency Instrument (HFI) and Low Frequency Instrument (LFI), which require cooling the sensors to one-tenth of a degree above absolute zero (minus 273.05 °C) in order to detect the faint CMB. As expected, the HFI's coolant ran out around 14 Jan. 2012, ending the instrument's ability to detect faint CMB. The LFI will continue to gather calibration data throughout 2012 until its

[161] Brinton, Turner. "Raytheon Touts Progress on Hyperspectral Data Exploitation." SpaceNews 13 Jun. 2011: 14.

[162] De Selding, Peter. "ESA Controllers Buy Time To Fix Glitches on Comet Chaser." SpaceNews 20 Jun. 2011: 10.

coolant reservoir is likewise depleted. Planck findings about the Big Bang and CMB will be released in two stages between 2013 and 2014; i.e. observations from the first 15.5 months of operation will be published in early 2013, with a full data release of the entire mission occurring in the following year.[163]

In the last reporting period, the British imaging solutions company E2v had been developing image sensors for Euclid, a dark-energy mapping mission currently under development for ESA.[164] E2v was also awarded a contract with ESA to develop image sensors for Plato, a space telescope that searches for planets; the Plato mission competed with the Euclid and Solar Orbiter space science missions for two launch opportunities under ESA's Cosmic Vision program.[165] Euclid and Solar Orbiter were ultimately selected; however, if Plato had been selected, it would have been equipped with 34 mini-telescopes, containing four back-illuminated charge-coupled devices on each of them. Plato would have orbited the Sun 1.5 million km beyond Earth for 6–8 years, and surveyed up to 40 % of the sky. Euclid is set for a launch in 2019 aboard a Europeanized version of Russia's medium-lift Soyuz rocket. Equipped with a 1.2 m-aperture telescope and other observing instruments furnished mainly by European laboratories, this spacecraft will study the universe's quickening expansion and its dark energy catalyst. Solar Orbiter will study solar phenomena from an unprecedented orbit altitude of 41.9 million km beyond the sun's surface. Slated to launch in 2017, ESA will develop the spacecraft as a bilateral endeavour with NASA; where the former will provide a majority of the spacecraft's payload and instruments, and the latter will provide part of the satellite's payload and the use of an Atlas rocket.[166]

Across the Atlantic, since the retirement of the U.S. Space Shuttle Fleet in July 2011, NASA has focused on destinations that extend deeper in space. NASA's Technology Demonstration Missions programme selected three pioneering technology demonstrations to aid in this pursuit, i.e. a deep space atomic clock project, a solar sail demonstration mission, and an optical in-space communication system. The first mission will launch and validate a miniature mercury-ion atomic clock that is ten times more accurate than today's systems; its precision timing will be vital to the navigation and performance of any deep space mission. The second technology is an expansion of the NanoSail-D's concept project; here, the Solar Sail demonstration mission will deploy and operate a sail that is seven times larger than any ever flown in space (38-by-38 m in area). In addition to its use in deep space missions that travel through space without the need of traditional propellant, solar sails may help set up systems to monitor space weather and clean up orbital debris. The third sponsored demonstration is the optical in-space communication system

[163] SpaceNews Staff. "Key Sensor Aboard Planck Runs Out of Vital Coolant." SpaceNews 23 Jan. 2012: 9.

[164] "E2v Tapped To Develop Image Sensors for Euclid." Space News 18 Oct. 2010:8.

[165] "E2v Picked To Develop Image Sensors for Plato." Space News 25 Oct. 2010: 8.

[166] De Selding, Peter. "Solar Orbiter, Euclid Chosen as ESA's Next Medium-class Missions." SpaceNews 5 Oct. 2011: 11.

that sends information with the use of encoded laser beams, thus allowing a spacecraft to transmit data at speeds up to 100 times faster than current capabilities. Whereas this latter project should be developed sometime in 2015, its two former projects are planned to debut 1 year earlier. The synergistic results of the clock and laser duo should greatly enhance navigation and communication capabilities on future spacecraft, and the additional benefit of a solar sail when travelling vast distances should reduce the need for heavier alternative propellant sources.[167]

NASA planned to launch its PhoneSat mission onboard Orbital Science Corp.'s Antares medium-lift rocket on its inaugural launch in early 2013. The PhoneSat project, led by the Ames Research Center, intends to demonstrate that cubesats can be produced at a cost below $10,000 using only commercial off-the-shelf components. Three identical PhoneSat 1.0 cubesats were made at a cost of about $3,500 each. At 10 cm in length per side, each cubesat holds an HTC Corp. Nexus One smartphone running Google's Linux-based Android operating system, and does not have propulsion capability. When launched, these cubesats will be released by an Isipod cubesat deployer, developed by Innovative Solutions in Space of the Netherlands. The cubesats will then drift in low Earth orbit for about 2 weeks sending data to its ground controllers on the health of its system, batteries, and sensors, and will also take photos and attempt to send them through its transmission, prior to re-entering Earth's atmosphere.[168]

The space agency also wants to conduct an unmanned test flight of its Orion Multi-Purpose Crew Vehicle (MPCV) aboard a Delta 4 Heavy rocket by 2014. If the undertaking can be secured by funds in the MPCV budget, the test flight would be 3 years ahead of its intended carrier rocket, the U.S. Space Launch System. The test would occur following the completion of two orbits around the Earth, when Orion would then re-enter the atmosphere at speeds verging on those that would occur during a return from deep space. The key purpose of this launch would be to test whether Orion's heat shield can withstand the forces of atmospheric re-entry at about 32,000 km/h.[169]

In the private sector, following the permanent grounding of the NASA space shuttle fleet, Boeing is developing its own Commercial Space Transportation-100 (CST-100) crew capsule, designed to send astronauts to the ISS as early as 2015. Boeing intends flight-testing CST-100 on three expendable Altas 5 rockets; four tests are planned, the first of which is a launch pad abort test planned for 2014. In 2015, three additional tests will be conducted involving an unmanned orbital flight

[167] Wall, Mike. "NASA Picks 3 Pioneering Tech Missions for Deep Space." SpaceNews 22 Aug. 2011: 10.

[168] Leone, Dan. "Smartphone-powered Cubesats To Fly on Antares Maiden Launch." SpaceNews 6 Feb. 2012: 10.

[169] Leone, Dan. "NASA Proposes Orion Test Flight in 2014." 8 Nov. 2011. SpaceNews 12 Mar. 2013 http://www.spacenews.com/article/nasa-proposes-orion-test-flight-2014>.

of the CST-100, an in-flight test of the capsule's abort system, and the first manned flight of the CST-100 at the end of the year.[170] During October 2011, Boeing applied hundreds of small sensors on a 7 % scale aluminium model of the CST-100 space capsule, and subjected it to ultra-high-speed winds. Over 20 different positions will be used when measuring the airflow across the model, to simulate aborted landing stages, with the goal to ensure structural soundness and safe operation.[171] Drop tests of the real CST-100 capsule were conducted in April and May 2012, whereupon after release from an altitude of 3.3 and 4.2 km, the capsule's three main parachutes deployed to slow its descent before six airbags inflated to cushion the landing. The second drop test added a drogue parachute which deploys first in order to pull the main parachute into full deployment, to demonstrate the full normal parachute deployment sequence.[172] Measuring 4.5 m across at its widest point, this seven-seat gumdrop-shaped capsule will be reusable for up to ten flights.[173]

Notwithstanding Boeing's progress in the development of the CST-100 under NASA's Commercial Crew Development program; as of 26 April 2012, Blue Origin had completed a series of wind tunnel tests to validate the shape of its reusable spacecraft under development. Blue Origin's spacecraft has a bi-conic shape to provide greater cross-range and interior volume than traditional gumdrop-shaped capsules without the need to use winged spacecraft. Over 180 wind tunnel tests were conducted in March and April 2012 to validate the spacecraft's aerodynamics during re-entry, including its ability to change its flight path to increase the number of landing opportunities available to the vehicle. Blue Origin is also testing its thrust chamber assembly for the BE-3, a liquid oxygen-liquid hydrogen engine, capable of producing 100,000 lb of thrust; and the company will also conduct a test of a pusher escape system, that will control the flight path of the capsule by use of a new thrust vector control system, later in 2012.[174]

Finally, in the previous reporting period, Space Exploration Technologies (SpaceX) had three Commercial Orbital Transportation Services (COTS) capsules, named Dragons, in varying stages of assembly. Despite expected cost increases and delays, as of 25 May 2012, following the successful launch and birthing of its cargo-carrying Dragon capsule to the International Space Station, SpaceX could now be cleared to begin commercial cargo operations to the station. Upon clearance, SpaceX could make its ISS deliveries under a fixed-price NASA contract

[170] SpaceNews Staff. "Boeing Picks Atlas 5 for CST-100 Test Flights." SpaceNews 8 Aug. 2011: 3.

[171] SpaceNews Staff. "Boeing CST-100 Model Undergoes Wind Tunnel Testing." SpaceNews 24 Oct. 2011: 9.

[172] SpaceNews Staff. "Boeing Drop Tests CST-100 Over the Nevada Desert." SpaceNews 9 Apr. 2012: 3; *see also* SpaceNews Staff. "Boeing Conducts Second CST-100 Parachute Test." SpaceNews 7 May 2012: 8.

[173] Chow, Denise. "Boeing's CST-100 Capsule Shooting for 2015 Debut." SpaceNews 30 Apr. 2012: 14.

[174] SpaceNews Staff. "Blue Origin's Space Vehicle Completes Wind Tunnel Tests." SpaceNews 7 May 2012: 9.

worth $1.6 billion. Having already received $336.7 million as an advance of the contract award, upon the completion of development and demonstration of the milestones, SpaceX will receive the final instalment of its $396 million COTS portion of its contract agreement with NASA. SpaceX launched Dragon aboard its Falcon 9 rocket on 22 May 2012, to which over the course of just over 3 days and 6 h, the capsule conducted a series of automatic orbital manoeuvres adjacent to the ISS prior to gradually entering within the range of capture of the station.[175]

2.5.4 Suborbital Activities

Virgin Galactic and similar US firms have been given a regulatory grace period extension on developing suborbital spacecraft without Federal Aviation Administration (FAA) imposed passenger and crew safety rules. The 2004 Commercial Space Launch Amendments Act barred the FAA from imposing those rules for a period of 8 years, unless an operator experienced a serious accident or an especially dangerous close call. In either circumstance, the FAA would be limited to restricting or prohibiting the craft's design features or operating practices resulting in those events. Whereas this grace period was due to finish by the end of 2012, with the expectation that commercial suborbital spaceflight had become established; this extension to September 2015 allows the fledgling commercial human spaceflight industry to develop without being weighed down with regulation, and allows operators to establish a base of safety-related best practices that the FAA could later-on convert into regulations.[176]

On a related note, Virgin Galactic planned to begin powered test flights of its SpaceShip Two suborbital launch vehicle in early 2012. While it previously planned to begin in 2011, the suborbital spacecraft had experienced a problem following a clean release from its carrier during its 16th glide test, held on 16 Sept. 2011. Experiencing a tail-stall due to an excessive downward pitch, the pilot was able to stabilize the craft at an altitude above 6 km and make a safe runway landing.[177] The company cleared an important regulatory hurdle in 2012, when the U.S. government granted Virgin Galactic a favourable EAR99 ruling, removing its suborbital operations from ITAR control. Without this ruling, the training and/or launching of non-U.S. citizens on Virgin Galactic suborbital flights would have constituted an export activity requiring federal approval, and thus the need for export licenses which could take months for each case. However, as the spacecraft's

---

[175] Ferster, Warren and Dan Leone. "SpaceX's Successful Mission Boosts Commercial Credibility." SpaceNews 25 May 2012: 1 + .

[176] Leone, Dan. "Private Spaceflight 'Learning Curve' Extension Approved." SpaceNews 13 Feb. 2012: 4.

[177] Leonard David. "Virgin Galactic's SpaceShipTwo Powered Flights Expected To Begin in 2012." SpaceNews 31 Oct. 2011: 13.

flight hardware is still under ITAR's export control, any disclosure of controlled technical data to a foreign national, regardless of whether disclosed within the U.S., is treated as an 'export' of the technical data. The company expects to fly its first paying customers in 2013.[178]

Other groups have seen mixed results in their suborbital spacecraft development. Both Masten Space Systems and Armadillo Aerospace are developing spacecraft to rival Virgin Galactic's SpaceShip Two. Funded in-part with seed money from NASA's Commercial Reusable Suborbital Research program – a programme intended to mature experimental suborbital launch technology and help create a commercial suborbital launch industry – during agency-funded flights, these groups are required to carry NASA payloads within near-space altitudes. Near-space is the grey-zone where the delimitation between airspace and outer space is uncertain, i.e. between 20 and 107 km in altitude. Masten is developing the Xaero; a vertical-takeoff, vertical-landing vehicle, that uses the firm's isopropyl alcohol- and liquid oxygen-burning Cyclops-AL-3 engine. Its successful first flight test, held on 29 June 2011, had the vehicle ascend to 1.8 m above ground level (its maximum altitude is intended to reach 30 km), hover briefly, and then descend to land. Armadillo Aerospace is developing the SuperMod suborbital rocket; another vertical-takeoff, vertical-landing rocket, derived from Armadillo's Module 1 (Mod) vehicle. Eleven seconds into its 11 June 2011 launch, the rocket veered sharply off course and thus an abort was triggered at 1.5 km in altitude. The failure was likely caused by a cracked standpipe that broke off inside the rocket's fuel tank. The companies shared a NASA Commercial Reusable Suborbital Research award in August 2010.[179]

And in an effort to reduce launch costs, NASA and the Pentagon's Operationally Responsive Space Office have developed and are testing an inexpensive microsatellite platform aboard a Terrier-Improved Orion suborbital rocket. Measuring 40 cm in diameter, the primary payload is a modular reconfigurable microsatellite bus that can be integrated and readied for launch in as few as 7 days at a cost of less than $1 million. The Small Rocket/Spacecraft Technology microsatellite bus, dubbed SMART, can be used for a variety of missions, e.g. optical imaging and radio-frequency applications. Following the launches, the payload may be recovered for data analysis.[180]

---

[178] Leone, Dan. "Virgin Galactic Granted License Exemption for Spaceflight Experience." SpaceNews 16 Apr. 2012: 22.

[179] SpaceNews Staff. "Suborbital Firms Have Mixed Results in Tests." SpaceNews 5 Jul. 2011: 3.

[180] SpaceNews Staff. "NASA Sounding Rocket Tests New Technology." SpaceNews 13 Jun. 2011: 3.

2.5.5 Other Technologies

Developments in technology and science continued to advance within the 2011–2012 period, with implications both within and outside the space sector.

Sweden's ACC Microtec, in association with the Swedish Defense Materiel Administration and the U.S. Air Force Research Laboratory, has developed plug-and-play avionics meant for use in the University of Tokyo's nanosatellite research programme. With technology on par with the U.S. Space Plug-and-play Avionics standard, the spinoff from Uppsala University's Angstrom Laboratory has developed remote terminal units and distributed power control units, which are designed to permit satellite builders to integrate their hardware with increased speed, and at a decreased cost, allowing for cost-effective development of the technology.[181]

NASA planned to demonstrate on-orbit satellite refuelling and repair technology with its Robotic Refuelling Mission (RRM) satellite mock-up in November 2011. With the aide of the Special Purpose Dexterous Manipulator (Dextre), a twin-armed Canadian-built robot that has been onboard the ISS since 2008, and with specially updated software, Dextre will use a set of satellite-servicing tools to perform simulated refuelling tasks on the RRM. The mock-up was lifted to the ISS on the final U.S. Space Shuttle launch on 8 July 2011, and it is mounted outside of the station. As the RRM is fitted with the same valves, caps and connectors found on many commercial and government satellites, Dextre will be used to cut away protective thermal blankets, unscrew fuel caps and transfer simulated fuel from one reservoir to another. The aim of the mission is to demonstrate that existing technology can be used to perform such work. Moreover, the RRM might also spark a private satellite-servicing industry, which might also encourage satellite owners to put additional sensors, electronics and fuel-carrying capacity onto future spacecraft, as the potential to extend operational life would be seen as sufficiently worthwhile to make up for a higher up-front investment. However, the RRM is only meant to be a demonstrator, and it will be left to companies like MacDonald Dettwiler and Associates to conduct commercial refuelling operations on satellites.[182]

The humanoid robot, Robonaut 2, has been activated on the International Space Station. The product of a joint project between NASA and car manufacturer General Motors, this $2.5 million robot is designed to assist humans in complex tasks, either in on the space station or on Earth. Weighing about 150 kg on Earth, and with an upright height of 1 m, Robonaut 2 has the configuration of a head, torso, arms and super-dexterous hands; and as the need for legs and feet onboard the ISS is superfluous, an anchor-like pole is used to attach the robot to the wall of the U.S. Destiny module. A second Robotnaut 2 is undergoing tests in different terrains

---

[181] SpaceNews Staff. "Swedish Firm Tapped for Nanosatellite Components." SpaceNews 13 Jun. 2011: 9.

[182] Leone, Dan. "Space Station-bound Refueling Demo Won't Start Before November." SpaceNews 5 Jul. 2011: 6.

on Earth as part of a space exploration technology demonstration test program called Desert Research and Technology Studies.[183]

In an effort to assist astronauts in adjusting to weightlessness, the Draper Lab group in Cambridge, Massachusetts, is developing a concept for a gravity imitating spacesuit that will simulate the resistance experienced while moving against a gravitational acceleration – applicable for use on the ISS, asteroids, Mars and beyond. This action could be done with the use of an inertial measurement unit that would help to gauge movement, enabling flywheel gyroscopes to vary spacesuit resistance for every movement, by increasing speed or changing direction. Astronauts might also benefit from maintaining muscle coordination while operating in resistance-free microgravity; and spinoff uses with physical rehabilitation are also foreseeable. While still in development, Draper Lab is looking to make a prototype for a spacesuit arm by 2012, and a full-body wearable suit within a decade.[184]

High-radiation environments have a degrading effect on the susceptible and increasingly sophisticated microelectronics within spacecraft. To lessen the harm caused by this radiation, the U.S. Air Force Research Laboratory and BAE Systems are developing powerful radiation-hardened RAD750 microprocessors that are modified for better use in space applications. While the first generation of RAD750 microprocessors were radiation-tolerant with modest processing speeds, strategic radiation-hardened versions with higher processing speeds are now available.[185]

The Pentagon's Defense Advanced Research Projects Agency (DARPA) wants to recycle the still-functioning components of non-operational satellites to be refitted on cheaper-to-launch small- and nano-satellites that have entered geosynchronous orbit. Dubbed the Phoenix programme, the intent is to use a tender vehicle – equipped with grasping mechanical arms and remote vision systems – to act like a robotic satellite servicing station, attaching the salvaged antennas to the antenna-less satellites placed into orbit at a reduced launch cost. As existing hardware is usually molded or welded onto satellites, accomplishing this task requires advanced remote imaging and robotics technology and special tools to capture, detach, and modify complex systems. If successful, the Phoenix programme could have the dual benefit of reducing space debris and saving money by not having to launch satellites with large and bulky antennas.[186]

Draper Laboratory has also developed the Guidance Embedded Navigator Integration Environment (GENIE) for NASA's Autonomous Landing and Hazard

[183] SpaceNews Staff. "Humanoid Robot Powers Up on Space Station." SpaceNews 29 Aug. 2011: 9.

[184] SpaceNews Staff. "Spacesuit To Imitate Gravity on Long NASA Missions." SpaceNews 12 Sep. 2011: 9.

[185] SpaceNews Staff, "BAE and Air Force Fielding Hardened Microprocessors." SpaceNews 10 Oct. 2011: 8.

[186] SpaceNews Staff. "DARPA Project Aims To Turn Space Junk into Satellites." SpaceNews 24 Oct. 2011: 8.

Avoidance Technology (ALHAT) project, which has since been integrated into Masten Space Systems' Xombie rocket for initial testing prior to testing GENIE at high altitudes. The GENIE achieved Technology Readiness Level 6, i.e. achieving a successful demonstration in a relevant environment, by vertically flying 50 m, laterally moving 50 m while maintaining altitude, before landing within 12 cm of its target. The next test flight will be conducted on Masten's Xaero suborbital rocket in 2013, to further mature the technology intended for landers by imitating the landing trajectories that spacecraft would follow when landing on other celestial bodies.[187]

2.5.6 *Innovation Policy*

Near Earth LLC conducted a study for NASA on opportunities and obstacles inherent in attracting private capital to commercial space ventures. The study found that institutional investors desire adequate assurance that the risks and rewards of commercial space investments are in line with their overall investment goals.[188] As public-private partnerships (PPPs) help spur innovation, NASA has already taken important steps in establishing its Commercial Orbital Transportation Services and Commercial Resupply Services contracts, meant to support private industry efforts to develop space transportation services. However, in light of the unique and inherent risks in developing space technology, additional steps are needed to make commercial space ventures attractive to cautious investors; e.g. developing business plans that include incremental steps toward a goal. Another way that NASA could attract private capital and build the market is by serving as the first customer for a new space-based product, or by functioning as an anchor tenant ensuring enough purchases to make commercial businesses viable. Tier awards for achieving milestones could also be deemed to be more useful than winner-take-all competitions. Additional recommendations included: the continued use of seed money and Centennial Challenge programmes; the need for a roadmap identifying the types of jobs NASA would like to turn over to industry; the development of an innovative technology programme like the CIA's 'In-Q-Tel' non-profit programme, which could provide funding for commercial technology to meet the space community's needs; and loan guarantees to enable private space companies to borrow money at low interest rates for programs that have reached a level of maturity where the risk of default is low.[189]

---

[187] Leone, Dan. "Draper, Masten Test Guidance System." SpaceNews 13 Feb. 2012: 12.

[188] Werner, Debra. "Institutional Investors Seek Profit Potential in Commercial Space." SpaceNews 22 Aug. 2011: 7.

[189] Davidson, Hoyt. "Near Earth Report – Supporting Commercial Space Development." Near Earth LLC. Presentation. NewSpace 2011 Space Conference. NASA-Ames Research Center, San Jose, USA. 29 July 2011 http://www.youtube.com/watch?v = yScSEV-Mqh8.

NASA, desiring the long-term cryogen storage capability needed for deep-space exploration, will invest up to $600,000 a piece in four companies that study concepts that demonstrate the storage and uniform transfer of liquid oxygen with "zero boil-off" and liquid hydrogen with "minimal boil-off" in space. In managing storage difficulties associated with the very small size of the hydrogen molecule, the current state-of-the-art for liquid hydrogen storage only lasts for several hours prior to leakage; NASA would like the ability to store liquid hydrogen for several months. In addition to demonstrating the concept, awardees must ensure that the mission costs remain between $200 and $300 million.[190]

The U.S. Air Force (USAF) would like to find new launch service providers to compete in launching small- and medium-class military payloads. Having relied solely on Minotaur rockets, with repurposed ICBM solid-rocket motors, since 1997, the USAF's 2003 Orbital/Suborbital Program (OSP)-2 fixed-price contract will end earlier than expected due to higher-than-anticipated demand for the Minotaur 4 launch vehicle. The USAF planned to award multiple indefinite-delivery, indefinite-quantity contracts to several new U.S. launch providers by February 2012, for the cost-effective purchase of as many as 16 launches during the next 8-year period. With a desire to have at least two launch vehicles available for any given mission, contract eligibility requires launchers to have already undergone critical design review; while Minotaur rockets still remain a viable option in lower tier launches, potential options for the heaviest missions might be SpaceX's Falcon 9 rocket or Orbital Sciences' Taurus 2 rocket.[191]

[190] Leone, Dan. "NASA Invests in Cryogen Storage Mission Concepts." SpaceNews 22 Aug. 2011: 6.

[191] Brinton, Turner. "USAF To Allow Competition for Small- and Medium-lift Rockets." SpaceNews 13 Jun. 2011: 4.

Part II
Views and Insights

Chapter 3
The Effect of the Financial Crisis on Political Decision-Making

Per Tegnér

3.1 The Financial Crisis

We Europeans, who are today experiencing to different extents the effects of the financial crisis, tend to think that this is a worldwide phenomenon. That, however, is not the case and the global picture is mixed at least. Many space nations around the world, such as China, Russia, India and Brazil, are in a different situation. Russia has just announced that its space budget will increase substantially during the coming years. At the same time, there is no sign that China will reduce its ambitious efforts in space science, exploration and strategic capabilities. The United States, accounting for over half of the world's spending on space activities, is currently witnessing an enduring economic stagnation combined with political disagreement on its federal budget. Although politicians in Congress have agreed upon certain issues, many challenges need to be resolved in the near future. As long as the issue of federal spending remains unresolved there is uncertainty as to how defence and space budgets will be affected in the coming years. Over the last couple of years, the U.S. space budget has already diminished in relative terms vis-à-vis other nations. While in the early 2000s U.S. spending accounted for 75 % of global spending on space, this number has now dropped to approximately 60 % – a remarkable drop for a relatively short timeframe.

The situation in Japan is somewhat more difficult to judge, but it is nevertheless likely that space activities there will also be touched by the current political and economic turbulence.

This article will focus on the situation in a European context. In order to assess both the short and long term effects of the financial crisis on the European space sector, an introduction to the European institutional set-up for Space will be given first, followed by an analysis of the strengths and weaknesses of the sector.

P. Tegnér (✉)
Swedish National Space Board, Stockholm, Sweden
e-mail: per@tegner.net

C. Al-Ekabi et al. (eds.), *Yearbook on Space Policy 2011/2012*, Yearbook on Space Policy, DOI 10.1007/978-3-7091-1649-4_3, © Springer-Verlag Wien 2014

3.2 The Institutional Set-Up for Space in Europe

European states have for long realised that they have to cooperate in order to achieve an effective outcome in space activities. The European Space Agency (ESA), the European intergovernmental body for space activities founded in 1975, has been the main vehicle for this cooperation. Today, ESA has 20 Members States and all European countries, even those that are not members of ESA, are involved in space matters in one way or another. Although many of them are still very small players in space activities there is a clear understanding in Europe that space plays a vital role in economic performance, as well as in many other fields in society. This understanding is also reflected in the explicit interest and participation of EU Member States as 26 (of the 27) of them were present at the last ESA Ministerial Council in November 2012.

Some European nations have set up dedicated agencies to deal with space affairs. All the big spenders (France, Germany, Italy and UK) have done so, as well as some smaller states such as the Netherlands, Norway and Sweden. Some states have embedded space in an agency with wider scope such as Austria, Finland and Spain. Others, like Switzerland the Czech Republic, have formed a special space office inside a Ministry. In most other countries space is an integral part of a Ministry. Belgium, being also a big spender in space activities, has asked ESA to be the space agency for its national activities.

Many ESA Member States spend a considerable part of their space budget on national activities (France 59 %, Germany and Italy 48 % each), while others have smaller portions of their budget allocated to national activities. In terms of space governance, the differences between ESA Member-States in overall budgets and proportions spent on national activities, are remarkable. It is this variety that makes decision-making processes in Europe so particular, interesting and also different from anywhere else in the world. Moreover, it makes the eventual impact of the financial crisis on spending on space activities diffuse and hence relatively hard to predict.

Another factor of importance is that space activities of ESA Member-States are typically managed in at least two different types of Ministries; either the Ministry responsible for research or the Ministry responsible for economic activities or enterprise. Examples of the first category include Denmark, France, Italy, Spain and Sweden. The latter category includes Germany, the Netherlands, Norway, Switzerland and the UK. Why this organisational difference can be important in the decision-making process will be discussed below.

3.3 Why Space Activities Are So Peculiar in Terms of Decision-Making

There are several aspects of space that make it different from other fields.

First, space activities are characterised by their horizontal nature. By that is meant that the benefits of space can be attributed to so many different parts of public and private activities. Many Ministries, not to say most, benefit from space activities: foreign policy, development aid, security and defence, traffic and communications, research and education, media, industry, innovation, environment, urban planning, health and agriculture are all examples of Ministries that clearly have interests in the use of space. One can ask whether this is a real problem. In my view, it is indeed problematic in a certain way. Usually there is a very clear understanding in Ministries as to what kinds of issues they want to promote and support – these are the core issues with a political priority. In a Ministry there is often an understanding of what sort of benefits space activities can bring to the areas of responsibility of the particular Ministry. Often, however, it is on the fringes of the Ministry's interest, and so there is no willingness to support it financially. There might be support for the activity if somebody else pays. Ministries responsible for space typically encounter this attitude when trying to ask for financial support. Most Ministries responsible for space have set up a network of Ministries or Agencies with an interest in one or more existing space applications. How this works differs from country to country, but the most usual outcome is that there is no or very little financial support from other Ministries and often also very little support for the responsible Ministry when negotiating with the Ministry of Finance. In those negotiations the responsible Ministry usually has to show to the Finance Ministry the benefits for other ministries, weakening the strength of its arguments in the eyes of the Ministry of Finance. In many countries this tendency is less present when Ministries for Communication and or Traffic are involved, because of their wide support for the Galileo system.

Another issue that makes space so different is the lack of identifiable pressure groups other than industry. Although ultimate users are the major stakeholders of space activities, the user communities are not very well organised. As a result politicians do not feel pressure to deliver adequate space solutions and applications. In most countries the most powerful pressure group remains the space industry. Space activities and programmes have been of vital importance for industry to develop knowledge and skills also regarding products and services other than those procured by space agencies. There has been a clear understanding by European countries that space programmes are a way to promote development in sectors in addition to those relating to products and services for public space activities, and that is why the geographical return principle in ESA's industrial policy is so important for the ESA Member States. Obviously, spending on space should in the first instance be done to serve the functional needs of the user communities while support for industrial development can only be a concern of secondary importance. That is why too strong lobbying support from industry is problematic for politicians; it would create an imbalance because of the absence of inputs from strong and diverse user communities.

The most obvious pressure group for politicians is of course the general public, but what exactly is the interest of the general public in space activities and how is this possible interest communicated to politicians? If you ask the general public how it appreciates space activities – something that has been done in many countries by the way – a common answer is that other public activities are much more important. Of course people give priority to public activities or services that are closer to them such as employment, health, social security and education. Issues such as research, innovation and industrial development are rarely the focal point of political discussions nor are they a major concern in the mind of the general public, hence there is very little pressure from the general public for space activities. One factor of exception seems to be the presence of an astronaut, in any given country. It increases the interest of the general public to a very high level, however, not for a long time. The media also plays a vital role in keeping the general public's interest in space.

In other parts of the world space activities are looked upon as a national interest and even as a flagship of national pride and achievement. These sorts of feelings are not widespread in Europe. Up to now we have been so spoiled by high-class performance achievements in many fields that the pride argument does not play out as it does in other space nations such as China, India and Russia.

Another issue that makes Europe different from other space nations is the simple fact that we are not one nation. Never have I heard a Prime Minister in Europe talk about space as important unless he welcomes an astronaut back to his country.

The last point I want to highlight in this section is the typically extremely long duration characterising space activities. This is something that Ministries of Finance are allergic to. Commitment to financing long term programmes is in itself not unusual for Finance Ministries but what is unusual is the discretionary return of space investments; if a programme is terminated before its scheduled end the entire return on investment is at risk. This combined with the fact that long-term programmes are often made in a framework of international cooperation makes it difficult to withdraw if there is a desire to change priorities. These facts make Ministries of Finance hesitant to commit to space investments.

3.4 Decision-Making in ESA and the EU

ESA is responsible for the majority of European space activities and in this context decision making is very complex. Each Member State has one vote and while for some issues only a qualified majority is required, in the end the prevailing culture tends towards consensus decisions. For the sake of maintaining sufficient support and political continuity, there is a very strong tendency to avoid that one of the big contributors to a programme is down voted. This has created a complex negotiating process but, despite this, ESA has managed to find a very successful way forward in this process. For ESA executives and Member States' delegations this form of decision-making requires great skill. Therefore, the process is vulnerable if new people enter the playing field without the proper feeling for the subtleties of the process, regardless of whether they come from new or old Member States. Discussions are ongoing in ESA on whether the – so far unchanged – Convention should be amended so as to reform the decision making processes. It is often said that this should allow the organisation to incorporate into its working structure the multiple internal and external changes that have happened since its founding. The stability of ESA's legal framework reflects the workings of a very result-driven organisation in which strong support from its Member States is an indispensable requirement to guarantee the social cohesion and political impetus it needs to operate. This is of course necessary since ESA is an intergovernmental organization for which – to a certain extent – the objective is more important than the formalities even though ESA and its Member-States are keen to keep to the regulations that have been set up through consensus.

The European Union (EU), on the other hand, is a very different type of body. The most important difference is the position of the European Commission (EC). The EC has powers that the ESA executive totally lacks. These powers are strictly regulated, which means that it is not possible to violate the rules, even if everybody agreed to do so and was aware that sticking to the rules leads to suboptimal results. This does not mean the situation is bad per se. Given its size and variety of competences there is no way for a political body such as the EU other than to have strict rules and to keep to them; other options could even be dangerous. The result, however, is that the provisions are different for the EU and ESA, and that this institutional discrepancy makes cooperation troubled to say the least. It seems that Member States in the ESA and EU want to keep the foundations and the different sets of rules intact and so the key issue is to find the right ways and modalities for cooperation. Decisions on the way forward for future cooperation with the EU have been taken by Ministers at the ESA Council. ESA had its Council on Ministerial level in November 2012 and will have a new Council at Ministerial Level meeting in 2014.

The EU has now a shared competence in space matters with Member States but the activities will be heavily dependent on funding decisions regarding the activities in question.

Decision-making in ESA is done by the Council, which meets at least four times a year on a delegate level. So far meetings at Ministerial level have only taken place every third or fourth year.

Decisions in the EU are taken by Ministers in the Competitive Council who meet several times a year. In this Council the Commissioner for Enterprise is responsible for space. This means that Ministers who have the responsibility for enterprise issues will meet several times a year, while Ministers responsible for research will meet less often with this Commissioner.

Personal involvement from Ministers is crucial for the success of space affairs. It is necessary that all Ministers responsible for space meet more often and in this sense a more frequent gathering of the ESA Council at Ministerial level, as is currently happening, is a positive signal. Also, interagency representation should be reinforced, for example, by letting Ministers responsible for space also attend the Competitive Council when space is discussed.

3.5 The Present Situation in Europe

Everybody in Europe is aware of the financial crisis, although in some countries its effects are worse than in others. Certain countries are faced with such a severe debt situation that financial markets are not willing to lend them money anymore. Other institutions, such as the EU, the European Central Bank and the International Monetary Fund, offer financial support in exchange for far-reaching structural reform. This creates a situation where budget cuts are very dramatic.

It also creates a tense situation between states with a healthy and well-functioning economy and those who do not. As a result, solidarity and social cohesion between member states in the union has decreased. It is clear that eventually this will have an impact on the EU and the Eurozone countries, especially when it comes to the issue of the financial perspectives of the EU. The attention and energy devoted to space will be minor as there are many more important issues to be resolved during budget negotiations. Therefore, it is not unthinkable that the crisis will have a considerable impact on the EU budget for space. While it is very difficult to predict the outcome, one certainty is that no new big commitments will be made. The current EU flagship projects in space, Galileo and Copernicus (formerly known as GMES) will continue but with strict orders to cut spending wherever possible. If Copernicus' spending is cut, national reactions remain to be seen, but it seems unlikely that member states would be willing to let ESA take over the operation, as this is the responsibility of the Commission. The same is true for Galileo.

Predictions are at this point very hard to make. In the Horizon 2020 it is doubtful whether member states will be willing to let the European Commission spend money on space science that is complementary to what is going on in ESA.

The outcome of the negotiations on the financial perspectives is of course of vital importance for ESA and that is why ESA has called for a new Ministerial Council meeting in 2014, when the result of the EU decisions will be fully transparent. Coordination between what Ministers are doing in the EU Competitive Council and in the ESA Council at Ministerial Level is of utmost importance. Hopefully Ministers and their Ministries are fully aware of this by now. Although many countries have excellent internal cooperation, some countries still seem to have big difficulties in this respect. Granted, coordination is more difficult for big countries than for smaller ones, but it remains necessary for everyone to improve. Europe especially cannot afford to fail in its coordination, as this is an essential task of its functions.

The task of improving working relations between ESA and the EU must be taken seriously. Rules and regulations must be changed on both sides in order to have more effective cooperation, even if the objectives, as well as the rules, are built upon a different logic.

3.6 The Situation in Certain European Countries

As indicated above, the situation and challenges are very diverse for different European countries and, moreover, the gap between countries has widened in recent years. The outcome of the ESA Ministerial Council is very clear in this respect, as southern European countries with budgetary problems have withdrawn from some ESA programmes. Spain stopped funding all the ESA optional programmes and this will have serious effects on the Spanish space industry. After several years of investments in optional programmes and building up a competitive sector, the industry will face severe problems in terms of viability. On the other extreme is the UK which has increased its contribution to ESA programmes by 25 %. Germany and France remain the biggest contributors to ESA programmes. Italy has difficulties in maintaining its position and Italian industry will suffer as a result of this. Generally, the outcome of the ESA Ministerial was surprisingly good and a decision was even taken on new programmes. The majority of countries have given priority to ESA programmes even though total funding has not increased. This has been the case over the past decade. Overall, ESA's space activities have not increased if the funding that came from the EU is not taken into account.

It is difficult to judge whether Member States' contributions to ESA programmes will have a negative impact on national space activities. Countries with extensive national space programmes are still capable of maintaining their level of activity, at least France and Germany. There is a strong political will in these two countries to maintain a balance between them and so far that has resulted in overall high spending on both domestic and ESA programmes. Some of the smaller countries, like the Netherlands, have reduced their space budgets due to budgetary constraints. The overall balance in ESA is maintained because new countries have acceded as members. So far, even though this is a small fraction of the total, it is an important step from a political and budget stability point of view.

3.7 The Impact of the Crisis on Programmes

Even though the funding of ESA has been kept stable, there has been a change in the willingness to fund programmes, which has not always been officially announced but is a reality. Programmes that are of more direct benefit to the general public, such as meteorology and telecommunications, are generally more favoured than manned space activities, for example. In some countries there is a noticeable weariness of ISS funding. Reasons for ISS budget stability mainly include commitments to the agreement with international partners and the fact that 'payments' for the operations are made with European hardware. Funding of launchers has always been tricky in Europe. Even though the Ministerial Council decided to start new programmes in this field, it remains to be seen how the funding of the whole European launcher sector will evolve over time. The science programme has a constant difficulty because its missions tend to be more and more expensive and the costs of the instruments provided and funded by the Member States also tend to increase. Moreover increases in funding of the programme are not sufficient to cover its ambitions.

3.8 The European Space Industry

In recent years, the European space industry has transformed into a competitive industry benefitting from ESA programmes, ESA procurement and national initiatives. Many companies have expanded their market potential and are successful in markets beyond the European public market as well as in other, sometimes less space-oriented, products and services. As the financial crisis hits the commercial market, the public market becomes more important for the industry's survival. Despite the oligopoly market structure of large system integrators, competition in the public market is fierce. The European public market is too small for current industrial capacity and some companies are very vulnerable, even when they are a part of bigger groups. There is a risk of some companies closing if public procurement stops or drops too dramatically. As indicated above, this will be the case for industries in countries that have changed their contributions to ESA optional programmes. At the same time other companies, eager to get a bigger market share, will take their place, and so the overall effect of the financial crisis is very hard to predict on a company level. The only sure thing is that it will change the industrial landscape in Europe. One issue of particular concern that has been a long term problem for ESA and its Member States is the size of the French space industry, which is too big for the country's budget for space. This has resulted in a historic net positive geo-return for France in the ESA procurement system, which creates difficulties for ESA and its Member States. In the long term such a persistent imbalance is not sustainable.

3.9 Conclusions

This article has focussed on different elements that are crucial in determining the eventual impact of the financial crisis on political decision-making. This is very important since space activities are so very specific that one cannot draw conclusions from other fields of economic activity. To recapitulate, the specific factors are that European Space affairs are dependent on:

- The budgets for space in ESA Member States.
- The budget for space in the EU budget.
- The decision-making process in Member States, the ESA and the EU.
- Political relations between the abovementioned parties.

- The personal interest in space of decision makers.
- The various pressure groups impacting on decision making.

All these factors are affected by the economic crisis in Europe and while some of them will have only a short term effect, others might have long term impact on space activities.

Although the overall budget for space in ESA has not yet changed much because of the crisis, certain countries are heavily affected. This means that influence on space programmes will shift and that certain industries from now on will suffer from this change while others will benefit.

Some countries – often those with strong national space industries – will have a bigger role in the future. This may increase tension between ESA Member States in such way that a change of the voting system in ESA may become necessary. At the same time, these big countries also want to exert more influence on ESA decision-making. On one hand the tension between France and Germany is still very noticeable; on the other, the large countries have the same interests when it comes to having influence over ESA decisions.

There are also differences in the approach of Member States to EU financial perspectives on space. France is very positive towards a larger EU budget for space. The obvious reason for this position is the strength of its space industry and the fact that in an EU framework it can avoid ESA procurement rules with their limiting geo-return principle. The financial crisis will most likely increase tensions between ESA Member States. This is a long term issue and it will intensify the process of change within the ESA system. Decisions on ESA programmes and long term budgets are time-wise not correlated with decisions on budgets in Member States, which in a time of crisis may affect the willingness of at least Ministries of Finance to fund the ESA in the usual way. The personal interests in space of several European Ministers have played an important role in decision-making in ESA Member States and in ESA itself. The economic crisis may affect the possibilities for Ministers responsible for space to be able to exert influence on these issues. Ministers responsible for space in their governments often meet at Council meetings in the EU framework. It is important that Ministers also become more familiar with space issues within the ESA framework, so they can get a better understanding of the issues and regulations within ESA. Changes in this direction are being implemented. Pressure groups, especially space industry, are more active than ever. The public procurement of space related products and services is, in a longer perspective, too small for the current capacity. Competition is fierce. Many companies in the space business are also active in the defence business, so the situation for those companies is dependent on procurement policies in this field. This article has tried to point out some of the weaknesses in the European "space ecosystem", which in the longer timeframe will have an impact on states' willingness to keep space budgets at least at a constant level. These weaknesses make the "space ecosystem" more vulnerable in a time of financial crises than under more normal circumstances. Personally, I believe that if the European "space system" takes the opportunity that the financial crises has created to change itself, the long term effects will be beneficial. It will of course mean changes for a lot of actors. The short term effects, such as programme decreases and decreases in national space budgets and activities, have already occurred. This is of course sad, especially given that everyone agrees that European space efforts have been very successful over many years.

Chapter 4
The Psychology and Reality of the Financial Crisis in Terms of Space Cooperation

Peter Hulsroj

4.1 Introduction

Every economic sector within every country has been, and is, called upon to assess the consequences of the financial crisis for its specific domain. This is true also for space.

At the start of the financial crisis space tended to be very persuasive in explaining how investment in space has a sustaining effect: counter-cyclical when times are bad, but pro-cyclical when times are good. This sustaining effect space shares with most investment areas with long time horizons. The logic has meant that NASA's budgets did not decrease even at the height of crisis, and also in austerity-addicted Europe both the 2008 and 2012 ESA Ministerial Conferences brought new subscriptions at levels that maintained or exceeded those of non-crisis periods.

In view of the solid support for space in both good and bad times it could perhaps have been assumed that space activities and the attendant cooperation would have continued almost unaffected by the crisis, but this has not been the case for reasons that will be illustrated below.

4.2 The Psychology of the Financial Crisis and Its Effect on Cooperation

The near disaster of the Lehmann Brothers bankruptcy, the Greek debt crisis, the European bail-outs and the endless talk of austerity and diminishing wealth has moved the rich part of the world from a feeling of natural privilege to a feeling of

P. Hulsroj (✉)
ESPI (European Space Policy Institute), Vienna, Austria
e-mail: peter.hulsroj@espi.or.at

C. Al-Ekabi et al. (eds.), *Yearbook on Space Policy 2011/2012*, Yearbook on Space Policy, DOI 10.1007/978-3-7091-1649-4_4, © Springer-Verlag Wien 2014

gloom and incapacity in facing even the simplest challenges. The United States suddenly doubted whether it should service its debts; the Greek debt crisis, which would have been resolved by the IMF in 6 months had it been a Latin American or Asian country, has been allowed to become a EURO crisis with endless hand-wringing, little structural action, and a high price of unemployment to pay for the youth of periphery-Europe.

It seems entirely forgotten that even in the past decade the rich 'West' has been richer than at any time before in human history. Clearly that is not a reason for complacency, but it is a reason not to let ourselves be weighed down by burdens purely of our own making. And it is certainly not a reason to ignore the stark social consequences for us of having been led to believe that we are poor.

The space business might have reasons to look more than once at how it can enhance its beneficial effects for domestic populations, but, in addition, it has a special calling to make sure that it does not freeze at the controls in such a fashion that international cooperation is hindered or reduced at a time when it is of central importance that space plays its traditional bridge building role.

In another contribution to this Yearbook, Per Tegnér has explained how uneven the effect of the financial crisis has been both within Europe and globally. In itself, this is an important fact. However, the asymmetry in readiness and ability to fund space activities, and the general change in geopolitical dynamics, do not explain completely why emerging players in space appear to have so much wind in their sails and why existing players appear to be so hamstrung. The suspicion is that some of the new players might be blowing their trumpets a bit too hard, but particularly that existing powerhouses, despite the good funding situation, have convinced themselves that they are in rapid decline. On the international stage overreaction is as well-known as it is in the stock markets! Space has not had its Lehmann Brothers event, so why has it become so hard to create visionary, long-term, collaborative projects?

One explanation, of course, could be that actors are already in the midst of executing a number of highly demanding, highly cooperative programmes, and that the time is not ripe for new challenges before the existing ones are mastered. This might, perhaps, be part of the difficulty with new projects in human exploration. ISS is the measure of all things, and, at least NASA will be inclined to heed the lesson of the past, when it was found that resources did not allow the simultaneous development of the Constellation program whilst running the space shuttle and operating the ISS. Obviously a result where neither the shuttle nor another human launch capability is available in the United States is devastating, particularly since there is also no clear path from ISS to follow-on capability. But learning the lessons from this debacle does not actually lead to the logical conclusion that new activities should not start or that cooperation should be approached cautiously. Clearly the reliance on Soyuz for launching humans has stung, but at the same time US National Space Policy declares it necessary that projects include international partners on the critical path.

This last statement can, of course, be understood as a concession to the financial crisis, but that is not the whole story. The National Space Policy was defined at the start of the Obama presidency, at a time when President Obama was very keen to show that he represented a sea-change in terms of American readiness to engage in multilateralism. Has that resolve weakened? The NASA pull-out from Exomars seems to say: yes, and the unwillingness or inability to engage in a meaningful way with China seems to point in the same direction, as does the seeming insistence by the US to always be in control when discussing exploration strategies. But this is hardly the end of the story. It is not surprising that the superpower which has been so dominant for the last 25 years, and has been the measure of all things space for more than 40, finds it difficult to share the sandbox more equally and finds it painful to have to borrow a shovel from someone else once in a while. However, the latest development on the Orion spacecraft, which will be the vehicle to ferry astronauts beyond low earth orbit, is a significant step in the direction of true international cooperation since its indispensable service module will be provided by Europe and by European industry.

But let us back up a bit and look at the relationship psychology of the main players in the world, before we get too deeply into analysis of specific cooperation topics. And let us look at relationship psychology both in terms of self-perception and external perception, since both are key to understanding cooperation possibilities and realities.

The United States sees itself as the preeminent space power and has much evidence to back this up, starting with the space race, over Hubble to the shuttle and the leading role in ISS. Based on this self-confident self-perception, George Bush could declare in 2004 that the United States was going to Mars via the Moon. Little has come of that, however, and bold statements has been followed by re-directions, termination of, first, the shuttle program and thereafter the Constellation program, the quest for game-changing technologies, and vague plans for a visit to an asteroid as a precursor to finally going to Mars. The remarkable thing about all these plans is that they were created and abandoned virtually without the involvement of friends or partners from the outside. The critical path was not trodden by foreigners. The plans for the Orion spacecraft represents a significant shift towards interdependence, and hence a notable change of role definition by the United States; surely not one which puts American pre-eminence into question in the American self-understanding, but one which allows this self-understanding to acknowledge both the utility and political sense of working in an integrated fashion with partners that are in any event joined at the hip with the United States in the geopolitical power structure.

The perception of the United States by foreign partners and competitors is ambiguous. Partners have grown accustomed to the rapid and sometimes dramatic changes in US planning due to the high political significance of space in the United States, and hence close political scrutiny by both the executive and the legislative branches, much aided by the annual budget cycle which ultimately determines what can be pursued and what not. The evolution of space station Freedom to an ISS with five partners is the classic example of the swings and unpredictability of US planning. However, the Freedom to ISS evolution illustrates not only the difficulties of staying on firm planning, but also the flexibility in opening towards new partners offered by a planning process that must respond quickly to the political exigencies defined in Washington. The Orion inclusion of Europe as the provider of the service module could be seen a bit in the same light. The partnering situation with the United States is thus characterized by continuous possibility, within, it must be said, a playing field which is entirely controlled by the United States.

The fluidity in partnering makes it difficult to be a competitor of the United States, its economic muscle aside. China remains untouchable for the United States, which seems to seriously think that a country of 300 million in the mid-to long term can remain the lone agenda setter, in space and beyond, and that hence rapprochement with an ambitious country of 1.4 billion is not necessary. This leaves China no choice but to be a competitor, and fish in the partnership waters left undisturbed by the United States. Having been rebuffed in seeking a way in on the ISS, China has logically and, so far, successfully, pursued its own space station, much as the Soviet Union did in the 70s and early 80s. Such success builds credibility, which one would assume would be brought to bear first through APSCO, the Chinese led regional space organization for Asia, but remarkably with Iran, Peru and Turkey as members as well. APSCO might have its difficulties, but it is hard to imagine a clearer warning flag for the United States. For China the situation must appear to be that, even with a multilateralist President like Barrack Obama, serious cooperation cannot be established with the United States because of the influence of adverse forces in Congress. China cannot assume that any new Administration would be more approachable, and hence must conclude that if in a financial crisis, with no ability to launch humans itself, and with a constructive player in the White House, cooperation cannot be achieved, then China will always have to be the competitor.

It is difficult to know how China sees itself in space. Many plans are known and many successes speak for themselves. China generally has tended to see the end of the nineteenth century and the first part of the 20th as humiliating for an ancient and proud civilisation, the Middle Kingdom. The current leadership will seek to latch on to the proud tradition, but will probably also be operating with some resentment and well-hidden feelings of inferiority. This is what makes the attitude of the United States so frustrating. Deng Xiaoping reached out to the United States and Japan for the modernization of China, and now both doors are firmly shut, at a time when China needs them less and less. While there is thus not much evidence of space as a bridge builder between those three players, there is plenty of evidence that China is steaming ahead unaffected by the economic crisis. In both economic and psychological terms the game has moved in favour of China, and that has not been lost on the leadership.

How the rest of the world perceives China in space is fascinating. Some potential partners may believe that China will leverage it space capabilities in a similar way to how it operates its investments in Africa – with few strings attached. Others might believe that China will be the 800 lb gorilla, allowing cooperation but reserving the right to crush lesser animals, should they become troublesome. Others again might think that China will prefer to go it on its own in space, giving an imperial demonstration of capability of a power emerging from the shadows. While none of these perceptions may be correct and the approaches may be much subtler, nonetheless, the three options will to some extent determine how possible partners will react. Small partners on their own may well fear being crushed, so be careful Venezuela and Peru, bigger partners may find that they are too big for China to be comfortable with, and hence China will not open the door. Partners will have to think what they can bring to the table that will make them indispensable, but that is hard given China's culture of self-reliance.

Russia has, by design or opportunity, been the most successful in bringing its space technology into a meaningful relationship with broader goals. Russia's presence in the space endeavours of India, South Korea, Brazil, the US, Europe and China, bears witness to its position as a linchpin of most current space architecture. Nevertheless, Russia sees itself increasingly as a space power in crisis because of the lack of industrial reform and the consequential string of embarrassing launch failures. At the same time, Russia is clearly adamant about protecting its capabilities. Both President Putin and Prime Minister Medvedev are committed and the financial crisis has had no influence at all on their determination. The open question is whether the societal boundary conditions are such that Russian space industry can be successfully reformed, and this open question can be assumed to be very clear in the minds of the Russian leadership. Particularly in a financial crisis in Europe and the United States, and almost none in Russia, the normal remedy for the Russian malaise would be alliance building between Russian industry and European and American industry. In a sense this was how Soyuz was saved when the Soviet Union collapsed. That so little is happening in alliance building has both Russian and non-Russian causes, but Russia as a teacher in India, South Korea, Brazil and China will not be sustainable without rapid renewal of its industry, and it is virtually impossible to see how this will come about. Where Chinese credibility is on the ascendant, Russian credibility is on the descendant.

This is well understood by Russia's partners and possible partners. Ironically, this might give Russia some competitive advantages, alongside the obvious negatives. Russia is possibly seen as a less threatening partner than China, and Russian technology remains state-of-the-art, meaning that the aim of Russian partners might be to get to licensing deals allowing domestic production of Russian technology, rather than relying on engineering implementation by the questionable Russian industry. This is hardly ideal for Russia, but perhaps one of the ways forward. Unfortunately not a way forward for those relying on Russian launchers, however, since the recreation of launcher production lines is neither feasible, nor would they be allowed by Russia, notwithstanding the precedent of having granted access to the RD-180 launcher engine technology to the U.S. firm Pratt&Whitney in the 90s.

In a sense, Russia's industrial difficulties make Russia an ideal partner for China, if historic rivalries can be put aside. China is, of course, a master of industrial production, albeit not yet of the kind of advanced manufacturing required for space, and China would be likely to be keen to get access to Russian space technology for domestic production, something which seems to have happened quite a bit in China's current human spaceflight programme. Russia certainly has a very realistic view of China, and vice-versa; both are keen to leverage space, and neither is clearly inferior to the other in the overall scheme of things. From a partnership perspective, this is an auspicious configuration.

Europe sees itself as an established player, with a prominent role in commercial space, be it launch services, satellites, satellite equipment or as service provider, and as a proven institutional actor. Europe takes pride in its ability to partner with very different partners and to create cooperation frameworks fitting this variety. The diversity within Europe itself is projected to the approach towards the rest of the world. With ESA and the EU, national space agencies, established space faring countries and newcomers, with small and big countries and great heterogeneity of capability, Europe sees its diversity as both a difficulty and a strength. The financial crisis has not incapacitated Europe from being able to be an active international partner, as funding has remained stable, but the psychology of crisis has meant that the readiness to dream big dreams involving international cooperation has become markedly less, although the readiness logically should increase.

The rest of the world tends to see Europe as increasingly complex. The unresolved issues on authority stemming from the Lisbon Treaty, and the difficulties in agreeing on launcher strategies, ISS extension and GMES operational funding, play into the hands of those seeing emasculation rather than strengthening. Little credit is given to the great stability of programmes once decided, i.a. because of the funding modalities, and little acknowledgement flows from the fact that European space, merely understood as space activities undertaken within the zone of EU and ESA Member States, represents a population of 450 million people with potential for a steeply increased curve of activity once the 'new' EU Member States become fully engaged in space and have built up proper industrial capabilities. Within and without Europe there is a tendency to believe that more and more players are fighting for their slices of a cake of unchanged size, where surely the tendency should be to assume that in the medium term the cake will grow very significantly. As it is, there might be a perception that Europe is so focused on its internal cuisine that there is little appetite for an intensification of international cooperation, and that the need to find 'noble work' for new European entrants means that there is less 'noble work' to be shared through international cooperation. There is little understanding that increased diversity, flowing from more European countries becoming directly involved in space, means that there are also more diverse possibilities for international cooperation. The rest of the world sees how difficult it is for Europe to make headway against the financial crisis, and assume, wrongly it would seem, that the same kind of reluctance will exist in terms of establishing new international cooperation in space.

There is a large number of smaller actors in space, who together certainly represent critical mass, but it goes beyond the confines of this contribution to analyse their positions one by one. What can be said is that space has become a commodity and this means that many countries will seek active involvement with space as a demonstration of technological prowess. And it means that many countries will seek to not be only consumers of space services, but suppliers as well.

These countries see themselves as aspirational, and are also perceived by the rest of the world as aspirational. The question is, however, whether potential partners, beyond commercial suppliers, correctly factor in the potential of these new players, and the importance of being supportive early in the process of moving towards space, rather than just chase business when a country has already become a lucrative market. In fact, Russia has been uniquely good at this game, but is hampered in turning this into commercial advantage by current industrial inadequacies and the lack of commercial *nous* of Russia's de facto state controlled industry.

Japan has also sought to leverage its space capabilities with emerging space players though APRSAF, a looser instrument for cooperation than APSCO led by China. For obvious reasons Japan's focus is increasingly Asia, although Japan has a broad range of international partners, and, of course, is a partner in the ISS. Analysing Japan's self-perception is not easy, despite Japan being an open society. The biggest technical and cultural shift, however, has clearly been Japan's recent move into defence related applications of space, reflecting the uneasy neighbourhood it lives in.

History, and the much higher living standards in Japan compared to the rest of Asia, introduces centrifugal forces that are also reflected in Japan's self-perception in space. Japan's path to space has not always been easy, yet it now has a proud tradition of accomplishment. There are no signs that the global financial crisis has dented Japan's commitment to space in the least, but it is probably true to say that the unique geopolitical position of Japan makes caution the keyword in how Japan approaches international cooperation.

Depending on where a country is positioned in the intricate web of Asian relations, Japan appears as a highly attractive partner for cooperation. Japan has cultivated a reputation for being extremely reliable in fulfilling commitments, and, as mentioned above, Japan has the financial wherewithal to cooperate widely. The challenge for the external world is whether it can offer something to Japan which will fit Japan's need profile, with the specificities this entails. Japan has until now shown little appetite for African or Latin American cooperation but also this may change as the space game globalizes far faster and far more broadly than generally acknowledged. That Peru, with its significant ethnic Japanese population, is a member of APSCO, but not APRSAF, should give Japan reason to think!

4.3 The Financial Crisis and Opportunities Lost and Open

Rahm Emmanuel's statement in 2008, as Chief of Staff of President Obama, that an economic crisis is too good an opportunity to waste, has become a trivial bon mot. Yet, there is some truth to the saying also for cooperation in space because a crisis with its economic and political dislocations to some extent shuffles the cooperation cards anew. Paralysis for some becomes opportunity for others.

The neglect by the United States of its Latin-American neighbourhood in the space domain is clearly not a reflection of the financial crisis. However, the relative strengthening of many of the economies of the region relative to regions deeply affected by the financial crisis makes partnering with Latin American countries very attractive for both competitors of the United States and its friends. Brazil is, of course, the biggest prize, and, in fairness, NASA does have bilateral cooperation with Brazil in the context of the ISS. However, it is interesting to see that Brazil cooperates with China on Earth Observation and with Ukraine on launchers. Thus choice parts of the roast have gone intercontinental, which in the overall scheme of things might not be bad, but probably should not elicit screams of joy in Washington.

The competition between APSCO and APRSAF in Asia was mentioned above. In Latin America a similar kind of competition might play out to some extent in an intra-regional context with Brazil in one corner and Argentina in the other, but, again, it is interesting that the United States has not been a catalyst at all in an effort to unite Latin American countries behind common space projects. Latin America often complains about domination by the big Northern neighbour, but in space, where it should be seen as a benefit rather than the opposite, the United States might now have lost the window of opportunity.

Europe has made different forays into Latin America, mostly Brazil, but it is true to say that activities until now have not been headline grabbing nor have they demonstrated a clear European focus on the region.

An interesting question that arises is whether in space the currently very loose concept of BRICS can be turned into an operational reality, given the neglect by the United States of Latin America, the modesty of the European effort, Russia's nascent but determined relationship with Brazil, Russia's strong links to India and significant involvement in China, and South African leadership ambitions. Complementarities abound, but so do, of course, overlaps and rivalries, not least between India and China. The logic of using space for full geopolitical effect in a BRICS context is strong however – an antidote to perceived American dominance would clearly be one argument, but, more convincingly for a joint effort, a BRICS space agency or a BRICS space exploration agency would signal emerging-world solidarity, aided by Russia operating in the twilight zone of an arrived and emerging economy.

Many people inside and outside the BRICS countries ask whether, in the final analysis, the concept is more than a smart acronym or a propagandistic talk-shop. The beauty of cooperation in a field like space is that cooperation becomes substantive and productive, as cooperation on the ISS has so impressively shown. A BRICS space station could be the emerging world equivalent, and a tremendous mobiliser of separate and joint capabilities. In some senses it could also be an answer to Russia's woes, since such an endeavour would give good opportunity, and pressure, to rethink the Russian industrial set-up and let its industry cross-fertilize with other partners.

Would a BRICS space station be bad news for the remaining ISS partners, or the established space players? Not necessarily. On the assumption that the space cake can be grown and that the space effort is not a zero sum game, such an alternative construct could make more resources available for space and could become a powerful and effective building block in trying to reach agreement on a roadmap for humanity's exploration of the Moon, asteroids and Mars through instruments like the International Space Exploration Coordination Group. A BRICS space agency could become a tool to break American isolationism, and a tool to finally make Kennedy's dream of humanity's conquest of the universe a reality 50 years after he mentioned this better option to the space race in an often-ignored speech to the General Assembly of the United Nations.

Tantalisingly, a BRICS space agency could also become a precursor for a 'G77 +China space agency'. This might appear too unwieldy, and might ultimately be so. However, aspiring states need practical ways to cooperate, and custom-made cooperation frameworks every time an international engagement is sought are ineffective and transaction hindering. If a G77+China space agency deployed an optional programme structure similar to that of ESA, states looking for partners for a specific project could use the menu-driven approach of ESA to achieve this. Clearly no programme could become mandatory in a G77+China space agency, but it would be a great step forward if states with ambitions in space could have a pre-defined and flexible framework which would allow them everything from bilateral programmes to huge multilateral undertakings.

The current predicament of G77+China and of the BRICS is that it is both too easy to agree and to disagree when they are only talk-shops. Project cooperation within a framework such as a space agency has the great advantage that it makes it difficult both to agree and to disagree!

4.4 States Move Aside!

The United States has been leading a crusade to let private players play a greater role even in domains that used to be strongly led by institutional actors, such as the transport of provisions and crew to the ISS. Interestingly, the United States has been leading this charge, although Europe can be said to have been the most successful commercial actor in space, with strong penetration of launch services, satellite development and operations businesses. It should be recalled that Arianespace preceded all other commercial providers of launch services by a long time.

Elon Musk of Space-X, in particular, has been spectacularly successful not only technically and managerially, but also in terms of influencing the public debate. That the US government underwrote the development of the Falcon rockets has not featured prominently in these debates, but his latest foray is the suggestion that private players should lead the way to Mars. What the argument for this is, if the taxpayer is still to underwrite and pay, is not so obvious, and all the analogies to the start of aviation are as misleading as analogies to the start of the automotive age.

However, with respect to the financial crisis and international cooperation, the point is a different one. In the past industry has not been particularly agenda setting in the domain of international cooperation. Yes, there are some examples, the use of the Russian RD-180 by Atlas is perhaps one, Atrium-Antrix cooperation on Hylas another, together with some instances of direct foreign investment in space companies. Nonetheless, it remains true that institutional players have defined the playing field for international cooperation in all but telecommunication services, where there is a lot of cross-boundary cooperation and investment.

In the launcher field it might not make business sense for Space-X to license its technology abroad, and it might be difficult to get around ITAR, but if it would make business sense then surely one could anticipate a major push from Elon Musk to allow him to make his own international cooperation decisions, based ultimately purely on business logic.

In a similar vein, could one then assume, hope for, an independent push from industrial players in the exploration domain, with Elon Musk buying minority shares in Russia's Energia and India's Antrix, and using these platforms for a push towards concerted action on exploration? The examples are, of course, too farfetched, and it is easy to see why, but the thought model perhaps is not. It would be much worthier if people like Elon Musk would create a 'bottom up' industrial push for Kennedy's dream, than for him to try to convince the US government that tax payers money for exploration would be better placed with him than with NASA!

4.5 Conclusions

The financial crisis has exacerbated dynamic change in the space world, with change being hailed as an opportunity by many, but with others close to paralysis because of gloomy economic news and the mere speed of change. A world in flux is not a reason to overreact, but it is a reason to analyse whether patterns that have grown organically can still serve societal and organizational goals. The highly uneven effect of the financial crisis and the highly heterogeneous response – stimulus in the United States, austerity in Europe, emerging economies still steaming ahead – mean that cooperation patterns must adjust or wilt, and that new kinds of partners and partnering must be seriously examined. In such an examination wishful thinking is dangerous, unless it is based on a cool-headed analysis of one's own position and how that position is perceived by others, and of what the options are for other partners in term of cooperation.

Even in a crisis cooperation will not disappear, but it might go into a default mode. When it does this it is sad, because it means that some of the possible positive effects of the crisis are not harvested. You can always be sure that the negative effects will find you even if you are not looking. To get positive effects you need to both look, and be ready for action!

Chapter 5
Space Industry and the Financial Crisis

Jean-Jacques Tortora

5.1 Brief Overview of the Recent Evolutions of the Global Space Activity

Assessing the potential impact of the financial crisis on the space industry shall first start with an overview of the evolution of the global space activity.

Fig. 5.1 shows the overall annual fluctuations of the number of launches realised worldwide, and in yellow of the number of satellites produced. The gap between both reflects the number of multiple launches performed.

The length of the period of observation makes Fig. 5.1 quite significant and some major historical events are clearly visible here:

- The strong decrease observed all along the 90s, linked to the discontinuation of major space programmes carried out at the time of the former Soviet Union,
- A noticeable peak at the end of the 90s, reflecting the intensity of the business generated by the deployment of the constellations Globalstar and Iridium,
- The relative slowdown observed in the middle of 2000s, which is the conjunction of several factors, including the end of the constellations and the consolidation of the Space telecommunications operators leading to an optimisation of the fleets in orbit. During this period, the business of space manufacturers active on the commercial market was strongly depleted.

Since then, and in particular over the period 2008–2011, the space activity worldwide, has been steadily increasing, confirming the recovery of the commercial market, but this trend must be analysed in greater detail.

The next level of analysis is to differentiate the various types of Space applications. Fig. 5.2 displays the evolution of the total mass launched, broken down by main kinds of missions.

J.-J. Tortora (✉)
Eurospace, Paris, France
e-mail: jean-jacques.tortora@eurospace.org

C. Al-Ekabi et al. (eds.), *Yearbook on Space Policy 2011/2012*, Yearbook on Space Policy, DOI 10.1007/978-3-7091-1649-4_5, © Springer-Verlag Wien 2014

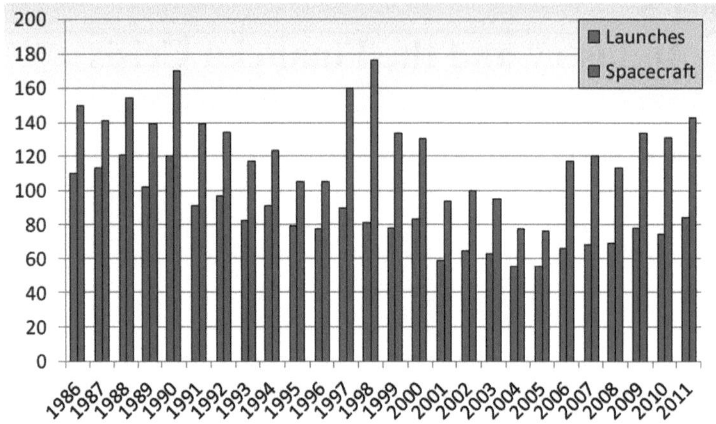

Fig. 5.1 Evolution of worldwide space activity (Source: Eurospace)

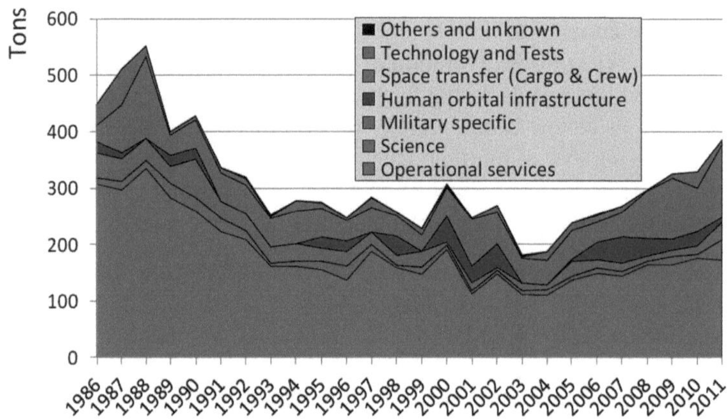

Fig. 5.2 Launched mass by mission categories (Source: Eurospace)

It shows that the major contributor to the recent increase is Space transfer to LEO and the associated manned flights. This is of course a consequence of the major effort made by NASA to terminate the deployment and assembly of the ISS. This has generated a lot of business for the space manufacturing industry, in particular in the US, but the situation will shortly stabilise with the retirement of the Shuttle and the end of the associated major operations.

Fig. 5.2 then shows slight and regular increase of launches carried out for space based operational services.

The other remaining categories of missions, including Science, Defence and Technology demonstration appear rather stable, with a limited impact on the overall

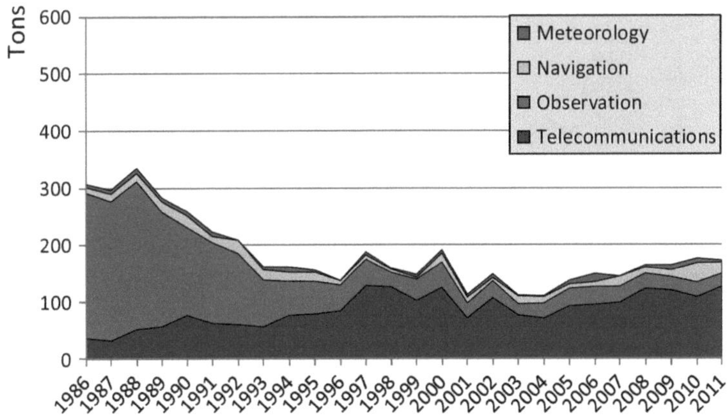

Fig. 5.3 Evolution of the launched mass for operational space-based services (Source: Eurospace)

activity when measured by the launched mass, which is not proportionate of course to the budgets involved.

The operational services themselves can be divided in four main categories, and the following histogram shows the evolution of the mass launched relatively to each of them.

The dramatic decrease of Earth Observation missions is once again linked to the end of the Soviet era. Since then, Space telecommunications are by far the most important domain of application, involving most of the mass of satellites produced and launched worldwide. After major fluctuations and the noticeable low mentioned above during the years 2003–2004, this activity seems to have stabilised so as to ensure in particular the continuity of operations of fleets in orbit.

The other areas must however not be disregarded as they are major stakes for industry as well:

– Navigation implies the production of a great number of satellites for the deployment and the replenishment of major constellations like GPS, Glonass, Beidou or Galileo. For European industry, it is expected to dramatically grow with the deployment of Galileo in the near future.
– Although not evidenced by Fig. 5.2, the business for industry related to Earth Observation has been growing in the recent years with a number of countries which have recently procured or developed domestic in-orbit infrastructures. The recent and constant progress in the associated technologies allows for smaller and lighter satellites with increased performances that are not correctly reflected when limiting the analysis the mass launched.
– Meteorology as well implies limited mass launched but represents important investments to develop innovative and more performing instruments.

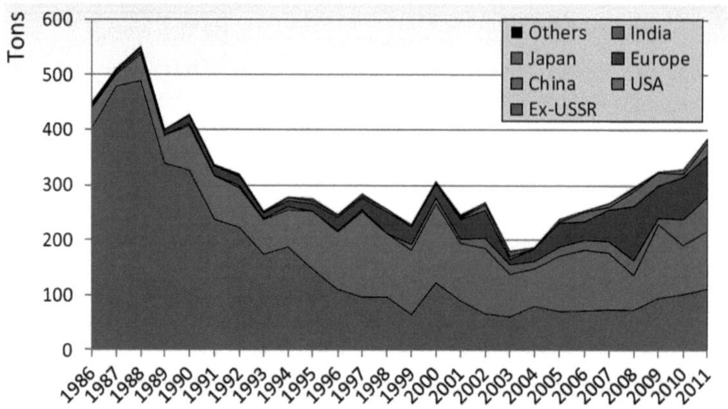

Fig. 5.4 Launched mass by manufacturers region (Source: Eurospace)

The analysis of the evolution of the global Space activity must be complemented by a regional approach. Fig. 5.4 shows the mass of satellites produced and launched by the major Space faring nations.

- Russia: Since the dramatic decrease of the 90s, the situation of Russia has stabilised, and the growth observed in the most recent years reflects major public investments made with the clear objective to reinstate Russia as a first rank Space power.
- USA: Strongly backed by massive public R&D driven by very ambitious defence oriented programmes, the US industry encompasses the whole spectrum of Space applications and is at the leading edge of each of them.
- China: Dramatic growth clearly visible in the last few years. This trend is expected to continue in the future, with great ambitions in all sectors and applications, including Defence.
- Europe: With relatively limited public investments as compared in particular to the United States, Europe has managed to become a very effective and highly regarded Space power. The activity measured in mass launched involves a major share devoted to the commercial market of GEO telecommunications satellites.
- In the rest of the world, Space activities are essentially concentrated in Japan and India. They appear as second rank players although they cover a broad spectrum of applications. However, they have, unlike USA and Russia, no Defence oriented Space programmes, and unlike Europe, very limited shares of the commercial markets.

A closer look at the GEO market, mostly driven by the commercial business of Space telecommunications operators, is proposed with Fig. 5.5 showing the number of GEO satellites annually produced worldwide broken down by region of production. Globally, this market remains overall steady, with strong variations, especially in the years 2003/2004 during which the consolidation of major private operators was a factor of reduction of the demand.

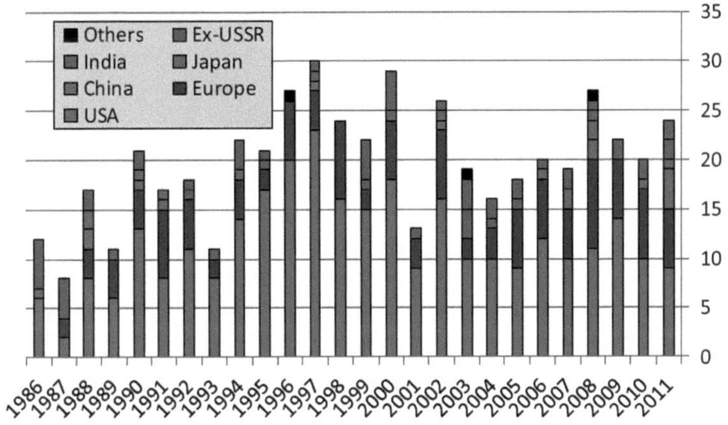

Fig. 5.5 Number of civil GEO satellites by supplier country (Source: Eurospace)

The regional distribution of the production shows that the US providers have been continuously leading this market.

However, it can be noticed the strong progression of European manufacturers on the open and accessible commercial market over the last decade, mostly at the expense of their US competitors.

It must also be stressed that beside US and European manufacturers, which strongly dominate the commercial market, the offer has been diversifying in the last few years, in particular – but not limited to – China, to the extent, that in 2011, third countries delivered an unprecedented share of almost one third of the global production.

5.2 The Case of the European Space Manufacturing Industry

Let us now focus on the case of European manufacturing industry. Fig. 5.6 shows the evolution of the total turnover of the sector in constant economic conditions, as well as the associated employment expressed in number of Equivalent Full Time Jobs.

With 6B€ per year and 35,000 employees, the European Space manufacturing sector remains small in size and the trend over the last 20 years shows that it is basically not growing.

The overall fluctuations are mostly due to the variation of the commercial markets, both for the delivery of telecommunications satellites or the associated launch services.

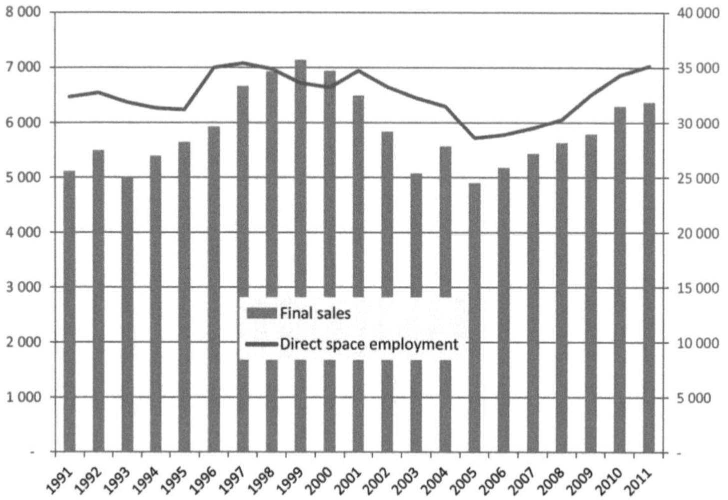

Fig. 5.6 Final sales (M€ constant e.c.) & employment (FTE) (Source: Eurospace)

The following chart focuses on the business of European manufacturers related to spacecraft, both in mass and in value. It presents separately the institutional and the commercial activities.

As previously mentioned, the commercial market has been dramatically variating over the last decade. The recovery from the major crisis of the 2003–2004 seems now complete, but no further growth is expected from the market analysts.

Expressed in current economic conditions, the institutional business is steadily increasing, which is consistent with the evolution of public budgets that have been kept constant – or slightly depleting – in constant economic conditions over the period.

The capacity of some member States to continue their support at the same level might be impacted by the financial crisis.

The total turnover made by the European industry for spacecraft development and production peaked in 2011 to almost 5 B€, for more than half on the institutional market.

On the launchers side, the shift from Ariane 4 to Ariane 5, along with the crisis of the telecommunications sector in the same period, generated a gap in the early 2000s. The production is now stabilised around 6 launchers per year for an associated turnover of about 700 M€ this is expressed in Fig. 5.8.

The European launchers business is almost exclusively driven by the commercial market.

If we now analyse the customers of the European Space industry, it appears that the export sales represent in 2011 about 19 % of the turnover as shown in Fig. 5.9. From a manufacturing standpoint, it must be reminded the Ariane related business

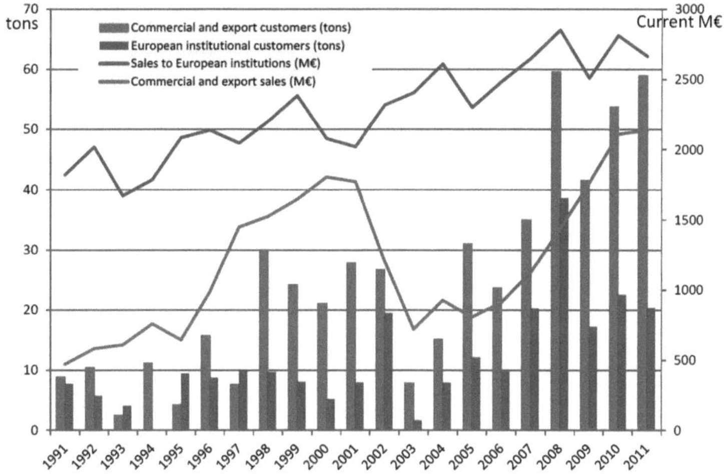

Fig. 5.7 Spacecraft sales (Source: Eurospace)

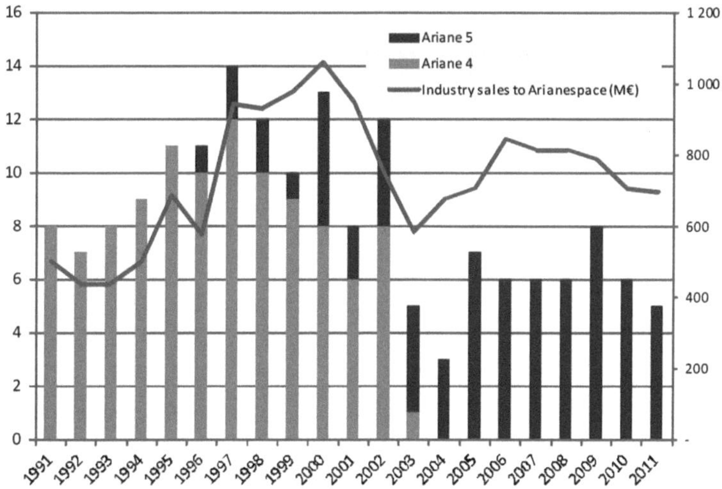

Fig. 5.8 Launchers sales (Source: Eurospace)

is not here considered as export since the final customer Arianespace is based in Europe.

The export sales included in the first pie-chart are thus essentially related to the telecommunications satellites business.

The second pie-chart shows the breakdown between private and public customers. It shows that the European Space manufacturing industry relied in 2011 for about 40 % on the commercial markets. Such level of exposure is a specific feature of Europe.

Fig. 5.9 European space industry customers in 2011 (Source: Eurospace)

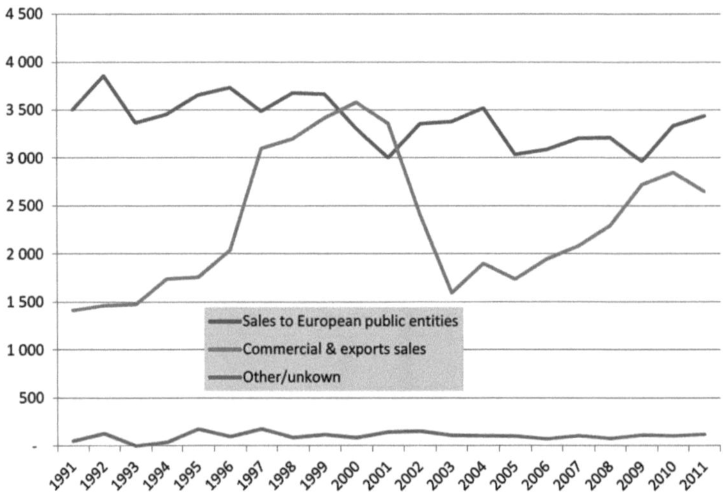

Fig. 5.10 Turnover with public/commercial entities (M€ constant economic conditions) (Source: Eurospace)

This ratio is of course highly variable and Fig. 5.10 illustrates the relative fluctuations of both markets.

Expressed in Constant Economic Conditions, the institutional business made by the manufacturers has been slightly receding. This is another key feature of the European Space industry while in the same period, all Space faring nations have dramatically increased their Space budgets.

With the fluctuations of the commercial market, the ratio of reliance on public budgets varies from more than 50 % to less than one third. However, the situation in 2011 seems quite typical.

Fig. 5.11 presents a closer look at the breakdown of the turnover of the European Space manufacturing industry in 2011.

It appears that ESA is by far its biggest public customer which represents some 43 % of its total turnover.

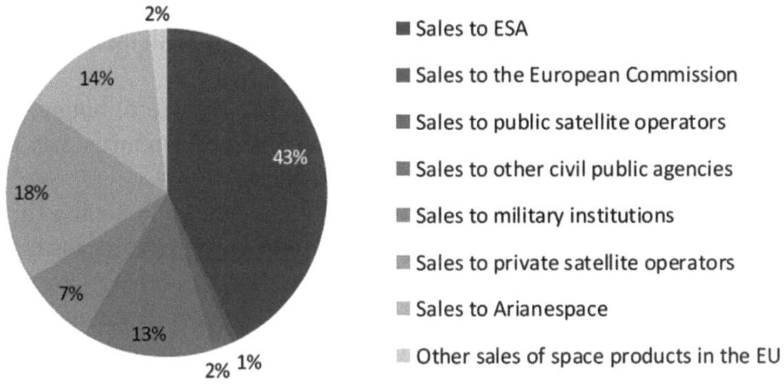

Fig. 5.11 Final sales by customers (Source: Eurospace)

To be noticed that the European Commission appears here as a customer only for the FP7 contracts awarded to industry. The most significant EC space-related activities – mostly Galileo and GMES – are delegated to ESA, which thus appears as the final customer for the manufacturers.

5.3 Impact of the Financial Crisis on Space Manufacturers

This brief overview shows that by the end of 2011, the global Space activity presents no sign of having been affected by the financial crisis yet. However, pressure is applying on public budgets to be voted by governments all over the world for the period to come. This will impact the direct institutional business of industry, but it will also have consequences on the commercial markets where the competition might be fiercer.

In particular, we have noted recently that US manufacturers, which are the primary competitors of European companies, were much more aggressive on the commercial markets, probably because they are anticipating major cuts in the federal budgets, especially from the DoD.

As a clear sign of this strong return, to date, the backlog of Boeing for commercial GEO spacecraft is, for instance, comparable to those of European manufacturers.

Boeing has also recently proposed a new offer for a full-electric platform, coupled with Space-X launcher in a dual-launch configuration. Despite the several month long launch process, this announcement generated a lot of interest from private operators as it might be proposed at a very competitive price.

Actually, the competitive relationship with the US is complex. Backed by the massive DoD Space R&D budgets, it is very difficult for European companies to challenge their technological leadership.

As a matter of fact, when the US administration increases its Space budgets, European industry has to make additional efforts to keep up with their technological developments.

On the other hand, in the current situation, the cut on federal budgets revives their motivation on the commercial markets, and European industry has again to make additional efforts to increase its competitiveness.

As compared to the US, Europe needs to elaborate and implement a successful strategy of follower. This implies to be proactive and to have the means to quickly adapt to changing situations.

But the return of the US providers is not the only risk of additional tension on the commercial markets. As highlighted in the analysis above, the emergence of new comers, especially China, in the Space sector is spectacular, and it is not expected that the financial crisis affects their Space budgets in the next few years. They are fuelled by very ambitious objectives embracing the full spectrum of Space applications and aim at achieving full technological autonomy.

For the time being, they have expressed limited interest in commercial business, first because they are not yet at the right technological level to meet the expectations of major telecommunications operators, but also because they have been assigned challenging governmental priorities.

No doubt that some time in the future they will enter the game, first addressing the lower end of the market with very aggressive pricing, and progressively getting to the upper end according to their technological progress.

5.4 Impact of the Financial Crisis on the Customers

Space manufacturing industry will be impacted by the financial crisis through its customers. As mentioned above, they are of two different kinds – private and public – and if no evidence of major impact is visible on the global Space activity yet, they will both be potentially affected, although in two different ways:

- Impact on the Private Customers

  There are today 38 private Space operators worldwide. Eighty percent of them have published results which have grown in 2011 as compared to 2010. Among those, 13 have even enjoyed an annual growth of more than 10 %.

  However, one should not hastily conclude that the financial crisis has no impact on the sector. Actually, if the growth is continuing, it is at a lower pace. Only 23–25 orders for big GEO telecommunications satellites are expected worldwide in 2012, and the market, which comprises a cyclical share linked to the renewal of assets in orbit, is expected to hit a low in 2014–2016.

  On top of that, the specific market for mobile telecommunications is developing below expectations. It reached 1.6 B$ in 2011 up from 1.4 B$ in 2010, which remains disappointingly low given the profusion of offers provided through multiple systems: Lightsquared, Globalstar, Iridium, Orbcomm, O3B.

But the reality of the business is that Space telecommunications systems are primarily in an endless competition with ground infrastructures. In this respect, we can regret that European public policies are mostly focused on the deployment of optic fibres.

Space operators on their side have developed several offers (Tooway, Astra2connect, KaSat, Athena-Fidus, THD-Sat), whose use should be encouraged.

- Impact on the Public Customers

On the public side of the demand, military and civil activities should be considered separately.

Defence programmes represent about 12 % of the turnover of European Space manufacturers. Although limited in size, this business is important to them as it generally implies the development of cutting edge technologies which is a great support to improve their competitiveness on the commercial markets.

The Space Defence budget is structured at national level and activities are concentrated in just four member States, namely France, Germany, Italy and UK. There is little doubt that the financial crisis will put a major pressure on Space Defence programmes in the next few years, and this share of the business of European Space manufacturers is not expected to grow in the next few years.

European civil Space programmes are funded through two different channels; EU and ESA:

- Provided that the funding of GMES is confirmed within the next Multi-annual Framework Programme, the overall Space budget of the Commission should increase in the next few years with the deployment of Galileo and the continuation of GMES. These are operational programmes which comprise a lot of recurring activities.
- The major contribution of the EC to Space R&T is expected through the Horizon 2020 programme, in strong progression as compared to the FP7, provided that the budget proposed by the Commission is approved by member States and the Parliament. However, the way this programme will be implemented is a major stake. To be useful and effective, it shall be based on a clear assessment of the needs of users and stakeholders, including downstream service providers, agencies, the scientific community, and of course industry. Furthermore, it shall also address maturation of technologies and accompany developments to high TRL (Technology Readiness Level), which is a major shift from FP7, implying a technological expertise that the Commission doesn't have in-house. EC should thus devise the right way to get support from ESA or other bodies where such expertise is available. This is still totally unclear at the moment.
- As far as ESA is concerned, the 2012 Council at Ministerial level is the next milestone and difficult discussions are going on among member States. As an outcome, industry wants to be challenged ambitious

programmes and looks forward to a reinforcement of the technology policy of the Agency to give it the means to keep up with its international competitors.
- It is expected that the financial crisis will impact the capacity of some member States to contribute to new programmes to be decided. This will directly affect the continuation of industrial activities in those countries.
- In the same time, the enlargement of ESA to new member States is politically most welcome. However, such enlargement should come along with an increasing overall budget allowing additional resources to make room to companies from these countries. Otherwise, the consequence is a reduction of the share accessible to established players with has negative implications on the integration of newcomers.

5.5 Conclusion

European Space industry is relatively small. It accounts for 35,000 employees in total, which cannot compare to the United States (250,000), Russia (280,000) or China (at least 300,000), etc.. . .

Despite its small size, it manages to be at the forefront on the international scene. This is the positive outcome of long-lasting and effective strategy of public investment of European member States, mostly through ESA. It should be continued and the financial crisis should not impair the capacity of reaction of Europe to adapt to a fast-evolving global context.

Despite the very effective positioning of its industry, Europe makes little use of Space capacities for its own needs.

One of the reasons is the absence of Europe-wide ambitious Space military programme which is the major driver for most of the other Space powers. There are no expectations that it might change in the near future.

On the civil side, according to the text of the European Space Policy, EU is in charge of aggregating the demand for space-based services. This is a major challenge and certainly a condition for the development of the sector. Potential contributions of Space to several public policies have been clearly identified, would it be for Environment, monitoring, security, transportation, maritime surveillance, etc.. . . There are strong expectations from the EC in this matter.

The consequence for industry of this weak demand from European public bodies is a uniquely high reliance on commercial markets, from 40 % to 50 % of its total turnover of industry. In order to preserve its share of the commercial markets, European industry must relentlessly improve its competitiveness, in particular in the field of telecommunications.

In the past, the improvement of the competitiveness of Space telecommunications systems was based on the augmentation of bandwidth and of the number of transceivers, at stable overall conditions (mass, size, consumption, platform). Major improvements have been made at system and integration level.

The emergence of newcomers at the lower end of the market will put more pressure on prices at overall constant technical performance conditions. This might impact equipment suppliers.

Today, the US competitors are back at the upper end of the commercial market with a strong technological basis, coming up with innovative solutions developed thanks to massive R&T support in particular from the DoD.

Europe is in the middle and must make efforts in order to be cheaper than the US and better than the new comers. This implies to come up with innovative solutions in order to meet the expectations of the operators for diversified services.

Chapter 6
Space: the Last Frontier for Socio-economic Impacts Evaluation?

Jean Bruston

6.1 Introduction: A Growing Need for Evaluation of the Impact of Space Activities

The last meeting of the ESA Council at Ministerial level, in November 2012 (CM12) in Napoli, Italy, was a clear and visible step forward to move space closer to mainstream public policies. With its political affidavit "Space for Competitiveness and Growth" the CM12 meeting, at which time ESA Member States were required to make important public investment decisions, emphasised how clearly the space sector is now required to demonstrate that it will deliver value for money and that the proposed public investment will benefit the economy and society.

Indisputably, the space sector is maturing towards being increasingly interconnected with the general economy – in particular through services, and technology and knowledge spill-overs. Recurring infrastructure and services – in telecoms and navigation systems essentially, but also increasingly in Earth observation, constitute a growing share of the space sector, and part of the economy of space is moving towards a market economy. However, the development and deployment of a large part of the infrastructure, access to space and enabling technologies remain publicly enabled.[1] Public investment still accounts for the main share of the space sector economy,[2] and the space sector is competing with many other public policies for limited – and now decreasing –funds.

The rationale for funding space activities is now expanding. Historically, it was mostly driven by sovereignty, prestige and strategic technology development

[1] The European Union's Galileo and Copernicus programmes are publicly funded infrastructures which are hoped to lead to applications markets worth several times the investment.

[2] Total European Public investment in space represent more than 4B€ annually for an industry turnover of 6B€ (ASD-Eurospace, The European Space Industry in 2010, Facts & Figures, June 2011).

J. Bruston (✉)
DG's Office for EU Relations, ESA (European Space Agency), Paris, France
e-mail: jean.bruston@esa.int

C. Al-Ekabi et al. (eds.), *Yearbook on Space Policy 2011/2012*, Yearbook on Space Policy, DOI 10.1007/978-3-7091-1649-4_6, © Springer-Verlag Wien 2014

ambitions. While these factors continue to influence policy, they are not sufficient to continue to convince decisions makers to invest in the space sector at a time when austerity has hit defence budgets as well as public services and infrastructure investments. Decision makers thus carefully assess the investment rationale when deciding between space and other important public policies. Furthermore, the need for accountability in government funded programmes is growing with tax payers requiring increased transparency. The space sector therefore needs to be evaluated according to public policy evaluation procedures that will provide decision makers the confidence and justification for investing public funds in the space sector.

The space sector is of course not new to evaluation.[3] Measuring all of its impacts and relating them back to the original investment in a structured and rigorous way is a challenge faced by all nations that invest in space. Many studies have attempted to measure the impact on the economy and society of either specific programmes or programmes at a national or regional level. However, recent activities at the European Space Agency have led us to conclude that we cannot identify a coherent, global evaluation of the socio-economic impacts of space activities in Europe. In particular, two studies[4] carried out in parallel, aimed at identifying the most suitable methodology for ESA to evaluate the socio-economic impacts of European public investment in space, concluded that while many evaluation studies were performed over the years, they used many different methods and were generally not based on comparable data. As a consequence, existing studies cannot be aggregated to provide a coherent, overall picture of the impact of the space sector. No single methodology alone could evaluate the impacts of global space activities. This paper will look at the need for coordinated, coherent parallel evaluation activities that together would enable global evaluation of the socio-economic impacts of space activities in Europe. It will discuss the current status of socio-economic evaluation of space activities, an approach for the space community to achieve a global assessment, the results to be expected and the use to be made of them.

6.2 Evaluating Space Activities: The Current Situation

As is well reported,[5] there are some structural difficulties, inherent to the space sector, which make its evaluation difficult. Although they are not unique to space, their combination has proven to be a challenge for academics and practitioners who

[3] Many references exist, see for example: Brendle P, Cohendet P, Larue R (1986) The economic impact of European Space Projects. Futures. April 1986. Chapman RL, Lohman LC, Chapman MJ (1989) An exploration of benefits from NASA Spinoff. Contract 88–01 with NERAC, Inc. Mathematica, Inc., Economic Analysis of New Space Transportation Systems, Princeton, NJ, May 1971.

[4] Reports by Technopolis (ESA Contract Reference 4000103623/11/F/MOS) and SEO-NLR (4000103624/11/F/MOS) (May 2012), Design of a Methodology to Evaluate the Direct and Indirect Economic and Social Benefits of Public Investments in Space.

[5] Hertzfeld HR (2012) A comment on economic theory, law, and policy in the space economy. 63rd International Astronautical Congress, Naples, Italy.

have been engaged in assessing the socio-economic impacts of space activities. The first issue is linked to the very nature of the space sector which, on the one hand, leverages numerous technologies and, on the other, supports manifold applications. For the space manufacturing sector, the economic data is lost within much larger sectors of aerospace and electronic equipment. And drawing the line between space and non-space for downstream services almost requires a case by case analysis. A well-known consequence is that space is not a recognised category in international standards of industrial classification, which results in a lack of consolidated economic data that is comparable between countries and across time.

The second structural difficulty linked to the space sector relates to the wide diffusion of its impacts. Space infrastructures support manifold applications with wide-ranging and widely disseminated economic and social benefits throughout many sectoral policies of very different natures. Furthermore, space activities are rich in R&D, the impact of which unfolds over decades, making the link between an impact and the original investment difficult to trace.

Finally, a third type of difficulty is related to the mind-set of the space community, for whom the idea that evaluation is needed for its activity has only recently and slowly emerged following the evolution of its justification from purely strategic to cost effectiveness criteria. The space sector is quite new to impact evaluation and does not routinely collect and report data that could support the evaluation of social and economic impacts.

This state of affairs means that the many evaluations performed in the past had very variable scope, used different methods and drew on non-comparable data. For example, the OECD[6] has assessed past macro-economic studies that for the main part attempted to measure the "multiplier effect" i.e. the direct effect on the space sector economy of space industry income from public investment. These national studies that were carried out in Belgium, the U.K., Denmark, Norway and the U.S. are not fully comparable. They nonetheless do provide a good first order indication, as the multiplier effect is found to converge somewhere between 2 and 5. While this indication is valuable, it is not sufficient to fully characterise the impact of the space sector, in particular its manifold indirect impacts that are often unquantifiable, on the economy and society through services and spill-overs. Furthermore, this multiplier effect is in the same range as the multiplier effect for public investment in any R&D sector,[7] and as such does not indicate where space can have more benefits than other sectors.

Other analyses have used a case study[8] approach to evaluate the micro-economic and social impacts of specific programmes, projects or sectors such as GMES,

[6] OECD (2011) The space economy at a glance 2011. OECD Publishing.

[7] Coe D, Helpman E (1995) International R&D Spillovers. Eur Econ Rev 39:859–887.

[8] PricewaterhouseCoopers (July 2006), Socio-Economic Benefits Analysis of GMES; Booz & Company (September 2011), Cost-Benefit Analysis for GMES; PricewaterhouseCoopers (November 2001), Inception Study to Support the Development of a Business Plan for the GALILEO Programme, TREN/B5/23-2001; Technolopolis (October 2010), Space Exploration and Innovation.

Galileo, Exploration, etc.. However their scope has usually been very narrow, using different definitions, with different goals, and not always transparent methodologies, and they cannot be aggregated to provide a coherent, overall picture of the impact of the space sector on society, although this is where space could be making a real difference. In addition, comparing two assessments using the same methodology is almost impossible due to the lack of standard classification of space related activities, coherent definitions of their impact and agreement on their metrics and, finally, inconsistent data gathering, formatting and analysis. As the OECD puts it:

> many space-based services have positive impacts on society, but issues concerning economic data definitions and methodologies have to be resolved to allow the benefits to be identified and quantified more precisely[9]

All these elements converge to show that the space sector needs to establish a recognised, satisfactory proven methodology for measuring its effects in a comprehensive, coherent manner.

6.3 Some Considerations About Evaluation Methodologies

Looking more into the details of what public policy evaluation methodologies exist and how they have been used for measuring the impact of space activities, we find[10] that most of the recognised monetary and non-monetary methods, from Financial Analysis to Impact Analysis, have been used for evaluating space activities. Those most frequently used are Cost Effectiveness Analysis and Social Cost Benefit Analysis for monetary approaches, and Multi-Criteria Analysis for non-monetary ones. Different methodologies can be applied, depending on the context and the exact objectives of the evaluation, and they can provide solid results when rigorously undertaken. Furthermore, different methodologies must be adapted depending on the nature of programmes as their impacts differ tremendously, varying for example between close-to-market Communications satellites project with direct economic impacts to R&D rich Exploration projects with long term knowledge spill-overs. It is clear that no single methodology can encompass the whole of the impact of space activities. As a very minimum, macro and micro economic studies and non-monetary ones should be considered. However, a sound, effective methodology can be found for conducting most types of evaluation, whatever the programme or the nature of its impacts.

Therefore, the issue of evaluating the impact of public investment in space in a global and coherent manner is not *stricto sensu* an issue of the socio-economic methodology to be used, but rather of the framework conditions to implement it.

[9] OECD (2012) OECD handbook on measuring the space economy. OECD Publishing.

[10] Reports by Technopolis (ESA Contract Reference 4000103623/11/F/MOS) and SEO-NLR (4000103624/11/F/MOS) (May 2012), Design of a Methodology to Evaluate the Direct and Indirect Economic and Social Benefits of Public Investments in Space.

Sound and effective socio-economic methods can be readily applied to the space sector, the limitation of their use being currently linked to what are called the boundary conditions (definition of the boundaries of the sector, macro-economic data availability) and the evaluation habit in the space sector or lack thereof. The search for solid socio-economic evaluation of the impact of public investment in space activities should therefore focus on changing these boundary conditions and the evaluation culture of the space community, towards performing evaluations in a systematic, coherent manner, making data available and the results of analysis meaningful and useful. What we are lacking is indeed a structured approach that would enable analyses to encompass the different nature of space activities, the diversity of the impacts, and ensure solid and consistent data availability. In order to structure such an approach, we first have to consider what exactly we want to evaluate and with what objective.

6.3.1 Different Motives

There are several reasons for which we may wish to evaluate the impact of space activities, depending both on the stand point of the targeted stakeholder (the "client") and the temporality of the evaluation. For whom are we conducting the evaluation and what is to be brought to light?

As actors from the space community, we want to demonstrate the value that can be achieved from investing in space. Our motives can be multi-fold, from genuine passion for space to a simple desire to get value for money for our ultimate employers, the taxpayers. Project managers will want to ensure that they deliver the "promised" benefits and have at their disposal an efficient monitoring tool. The policy maker needs to determine priorities for investment, i.e. identify areas with best return for consistency. Finally, the taxpayer needs reassurance that his money is well spent (with all the different interpretations of what "well" might really mean).

Another way of describing it is to consider the temporality of the evaluation: *ex-post* evaluation of space activities, in a global or in a programme specific manner will aim to provide a snapshot (at European/ESA or national level) of the impact of programmes or selected projects, related to specific past investments. *Ex-ante* evaluation of specific programmes or projects will aim to provide *e.g.* ESA, for its internal use, an impact assessment of specific programmes or projects during their definition phase in order to optimise their definition, and/or to provide decision makers (delegations, for example at the time of Ministerial Council meetings) with the means to make informed decisions between competing priorities. Finally, following the decision to implement a programme, evaluation as a monitoring capacity aims to provide the ability to measure delivery, throughout its development and exploitation, with respect to pre-selected objectives and to identify/remedy unforeseen impacts.

6.3.2 The Need for Aggregation

Depending on the "client" and on the "timing" of an evaluation, one must define different objectives, which might require different methodologies and processes. *Ex-ante*, *post-ante*, monitoring? Economic or social impacts, or both? Direct economic benefits are easier to measure but less relevant than the difficult to measure indirect economic and social impacts, where the evidence suggests space holds most of its added value compared to other publicly supported sectors.

Project, programme or sector? Impacts will usually be measured at project or programme level, which from one to another will have impacts of a very different nature: how can one compare the short-term, quantifiable economic impacts of a satcom project to the long-term, unquantifiable knowledge transfer from an exploration project?

Most important is that many space programmes cannot exist without other programmes: satellites need launch services, operational services need scientific knowledge, scientific progress needs new technologies etc. This suggests that socio-economic benefits should be evaluated globally, with the impacts of space activities being seen as a portfolio, with value to invest in space for knowledge, services and competitiveness. However, from a socio-economic theory perspective, space activities can be evaluated mostly at programme/project/sub-region level, and some would argue that each investment decision has to be justified on the basis of the specific programme that will result from it – will the marginal investment provide a sufficient marginal return on the investment?

A planned schedule of evaluation should attempt to provide the picture for the overall sector, while individual evaluations for each programme or project should focus on more immediately identifiable results. What must be done is to ensure that there is a consistent and coherent methodology that will permit the specific studies to provide inputs to those taking a broader view.

This underlines the necessity of aggregating the measure of a multitude of impacts into a global analysis. Beyond choosing adequate methodologies for measuring the specific impacts of specific programmes, the question is how to make sense of these measured impacts together, with regards to the *global funding* of space activities?

6.4 What Ambitions for the Space Sector?

The ambition to measure the socio economic impact of public investment in space is not new and has been addressed at several levels, yet we do not have an aggregated view of the impact of the space sector in a global, coherent manner, which would make sense over time. Beyond the inherent difficulties of achieving this in the space sector, and the need to structure evaluation practices to overcome the inherent difficulties, one should consider what we can really achieve, i.e. what are the limitations we should accept, what results are desirable and at what cost.

The first limitations are not specific to the space sector and are intrinsic to any socio-economic evaluation. We discussed above that to make sense, impact evaluations in the space sector should be global, i.e. assess the impact of the whole space sector (at least for the perimeter of interest, i.e. Europe in our case). However this should not be confused with a holistic evaluation that would attempt to encompass all impacts of all activities. Evaluations will always address a selected subset of impacts, programmes and investments and the motivations of both the evaluator and the "client" will introduce biases. The evaluator may be inclined to use either the most favourable metrics or, which may be the bigger danger, those he can most easily measure. He may also be pressured to respond to the expectations of the client, resulting in partial or overly simplistic results. In particular, the need of decision makers for short, easily communicated information may lead them to seek out a single figure, usually the direct macro-economic multiplier effect, which in the case of space can undermine its real impact as discussed above. A global evaluation will induce complex output, articulating both quantitative and qualitative metrics, which will require educating the client in how to read them. The results need to be communicated clearly but that is not the same as communicating simplistically.

Bias may also come from data gathering and interpretation, both for qualitative and quantitative metrics, but also in the presentation and explanation of the results, which all require sound definitions to be largely exploitable. Sound, consistent and largely agreed and shared definitions are essential to enable results from different evaluations to be comparable, between places, sectors, over time and between different evaluators. One-shot evaluations have limited value. It is in particular the time dimension that will indicate trends that will give the most insight into impacts. This implies that standards need to be defined so that the efforts of the whole community converge towards a possible aggregation over time into trends and throughout impacts.

Results of an evaluation are legitimated through indisputable impact definitions and methodology for metrics measurement, which can be interpreted in a stable manner, repeated, and compared. The space community therefore needs to agree on a set of definitions, metrics, categories, and recognised methods to establish a stable evaluation framework. In addition to this framework, data availability must be pursued through a systematic and sustainable effort that will impact the very culture of the space community.

6.5 Next Steps: A Space Community Collective Effort to Measure Its Impact

At this stage we believe that the issue of providing solid, indisputable and meaningful measures of the socio-economic impact of space activities is out of the reach of any single actor as it would require massive resources, and most importantly, modification of both the boundary conditions and the culture of the space community for doing so. It is the responsibility of the whole community, probably led by space agencies and economic statistics organisations, to establish the framework for solid, coherent and global evaluations. The community needs to organise itself to agree on the relevant impacts to measure and their metrics and to set definitions and evaluation standards so that every evaluation is made using elements of a well-defined, limited “tool box”. It needs to ensure data availability and organise its systematic collection, and to work with economic statistics organisations at national and international level to ensure that industrial classification standards identify the space sector as such. Finally it needs to change its culture so as to make the evaluation of the impact of space activities a systematic habit at all levels.

The Space Forum of the OECD, and its efforts to coordinate the space community and define standards,[11] is a very promising step forward in this respect but will fail if not fully supported by Space Agencies (and hence their political masters). They must indeed ensure that they also transition towards an evaluation culture, organise the collection of their internal data in a systematic manner with evaluation in mind, and enforce the agreed standards when procuring evaluations. The global evaluation of space activities will come from the aggregation of a multitude of project or national level (or other specific evaluations) evaluations, which must be *de facto* coordinated within a common framework, using agreed definitions and standards.

6.6 Conclusion: Some Considerations About the Results

Space activities are in competition with other sectors for public funding. Many past evaluations show that the direct economic impact of space activities, as measured by the multiplier effect, is not significantly different from public investment in other R&D sectors, perhaps even from any other public investment in general. We assume that space provides more global benefits than many other sectors and that they are most significant through indirect economic (spill-overs) and social benefits. Furthermore, because of the distributed nature of the space sector, those effects need to be evaluated globally, which requires aggregation of the measures of many different impacts of different natures, quantitative and qualitative. We believe that managing such global evaluation will require restructuring the space community and changing some boundary conditions and the evaluation culture, which would be a long term activity.

This effort is necessary – space can no longer be a sector dependant on public investment and not abide by the requirement of modern public policy for accountability and transparency. But the road is paved with risks, including the risk of failing to show that space programmes are closely intertwined and the very different

[11] OECD (2012) OECD handbook on measuring the space economy. OECD Publishing.

nature of the impacts that extend over many different sectors of the economy and society, which would thus put space programmes, or even individual investments, in competition with one another. Other risks include having to respond to the pressure of decision makers for a "magic number" (a comprehensive multiplier effect) when impacts are of very different nature and cannot be added up, and biases towards specific impacts. The packaging and communication of the results will be the moment for the space community to question its very *raison d'être*, its future perspectives. While transparency is indispensable, aggregating and presenting in a simple manner a multitude of impacts with metrics of different natures will necessitate that the space community and decision makers agree on where to place the focus, and finally why they are investing in space.

Beyond the methodology of evaluating the impact of public investment in space activities, the challenge ahead is the implementation of a solid structure, which will require that space agencies and their partners, economic statistics organisations and industry, acquire the 'evaluation habit' and change their isolated working culture.

Chapter 7
The Case for the Eumetsat Polar System (EPS)/Metop Second-Generation Programme: Cost Benefit Analysis

Stéphane Hallegatte, John Eyre, Tony McNally, Roland Potthast, and Robert Husband

7.1 Introduction

Observations from meteorological satellites are crucial inputs for the generation of weather forecasts by the National Meteorological Services of EUMETSAT Member and Cooperating States, with such forecasts being used to produce warnings and other tailored information in support of public and private decision-making.

The socio-economic benefits of the full information chain result from the impact of the decisions actually made to avoid costs, optimise implementation of public policies, or increase gains in private business. The realisation of these benefits depends in the first place on the quality of the input forecasts, but also on the capability to exploit this information in the decision-making processes, and on the ability to respond to the situation through appropriate measures. Weather forecasts also have significant value for individual citizens in their day-to-day lives.

Because of the complexity of the processes involved and the variety of benefit areas, from protection of life and property to weather sensitive sectors,

S. Hallegatte (✉)
World Bank, Washington DC, USA
e-mail: shallegatte@worldbank.org

J. Eyre
Met Office (UK), Exeter, UK
e-mail: john.eyre@metoffice.gov.uk

T. McNally
European Centre for Medium-Range Weather Forecasts (ECMWF), Reading, UK
e-mail: tony.mcnally@ecmwf.int

R. Potthast
Deutscher Wetterdienst (DWD), Offenbach, Germany
e-mail: r.w.e.potthast@reading.ac.uk

R. Husband
MSYS Ltd, Isle of Man
e-mail: robert.husband@eumetsat.int

C. Al-Ekabi et al. (eds.), *Yearbook on Space Policy 2011/2012*, Yearbook on Space Policy, DOI 10.1007/978-3-7091-1649-4_7, © Springer-Verlag Wien 2014

e.g. transport, energy, agriculture, tourism, or trading of consumer goods, the benefits of weather information are difficult to assess in precise quantitative terms, but a number of studies have met the challenge and demonstrated that they are substantial.

Assessing the benefit of observations from polar orbiting meteorological satellites, like the current generation of EUMETSAT Metop satellites, is even more complex, as it means allocating a fraction of the estimated benefit of weather information to one input of the meteorological information production chain.

However, if it is possible to assess the level of contribution of satellite data to the skill of forecasts, then a relevant fraction of the estimated benefits of forecasts can be attributed on this basis. This has now become possible, because advanced numerical methods are now available to evaluate in an objective, quantitative and cost-effective manner the level of contribution of any type of observations to the performance of numerical weather prediction (NWP) models[1] which provide the basic material for elaborating most forecasts; as long as these observations are actually ingested into models. In particular, the relative contributions of various existing satellite data inputs to forecast skill can be measured, and as a result, estimates of their respective contributions to the socio-economic benefits of forecasts can be extracted.

This methodology has been followed to assess the socio-economic benefits of the observations of the current EUMETSAT polar orbiting satellite, Metop, and to estimate by extrapolation the socio-economic benefits, and benefit-to-cost ratios, to be expected from the EPS/Metop-SG programme now proposed by ESA and EUMETSAT to replace this generation of satellites in the 2020–2040 timeframe, i.e.:

- Section 7.2 assesses the socio-economic benefits of forecasts in the European Union for a set of key benefit areas where a quantitative estimation is, in general, feasible;
- Section 7.3 provides an analysis of the contribution of observations from polar-orbiting satellites to numerical weather forecast skill;
- Section 7.4 overlays both analyses to extract the proportion of the benefits of forecasts that can be reasonably attributed to existing satellites such as Metop, and proposes an extrapolation to EPS/Metop-SG in the 2020–2040 timeframe, based on the most conservative assumptions.

As a complement, Sect. 7.5 proposes a qualitative assessment of the contribution to other benefit areas, which are not amenable to a quantitative analysis.

The main conclusions as regards the benefits, and benefit to cost ratio, of the EPS/Metop-SG programme proposed by EUMETSAT and ESA are presented in Sect. 7.6.

[1] Joo S, Eyre JR, Marriott RT (2012) The impact of Metop and other satellite data within the Met Office global NWP system, using an adjoint-based sensitivity method. Forecasting research technical report 562, Met Office, UK.

7.2 Socio-ecomomic Benefits of Forecast Information

Forecast information is used by all parts of modern society, and the resultant benefits are diverse and wide-ranging.

In order to provide an insight into some of these benefits, the usage and impact of forecast information is assessed in a number of key areas, including estimation of associated benefits from published scientific literature.

7.2.1 Safety of Life and Protection of Property[2]

A significant part of the benefits of weather forecasts are indirect in the form of lives saved and "costs avoided" in the prevention and management of natural disasters related to extremes of weather.

For example, short-range forecasting (up to 3 days in range) enables preparations to be made for extreme events such as heat waves, cold spells, gales, thunderstorms, storms, floods or forest fires. These forecasts then feed early warning systems which, if embedded within efficient prevention and civil protection policies and used by public and private operators, enable life and property to be protected in many ways.

7.2.1.1 Safety of Life

There are many weather-related threats to safety of life in Europe, such as floods (e.g. Elbe in Germany in 2002, le Gard in France in 2002 and 2005; Lower Silesia in 2005, Var in France in 2010, Vara river valley in Italy in 2011), winter storms (e.g. Lothar and Martin in 1999, Klaus in 2009, Xynthia in 2010, Kyrill in 2011), heat waves, (e.g. the 2003 summer) and cold spells (e.g. 1984–1985 in France; 2001 in Hungary; 2010 in Poland), avalanches. (e.g. in les Orres in France in 1998) and forest fires, e.g. in Greece, Spain and Portugal in recent years.

In Europe, severe winter-storms have a return period of about 10 years and often lead to dozens of casualties. The 2003 heat wave caused about 70,000 deaths in Europe[3] and the 2006 event led to about 2,000 deaths in France alone. Avalanches cause on average 32 deaths per year in France, compared to a total of 4.5 million ski

[2] Hallegatte S (2012) A cost effective solution to reduce disaster losses in developing countries: hydro-meteorological services, early warning, and evacuation, policy research working paper 6058, The World Bank.

[3] Robine J-M, Cheung SLK, Le Roy S, Van Oyen H, Griffiths C, Michel J-P, Herrmann FR (2008). C R Biol 331(2):171–178.

tourists.[4] Forest fires, due to the combination of dry vegetation and wind, have also been devastating, mainly in southern Europe.

It is difficult to assess how many lives prevention and early warning systems save each year, even though civil protection agencies and other policy makers consider these systems to be essential for ensuring the safety of the population. For example, when floods are forecast individuals can avoid travel, implement mitigation actions (e.g. sandbagging), whilst civil protection agencies, organisations and businesses can take tailored safety measures including, in extreme cases, evacuation.

Such warning services have a large audience and certainly help avoid hundreds of accidents each year, as demonstrated by studies and analysis of specific examples.

For instance, in the case of heat waves, studies by Inserm and InVS[5] have shown that the forecasts, warning and prevention system implemented in France after the August 2003 heat wave reduced the number of deaths from 6,400 to around 2,000 during the less intense but longer episode of July 2006.

The ability to prepare emergency services before an event occurs is also essential. During the few hours before an intense weather event, much can be done to increase the efficiency of the emergency services. For example, in 2002 during the floods in the Gard region and, as a result of the forecast information, 22 out of the 26 helicopters able to conduct rescue airlifts were pre-positioned in the flood area. According to local emergency services, this pre-positioning saved about 100 lives, compared with a situation in which it would have taken hours to move helicopters to the affected areas, and there are many and frequent similar examples across Europe. From a more general standpoint, the possibility to relay accurate forecasts is critical to the sizing and efficiency of civil protection, rescue and other large-scale safety systems, enabling their optimum deployment and targeting to the most exposed and most vulnerable areas. The benefit is that more lives can be saved, at a lower cost.

It is also important to stress that the benefits from forecasts depend largely, and nonlinearly, on their accuracy, and that threshold effects are important. For instance, the decision to evacuate due to a potential flood cannot be made if the probability of a false alarm is too high (or if the warning area is too large) as, after a few unnecessary evacuations, the trust into the warning system is likely to disappear, and the warning system becomes useless. One example is the case of New Orleans, which had been evacuated unnecessarily twice (for hurricanes George in 1998 and Ivan in 2004), making it more difficult to convince inhabitants to leave before hurricane Katrina. Conversely, if the risk of false alarm becomes low enough to create and maintain trust, and thereby allow significant prevention measures to be

[4] Dossier d'information "Avalanche" du Ministère de l'Aménagement du Territoire et de l'Environnement (2000). Available on http://www.prim.net.fr.

[5] Estimation de la surmortalité observée et attendue au cours de la vague de chaleur du mois de juillet (2006). http://www.invs.sante.fr/publications/2007/canicule_2006/canicule_2006.pdf.

taken before disasters, a limited improvement in forecast accuracy can lead to a disproportionately large increase in socio-economic benefit.

Forecast information is critical for the safety of maritime transport and air traffic[6] management. Indeed, to observe appropriate levels of safety, these operations need detailed weather information and, in the absence of such information, it is unlikely that passenger air travel could be safe enough to be commercially viable in its current form.

Other hazards also need to be considered, such as technological catastrophes (accidents in a chemical plant or a nuclear plant) or volcanic eruptions. In these cases, the capacity to forecast winds, and thus the trajectory of the contamination cloud or ash plumes, can save hundreds of lives.

Even in everyday life, forecast information plays a large safety role for people exposed to specific weather-related risks in their outdoor activities linked to the sea (e.g. sailing), in the mountains (e.g., hiking, skiing) or in the air.

So, although no comprehensive quantitative assessment is offered for the benefits associated with safety of life, it can be conservatively assumed based on the figures available, that on a European scale, many hundreds of lives are saved annually by forecast and warning information. Although this cannot be reasonably expressed in monetary terms, this is an invaluable benefit for the society and public policy makers.

7.2.1.2 Protection of Property and Infrastructure[7]

Forecasts and warnings also serve to limit as far as possible the economic losses due to severe weather and related hazards. Based on a number of published studies, the benefits in the form of avoided costs can be estimated at least for floods and storms.

For floods, according to Thieken et al.,[8] a study on the Elbe and Danube floods in 2002 shows that 31 % of the population in flooded areas implemented preventive measures aimed at protecting property, and it has been estimated[9] that the early warning and the subsequent prevention measures reduced flood costs by about 6 %.

---

[6] According to the "Bureau Enquête Accident", 7.5 % of plane accidents have meteorological causes (http://www.bea-fr.org/etudes/stat9798/stats1997-1998.htm).

[7] The value of hydro-meteorological information in Europe, Stéphane Hallegatte, paper in preparation.

[8] Thieken AH, Kreibich H, Muller M, Merz B (2007) Coping with floods: preparedness, response and recovery of flood-affected residents in Germany in 2002. Hydrol Sci 52(5):1016–1037.

Thieken AH, Petrow T, Kreibich H, Merz B (2006) Insurability and mitigation of flood losses in private households in Germany. Risk Anal 26(2):383–395.

Thieken AH, Muller M, Kreibich H, Merz B (2005) Flood damage and influencing factors: new insights from the August 2002 flood in Germany. Water Resour Res 41:1–16.

Kreibich H, Thieken AH, Petrow TH, Muller M, Merz B (2005) Flood loss reduction of private households due to building precautionary measures – lessons learned from the elbe flood in August 2002. Nat Hazard Earth Syst Sci 5:117–126.

[9] Tapsell et al (2008) Modelling the damage reducing effects of flood warnings, final report of the FLOODsite project. Available on www.floodsite.net.

According to Carsell,[10] a warning emitted 48 h before a flood enables the overall damage to be reduced by more than 50 %. According to Barredo,[11] European floods cost on average €4 billion per year (normalised costs calculated over the 1970–2006 period). Assuming that that only half of the floods can actually be forecast and that warnings reduce losses by only 20 %, the benefits from early warnings would be €400 million per year. Using Carsell's estimate, and assuming that 75 % of the floods can be forecast, the benefits would reach €1.5 billion per year.[12]

For storms, according to Swiss Re,[13] the average cost is about €2.6 billion per year in Europe. If weather forecasts help reduce these losses by between 10 % and 50 %, due to similar preventive actions taken for floods, the corresponding gains lie in the range from €0.26 billion to €1.2 billion per year.

Overall, for floods and storms alone, the total benefit from forecasting could lie between €0.66 billion and €2.7 billion per year. In the same area of safety of protection of property and infrastructure, it is expected that forecasts of other severe phenomena (risks of forest fires, snow, heat waves, cold spells, etc.) bring benefits in the same order, leading to an overall estimate between €1.32 billion and €5.4 billion.

7.2.2 Direct Benefits to the European Economy in the Form of Added Value[14]

Forecast information is widely used by industry and businesses to optimise their activities.

For example, in the transport sector forecast information is used to assist in road network management, route planning, rail network management and air traffic management. In the latter case, EUROCONTROL's Central Office for Delay Analysis (CODA) collects operational data from airlines operating within Europe and in 2008 it was estimated that the total cost of delays greater than 15 min, from all causes, was more than €7 billion.[15] 10 % of these delays was attributed to

---

[10] Carsell KM, Pingel ND, Ford DT (2004) Quantifying the benefit of a flood warning system. Nat Hazards Rev 5(3):131–140.

[11] Barredo (2009) Normalised flood losses in Europe: 1970–2006. Nat Hazards Earth Syst Sci 9:97–104.

[12] It is assumed that false alarms have no cost, which is not the case, especially in case of large scale evacuations.

[13] Swiss Re (2006) The effect of climate change: storm damage in Europe on the rise. http://www.preventionweb.net/files/20629_publ06klimaveraenderungen1.pdf.

[14] The value of hydro-meteorological information in Europe, Stéphane Hallegatte, paper in preparation.

[15] Planning for Delay: influence of flight scheduling on airline punctuality, EUROCONTROL Trends in Air Traffic, Volume 7.

weather and, assuming that these weather delays linearly affect the delay costs, then the cost of weather-induced delays within Europe was over €700 million in 2008 economic conditions. Forecast information contributes directly to the minimisation of these delay costs attributed directly to weather; with the capability to minimise such costs being directly correlated to the forecast accuracy.

Also due to capacity and safety considerations, as noted previously, it is unlikely that the sector could exist in its current form without reliable forecast information. Hence forecasting has an intrinsic, and very significant, socio-economic benefit as an enabling precondition for the sector, in addition to its role in reducing weather-related delays.

The benefits of forecasts are expected to further increase in the 2020–2040 timeframe, with the foreseen increase of traffic in the Single European Sky, and the improved use of weather information in the implementation of the trajectory-based air traffic management system under development within the SESAR[16] initiative.

Electricity production industry and grid management authorities use forecast information to anticipate demand as a function of weather conditions, as well as to forecast and adjust available production capacity and, if needed, purchase additional capacity at the most affordable price on the market.

On the demand side, air temperature and wind are normally the most important governing factors, as cold conditions imply heating, and warm conditions imply cooling. On the supply side hydro-electric power is, of course, dependent on precipitation (water and snow), and temperature (which controls snow-melt in spring). Other renewable sources, in particular wind and solar energy, fluctuate in accordance with atmospheric conditions.

From a global perspective, according to IEA's World Energy Outlook 2009, world electricity demand is projected to grow at an annual rate of 2.5 % to 2030. If coal remains the dominant fuel worldwide, the share of renewables (particularly wind and solar energy) is expected to grow from 18 % in 2007 to 22 % in 2030. In Europe the energy efficiency policy, agreed at EU level, targets an increase of the share of renewable energy to 20 % by 2020, as a proportion of overall EU consumption, along with a decrease of CO_2 emissions by 20 % with respect to 1990 levels.[17] The fluctuating nature of these renewable energy sources will therefore need to be addressed. Technological solutions, like energy storage, may provide a limited contribution; but more accurate forecasts of the expected production capability, and the associated uncertainties, will be crucial for the further development and optimum use of these renewable energy sources in the context of the EU energy policy.

---

[16] Council Regulation (EC) No 219/2007 of 27 February 2007 on the establishment of a Joint Undertaking to develop the new generation European air traffic management system (SESAR).

[17] Directive 2012/27/EU of the European Parliament and of the Council of 25 October 2012 on energy efficiency.

Farmers and agro-businesses make use of weather forecast information to make decisions about planting and harvest dates, use of fertilizers and other inputs, and on preventive measures in case of floods or heavy precipitation. The agro-businesses also rely on meteorological information to forecast yield and production, and anticipate the market conditions.

Leisure businesses use weather forecasts to anticipate customer numbers and to adjust their workforce and capacities accordingly.

In all these and other sectors, weather forecasts increase productivity. Their use is growing worldwide, with many new specialised downstream businesses being created to assist in the tailored exploitation of this information.

Estimations of the value-added by weather forecasts to these various "weather-sensitive" sectors of the European economy are available within the published literature.

Some of the particularly weather-sensitive sectors, like transport and energy, have dual importance for the European economy. On the one hand aviation contributes over €120 billion[18] to European GDP in its own right and, on the other hand, it provides a vital component of European transport infrastructure on which many other sectors of the European economy depend.

In the energy sector, Roulston et al.[19] have estimated the value of weather information to optimise wind power production and have found a doubling in profits due to 1 and 2-day forecasts.

In the agriculture sector, some studies have assessed the productivity gains from short to medium term weather forecasts. For instance, Wills et Wolfe[20] look at the use of forecasts to optimize lettuce production in the state of New York, and they find a $900–$1,000 gain per hectare and per year, i.e. a 10 % increase in productivity.

In these few examples, a significant impact of weather information on productivity can be observed. Bearing in mind that about one third of the European GDP is sensitive to weather, and if one assumes 0.25 % value-added due to forecasts, which is a very conservative estimate in view of all figures published in the literature, the overall economic benefit to Europe of forecast information would be of the order of €10.23 billion per year, based on EU-27 GDP in 2010 of €12.28Tn, which does not include the GDP of Turkey, Croatia, Norway and Switzerland who are EUMETSAT Member States and Bulgaria, Iceland, Serbia who are Cooperating States. The published literature on the subject, and the high added value to particularly sensitive sectors like aviation, suggests that a 1 % added value is more likely, which would lead to a figure of €41 billion per year.

---

[18] http://ec.europa.eu/transport/air/internal_market/internal_market_en.htm

[19] Roulston MS, Kaplan DT, Hardenberg J, Smith LA (2003) Using medium-range weather forecasts to improve the value of wind energy production. Renew Energ 28:585–602.

[20] Wilks DS, Wolfe DW (1998) Optimal use and economic value of weather forecasts for lettuce irrigation in a humid climate. Agr Forest Meteorol 89:115–130.

The range of current benefits of forecasts is therefore estimated to be between a minimum of €10.23 billion per year and a more likely value of €41 billion per year, for the European Union alone.

7.2.3 *Private Use by Individual Citizens*[21]

Measuring the benefit of the private use of forecasts is not straightforward and involves measuring the willingness of users to pay (either directly or through taxation) for the service they get from meteorological information.

Lazo and colleagues conducted a survey of U.S. households to estimate their willingness to pay for the weather information that is currently provided to them, and for potential improvement of this information.[22] This survey concluded that 50 % of US households placed a value of at least US$280 per year on forecast information, with more than 80 % of households ready to pay in excess of $30.

Assuming that 50 % of European households are ready to pay at least €40 per year, again a conservative estimate, the societal benefit from weather information would be above €4 billion per year. With a more likely value of €160 per year, the estimate would exceed €15 billion.

Therefore, the current benefits of private use of forecasts by European citizens are estimated to be a minimum of €4 billion per year, with a more likely value being €15 billion per year.

7.2.4 *Summary of Socio-economic Benefits*

Assessing the socio-economic benefit of forecast information is not an exact science but, by using simple and approximate techniques, an estimate of the resultant minimum and likely socio-economic benefits[23] for these three benefit areas is provided in the following table (excluding any provision for safety of life).

So, with a focus on three benefit areas, excluding any provision for safety of life, and with a conservative estimation approach, approximate calculations suggest that for the EU27 economy the socio-economic benefits of forecast information exceed €15 billion per year, with a more likely benefit being €61 billion per year.

[21] The value of hydro-meteorological information in Europe, Stéphane Hallegatte, paper in preparation.

[22] Lazo JK, Morss RE, Demuth JL (2009) 300 billion served: sources, perceptions, uses, and values of weather forecasts. B Am Meteorol Soc 90(6):785–798.

[23] The benefits are independent and therefore additive.

7.3 Contribution of Polar-Orbiting Meteorological Satellite Observations to Forecast Skill

7.3.1 Background

Forecast models, which are the basis of modern forecasting, depend on having an accurate estimate of the initial state to function correctly.

This initial state is constructed from observations and, in the persistent absence of such observations, the model capability rapidly degenerates and the resultant forecast loses all its skill.

So the availability, quality and coverage of observations, via their impact on the initial state, is a pre-requisite for forecasting using models. Observations also have an important role to play in the design and improvement of forecast models, as these models are validated and improved via a systematic comparison with observations.

The importance of satellite observations for short-range forecasting, at centres across Europe, is reflected in the significant expansion in their use within NWP models, with the number of satellite observations ingested in such models typically increasing by more than an order of magnitude over the last decade.

Meteorological satellite observations mostly stem from the complementary combination of geostationary and polar-orbiting systems.

In comparison to geostationary meteorological satellites, polar-orbiting meteorological satellites deliver less frequent but global observations and, flying at a much lower altitude (850 km instead of 36,000 km) above the Earth, give access to a wealth of ocean, land and atmospheric parameters measured by instruments (e.g. microwave) that cannot be flown in geostationary orbit. This makes a much bigger impact on the accuracy of short-range forecasting, and this impact is now assessed from various perspectives.

7.3.2 Impact of Polar Observations on Short-Range Forecast Accuracy

The process of generating a short-range forecast starts with the running of a global model to produce a global forecast, and four centres within Europe maintain and run such global models:

- European Centre for Medium range Weather Forecasts (ECMWF, intergovernmental);
- Deutscher Wetterdienst (DWD, Germany);
- Météo-France (Fr ance);
- Met Office (UK).

For some applications, the resulting global forecast can be an end-product in itself,[24] but in all cases the global forecast output will be used at national level, by all National Meteorological services of the EUMETSAT Member States, to constrain the limited area models that they then run, with higher resolution, to generate short-range forecasts at a regional scale.

Hence for short-range forecasts, the overall quality of the forecast depends on the performance of both the global and limited area models. The limited area models can only add value if the forecasts from global models, used as boundary conditions, are good enough.

Global satellite data, such as that from EUMETSAT's Metop-A, is always assimilated in global models and, in a still limited but increasing number of cases, is also assimilated in limited area models as well, in this latter case resulting in an additional impact.

7.3.2.1 Positive Impact of Polar-Orbiting Observations

Three different perspectives on the impact of polar-orbiting observations on short range forecasting accuracy are presented in this section:

- Relative contributions of the various observation sources to forecast accuracy (Met Office, UK);
- Statistical impact on forecast accuracy estimated through data denial experiments (ECMWF);
- A case study on the impact of polar data on the forecasting of winter storms over Europe in February 2011 (DWD).

7.3.2.2 Relative Contributions of Observations to Forecast Accuracy

The Met Office (UK) has generated Forecast-Sensitivity-to-Observations (FSO) statistics[1] which illustrate, how much each observation type, contributes to the reduction in Day 1 numerical forecast error, relative to other observations. The forecast error is measured by a single figure of merit, a global moist energy norm from the surface to 150 hPa which takes into account a comprehensive set of forecast variables.

From Fig. 7.1, it is evident that the overall impact of satellite data dominates that of surface-based observations, and that the impact of polar orbiting satellites dominates contributions from geostationary satellites and accounts for more than 58 % of the total. More specifically, it can be concluded that one polar-orbiting satellite of the current generation, i.e. Metop-A, accounts for around 25 % of the impact of all observations on NWP forecasts for Day 1.

[24] E.g. in fulfillment of responsibilities to civil aviation as a World Area Forecast Centre (WAFC).

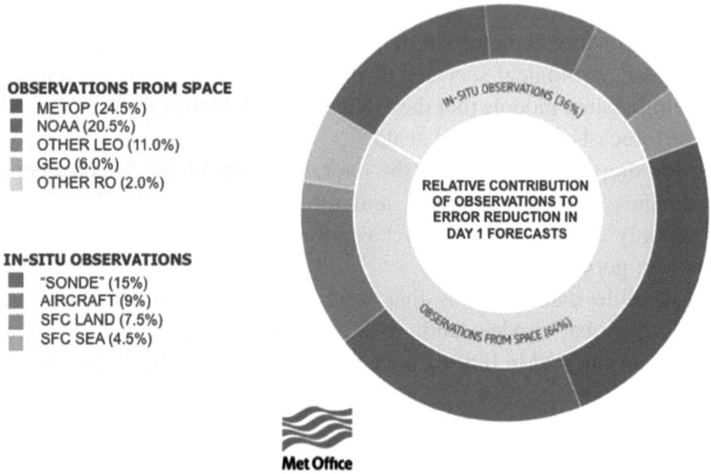

Fig. 7.1 Relative contributions of all observations to the reduction in the error of the day 1 forecast from the global NWP system of the Met Office (UK), estimated using the adjoint-based forecast sensitivity to observation method (Source: Met Office)

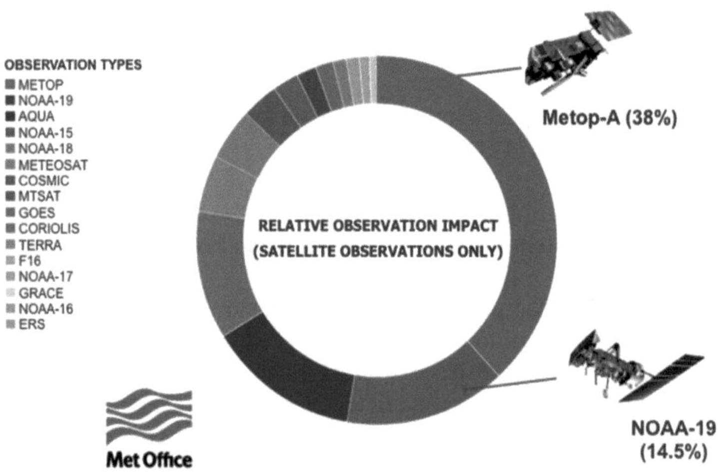

Fig. 7.2 Relative contributions of satellite observations to the reduction in the error of the day 1 forecast from the global NWP system of the Met Office (UK), estimated using the adjoint-based forecast sensitivity to observation method (Source: Met Office)

Figure 7.2 illustrates the relative impact of satellite data alone, and shows that Metop-A has the largest impact of any satellite with about 2.5 times more positive impact than a single polar orbiting satellite from the previous generation, i.e. NOAA 19 (38 % versus 14.5 % relative impact). This emphasises the benefits to be expected from investing in more capable satellite systems.

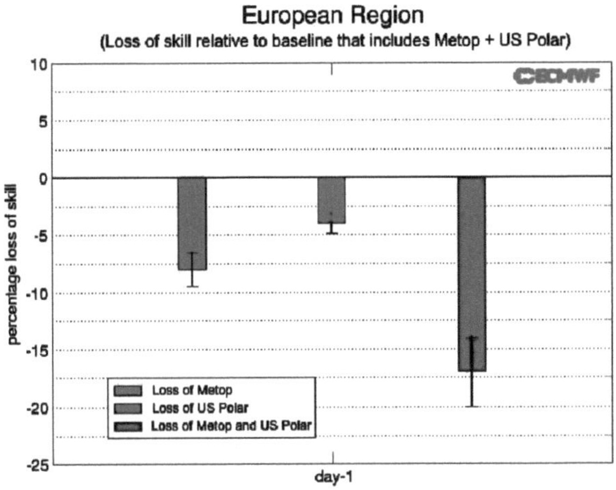

Fig. 7.3 Impact of loss of polar observations on day-1 forecast skill (Source: ECMWF)

7.3.2.3 Data Denial Experiments by ECMWF

A number of Observing System Experiments (OSEs) have been carried out by ECMWF to assess the impact of the loss of polar-orbiting observations on the skill of forecasts.[25] In this case, the skill of forecasts is measured by the root mean square error of the forecast 500 HPa geopotential height, a standard meteorological quantity involving a number of key parameters in the atmosphere.

From these experiments it can be seen (see Fig. 7.3) that the presence of the polar satellites has a clear and unambiguous positive impact on forecast accuracy. The simultaneous loss of both European and US polar satellites could be expected to degrade the 24-h forecast skill by 15–20 % over Europe.

Losing just one source of polar satellite observations produces a smaller, but still very significant, degradation of the short-range forecast accuracy; with around 8 % degradation for the loss of Metop observations, and just under 5 % degradation for the loss of US Polar observations.

Furthermore, and not explicitly shown in Fig. 7.3, if Metop were to be the only potential source of polar observations, its loss would have an even more pronounced impact at around 13 % (i.e. the 17 % degradation resulting from no polar observations – light grey bar, is reduced to a 4 % degradation – dark grey bar, leading to a net positive impact of around 13 %).

From these ECMWF experiments, one can conclude that the loss of Metop-A leads to a loss of 8 % in forecast skill over Europe (if the other components of the observing system remain in place to partially compensate for this loss) and a 13 %

[25] Observing system experiments to assess the impact of possible future degradation of the global satellite observing network, Tony McNally, ECMWF Technical Memo 672, November (2011) Final report in preparation.

loss of forecast skill if the missing Metop data source had previously been the only polar-orbiting satellite available. Finally a 15–20 % loss of forecast skill would result from the simultaneous unavailability of both Metop and the US satellite forming the Initial Joint Polar System shared with the U.S.

7.3.2.4 Case Study: Winter Storms Over Europe in February 2011

The German meteorological service (DWD) has investigated the impact of the assimilation of observational data from polar orbiting satellites on short-range forecasts using their operational global forecast model (GME).[26]

The evaluation period was selected so as to embrace winter storms over Northern Germany at the beginning of February 2011.

A verification of forecasts against surface observations showed that the assimilation of satellite data leads to better forecast scores at days 2 and 3 for surface pressure and wind gusts in the selected winter storm period, with the differences between the various forecasts and the verifying analysis (the best approximation of the ground truth) being particularly large for the winter storm "Nicolas", which hit northern Germany and Denmark during the night of 7 February 2011 – see Fig. 7.4. In this high impact weather event, without observations from Metop, the 45-h forecast would have totally missed the storm, both in terms of its intensity (minimum pressure over-estimated by 10 hPa) and position (missed by several hundred kilometres), leading to no significant warnings to the exposed populations.

With forecast lead times of 36–48 h, the experiments run with no, or less, satellite data ingestion show consistently larger errors in the severity (lower pressure gradients and wind speed), location and displacement and fast evolution of the trough, than the operational system using all satellite data. In summary, the assimilation of satellite data in general, and of Metop data in particular, has been found to increase the lead time of weather warnings for this severe weather event. As DWD's global data assimilation did not operationally use data from the IASI and the MHS instruments on-board Metop at that time, the potential benefit is expected to be even higher.

[26] The impact of satellite data on GME short-range forecasts – A case study, DWD internal report, January 20, 2012.

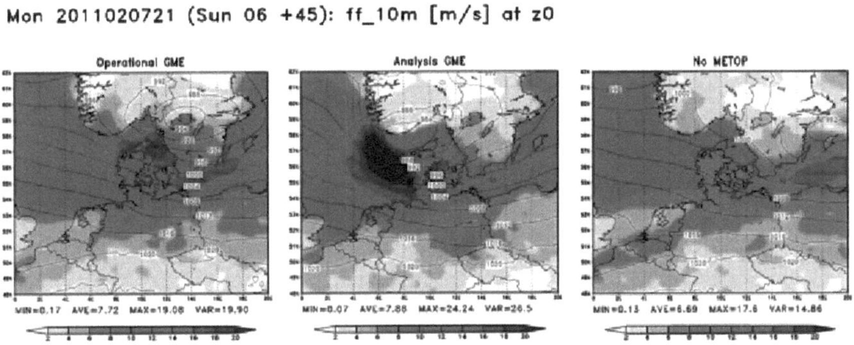

Fig. 7.4 Winter storm "Nicolas": 45 h forecasts with (*left*) and without (*right*) Metop observations compared to operational analysis (best approximation of ground truth, centre) of surface pressure (*contour lines*) and 10 m wind speed (shaded areas – units m/s) for 7 February 2011, 21:00 UTC (Source: DWD)

7.4 Socio-economic Benefits of EPS/Metop-SG Satellites Related to Forecasting

7.4.1 Apportionment Factors

The three approaches (Met Office, ECMWF and DWD) provide a complementary, and consistent, picture of the positive impact of EPS/Metop observations on short-range forecast accuracy.

Whilst the Met Office and ECMWF provide statistics that can be used to estimate, from two different perspectives, the contribution of polar-orbiting satellite data to forecasts skill, the case study demonstrates that the impact can be massive in the case of a high impact weather event, with potentially severe consequences. This illustrates that a poor forecast can, in such a case, lead to a dramatic loss of benefits to society.

From the Met Office statistics, one could conclude that, because one polar-orbiting satellite of the current generation, i.e. Metop-A, accounts for around 25 % of the impact of all observations on NWP forecasts for Day 1, 25 % of the estimated current benefits of forecasts could be attributed to this satellite.

However, for the purpose of assessing socio-economic benefits the most conservative, worst-case estimate of 8 % from the ECMWF statistics was adopted as the proportion of forecasting benefits attributable to Metop. On this basis, the resultant estimated annual benefits of the current Metop/EPS satellites range from a minimum of €1.2 billion to a more likely figure of €5 billion (obtained by applying the 8 % factor to the benefits given in Table 7.1).

The same assumption is also retained for EPS/Metop-SG in the 2020–2040 timeframe.

Table 7.1 Summary of estimated annual benefits of forecast information in the EU27

Benefit area	Minimum (billion/year)	Likely (billion/year)
Protection of property and infrastructure	€1.3	€5.5
Added value to the European economy	€10	€41
Private use by European citizens	€4	€15
TOTAL (rounded)	€15	€61

7.4.2 Estimated Socio-economic Benefits of the EPS/Metop-SG Programme in the Area of Forecasting

A rather conservative approach has been adopted to estimate the benefits to be expected from 20 years of EPS/Metop-SG observations in the 2020–2040 timeframe, based on the extrapolation of the current annual socio-economic benefit figures of forecast information given in Table 7.1 of Sect. 7.2.4, together with the following assumptions:

- An apportionment factor of 8 %, reflecting the worst case contribution to forecast skills of the current Metop satellite, with no improvement with EPS/Metop-SG;
- No allowance for progress in numerical weather prediction with respect to the current 'state-of-the-art' (against all expectations);
- An assumed 2 % annual increase in EU GDP for the benefit areas sensitive to GDP ("Protection of Property and Infrastructure[27]" and "Added Value to the European Economy");
- A 4 % annual discount rate.

The resultant estimated cumulative socio-economic benefits are given in Table 7.2 for the 20-year operational lifetime of EPS/Metop-SG from 2020 to 2040.

It is emphasised that the figures in Table 7.2 are the result of applying conservative apportionment assumptions and equally conservative estimates of the socio-economic benefits of forecasting.

A further aspect not reflected in these figures is the "leverage" accruing from the availability of high-quality EPS/Metop-SG observations to the worldwide user community which will, under the auspices of the World Meteorological Organisation (WMO), stimulate the development of comparable and complementary satellite systems by other space-faring nations, delivering additional observations that will be available for ingestion into European NWP models; thereby achieving additional positive impacts on forecast accuracy within Europe and, consequently, further socio-economic benefits. The high quality observations of the current Metop satellites are already a particularly attractive proposition for data exchange with other satellite operators.

[27] In the last 50 years flood and storm losses have generally increased, as fast or faster, than GDP.

Table 7.2 Estimated socio-economic benefits of EPS/Metop-SG observations due to their positive impact on forecasting

	Socio-economic benefit (over 20 years of EPS/Metop-SG)	
Benefit areas	Minimum (billion)	Likely (billion)
Protection of property and infrastructure	€1.5	€6.0
Added value to the European economy	€11.3	€45.2
Private use by European citizens	€3.0	€11.5
TOTAL (rounded)	€16	€63

Present value (2010) with a discount rate of 4 %
These benefits are extremely conservative, at an order of magnitude less than the benefits found in the US

Even with such conservative assumptions, when the estimated costs of the EPS/Metop-SG programme in the order of €3 billion are contrasted with the discounted benefits, the minimum benefit/cost ratio is over 5 and the likely benefit/cost ratio is over 20, with the understanding that these ratios would increase by a factor of 3, if a 25 % apportionment was assumed instead of the worst case 8 %.

Also, at the time of the operational programme, it is expected that the actual benefits will be further increased for reasons highlighted hereafter.

7.4.3 Further Evolution of Benefits in the EPS/Metop-SG Timeframe (2020–2040)

In the timeframe of EPS-SG (2020–2040) the actual benefits of forecasts, and thereby the benefit attributable to EPS/Metop-SG observations, are expected to be much higher than the worst-case figures used for the socio-economic benefit calculation, for a number of reasons:

(a) The fraction of benefits attributable to Metop-SG satellites may increase if its capabilities are significantly enhanced, compared to the current Metop, enabling Europe to keep its leadership in the provision of satellite observations for numerical weather prediction;
(b) Improvements in the absolute accuracy of forecasts due to advances in science, numerical weather prediction and better observations, are likely to disproportionately increase the benefit of the forecast information as its higher accuracy renders it suitable for new decision-making areas (e.g. flood evacuation decisions and the "threshold effect");
(c) Improvements in the response of decision-makers to forecasts and warnings;
(d) Increasing vulnerability of our society and economy to weather and related hazards;
(e) Changing climate and the expectation of more frequent high-impact weather events.

7.5 Other Benefit Areas

Polar meteorological satellites contribute to other important benefit areas or applications, not currently amenable to a quantitative assessment, and for which a qualitative description of the benefits is now provided, including defence and security, contribution to GMES services, climate monitoring and benefits to the European space industry.

7.5.1 Defence and Security

Most of meteorological satellite systems are civilian, but weather information is also important to support military and security operations such as peace-keeping, humanitarian or other operations undertaken under UN auspices. In such operations, and in particular where no in situ observations are available, meteorological satellites are an invaluable source of input information to produce the weather information required for safe operations. The related benefits cannot be assessed but are certainly significant.

7.5.2 Additional Benefits Through GMES Services

Additional benefits are expected from specialised forecast services already developed within the context of the Copernicus Initiative, which require weather forecasts as inputs. In the future, such services may be delivered by numerical models coupled to, or integrated with, numerical weather prediction systems.

This is the case for air quality forecasts generated by global, continental, regional and local scale prediction systems, which are used by public and private decision makers to maintain atmospheric pollution levels below harmful thresholds for health through emission regulation measures. It should be further noted that the EPS/Metop-SG satellites are expected to embark infrared and ultraviolet sounding instruments, including the Copernicus Sentinel five instrument, which, in combination, will provide a unique set of trace gases and aerosols observations for ingestion in air quality prediction models.

Some ocean and marine forecasts, in the open sea and in the coastal zone, are also highly dependent on accurate forecasts of surface pressure and winds. This is the case for the prediction of storm surges and the transport and dispersion of marine surface pollution, e.g. the likelihood of an oil spill affecting a vulnerable coastline.

It can therefore be assumed that a fraction of the benefits of relevant Copernicus services[28] can be attributed to the performance of weather forecasts and hence, to polar-orbiting meteorological satellites such as EPS/Metop-SG.

7.5.3 *Climate Monitoring*

In addition to their central role in weather forecasting, observations make a key contribution to the monitoring of climate variability and change and related applications; provided such observations span the long time periods associated with the slow dynamics of climate.

As observational continuity is equally essential for weather forecasting and climate monitoring, the instrument complement of EPS/Metop-SG has been designed to provide consistency and continuity with respect to the current EPS/Metop observations, together with some specific capability enhancements. As many of the EPS/Metop and EPS/Metop-SG observations combine global coverage with long-term continuity and are directly linked to Essential Climate Variables, they will provide a unique contribution to the generation of long series of consistent observations called Climate Data Records that are used in the context of the IPCC for assessing climate variability and change over several decades, and for validating the Earth system models used to deliver climate projections into the next decades. The necessary long-term consistency and coherence of such Climate Data Records is secured through a combination of cross-calibration of observations from overlapping missions (e.g. Metop and Metop-SG in the transition period) and reprocessing.

The financial impact of climate change is likely to be large, with many decisions having to be revisited based on climate change projections.[29] This is particularly true for sectors that are climate-sensitive and where decisions have consequences over many decades.

As an illustration, the infrastructure capital stock within Europe (including housing) is worth at least 200 % of GDP. If inappropriate climate information leads to the need to retrofit between 0.1 % and 1 % of this stock per year over the next century, the total cost would be a 0.2–2 % consumption decrease, i.e. a consumption loss of €20 billion to €200 billion per year at the current consumption level. Even if this retrofit need appears only in 2050, and with a 4 % discount rate and a 2 % economic growth, this sum represents a net present value of €500 billion to €5000 billion, equivalent to an annual cost of €10 billion to €100 billion per year henceforth.

---

[28] Socio-economic benefits analysis of GMES, PWC et al. 2006, Cost benefit analysis for GMES, Booz& co 2011.

[29] Hallegatte S (2009) Strategies to adapt to an uncertain climate change. Global Environ Chang 19:240–247.

So the financial amounts at stake for climate change mitigation and adaptation measures are large and could reach hundreds of billions of Euros per year in Europe during this century. It is likely that accurate information on future climate change, associated with efficient detection and attribution techniques, could result in large savings for the cost of adaptation.

However, as the development of climate services is at an early stage, under the Global Framework for Climate Services (GFCS) recently approved by the WMO Congress, the estimation of the associated benefits of long-term climate information is accordingly embryonic. So, at this stage, it is not possible to make a quantitative assessment of the benefits associated with availability of EPS and EPS/Metop-SG Climate Data Records.

7.5.4 Benefits to the European Space Industry

The EPS/Metop-SG programme will be a major development and will directly benefit European space industry and its contribution to EU GDP. As such it will be a driving force for innovation, competitiveness, growth, and the preservation and creation of highly-qualified jobs; thereby contributing to the objectives of the "Europe 2020 Strategy for smart, sustainable and inclusive growth". It will thus represent a major component of the implementation of the European Space Policy, being user-driven and delivering direct benefits to European citizens and, as a result, contributing to the objectives of competitiveness and innovation.

7.6 Conclusion: The Case for the EPS/Metop-SG Programme

The socio-economic benefits of weather forecasts to Europe are very large, with their contribution affecting many sectors of society.

The accuracy of these forecasts, which ultimately determines their benefit, depends on the availability of relevant and high quality observations, as these observations are used to constrain the forecast models that are the backbone of modern forecasting and provide critical information on the initial state. Without such observations the forecasts would have no skill.

Since the 1980s satellite observations have been increasingly used within forecast models. In this period forecasting accuracy has improved dramatically, typically with an increase of 1 day's predictability per decade in the medium range. The increased availability of high quality satellite observations during this period has been one decisive factor in this improvement.

By assessing the benefit of forecasts to the European economy in a number of key areas, and then combining this information with impact assessments and highly conservative apportionment techniques, it is possible to derive a socio-economic benefit range for the current EPS/Metop satellite system. The resultant estimated annual benefits range from a minimum of €1.2 billion to a more likely value of €5 billion.

When applied to the EPS/Metop-SG programme, over a period of 20 years ranging between 2020 and 2040, and taking a 2 % assumption on annual increase of EU GDP together with a 4 % discount rate, the 2010 Net Value of the benefits range from a "minimum" of €16 billion to a more "likely" value of €63 billion.

The conservative nature of these benefit estimates is underscored by the absence of provisions in these figures for the anticipated enhancements to the EPS/Metop-SG instruments the increasing value of forecasts due both to better observations and expected advancements in numerical modelling in the 2020–2040 timeframe, the increasing vulnerability of society and the economy to weather, and the expectation of more frequent high-impact weather events. This estimation also ignores a number of benefit areas, including contributions to protection of life, defence and security, climate monitoring, and direct benefits to the European space industry. The benefit estimates also take no account of the leverage accruing from the availability of high-quality EPS/Metop-SG observations to the worldwide user community, which stimulates the development of comparable and complementary satellite systems by other space-faring nations. As a result of these developments, additional observations will become available for ingestion into European NWP models; thereby achieving additional positive impacts on forecast accuracy within Europe and, consequently, further socio-economic benefits.

Furthermore, these figures do not include the benefits for those EUMETSAT Member and Cooperating States that are not EU Members (Croatia, Iceland, Norway, Serbia, Switzerland and Turkey).

When these highly conservative annual benefit estimates are contrasted with the estimated cost of the EPS-SG Programme in the order of €3 billion, **the minimum benefit to cost ratio is over 5 and the likely ratio exceeds 20**, with the understanding that these ratios would increase by a factor of 3, if a 25 % apportionment was assumed instead of the worst case 8 %.

In conclusion, there is a clear and compelling socio-economic benefit case for ESA and EUMETSAT embarking upon an ambitious EPS/Metop-SG programme that will deliver more capable satellites, secure the continuity of polar observations in the 2020–2040 timeframe and preserve Europe's leadership in the provision of satellite observations for numerical weather prediction.

Chapter 8
The Benefits of Human Space Flight; the Discourse at a Time of Financial Crisis

Cenan Al-Ekabi

In developing human space flight, having knowledge of the past, present, and future in all its permutations would be a gift that many would not refuse. With this idealised situation out-of-reach, in plotting the future we can still look to the hazy present and place reliance on what is known of the past, even with hindsight being subjective and incomplete. While we are aware of the manifold details of most of the historical moments in human space flight, the appreciation of societal benefits of such moments in the broader sense has been much like an after-thought. Thus, when considering the way forward from a future benefit perspective, we do so from an obscured vantage point, wondering "… one giant leap for mankind" to where?

The first half-century of human space flight stirred many scholars to reflect on its social impact in the sense of how society was affected, but they did not map consistently how society's interests were balanced to achieve such majestic feats at the time of decision.[1] John F. Kennedy's (JFK) enduring 1962 address to the American people, later to be adopted into the identity of all humankind, is one of the relatively few moments in our history where a speculative journey was truly planned, yet the perceived benefits of achieving such a goal were mainly geopolitical; the resulting impact of journeying to the Moon would affect the American and global society much more broadly.

> We choose to go to the Moon in this decade and do the other things, not because they are easy, but because they are hard, because that goal will serve to organise and measure the best of our energies and skills…

Exactly how this goal would bring benefit aside from outracing the Soviet Union was largely left to serendipity.

---

[1] See generally Dick SJ and Launius RD eds. Societal Impact of Human Spaceflight. Washington: National Aeronautics and Space Administration Office of External Relations History Division, 2009.

C. Al-Ekabi (✉)
ESPI (European Space Policy Institute), Vienna, Austria
e-mail: Cenan.Al-Ekabi@espi.or.at

C. Al-Ekabi et al. (eds.), *Yearbook on Space Policy 2011/2012*, Yearbook on Space Policy, DOI 10.1007/978-3-7091-1649-4_8, © Springer-Verlag Wien 2014

As with Columbus' exploratory voyages, the benefits derived through space exploration have often been harvested along the way. And like Newton's apocryphal apple, even with all facts exposed, an epiphany might be needed to achieve a different perspective toward our surrounding environment. Where we once looked outward to the stars to orient ourselves on Earth, we now look inward from space to do the same with GPS.

8.1 Prelude to Contemporary Space Exploration

At the beginning of the space age, the shadow of WWII and the resulting Cold War between the period's two super-powers fuelled the rapid development of space exploration with a devil-may-care attitude to funding, with both hegemons striving to outdo the other in this neutral arena. The intended benefits toward citizens, aside from organising the best of their energies and skills, may not have been contemplated beyond measurement as a badge of honour of being in first in the space race. What came out of the endeavour was an awareness that a new uncharted world exists, in some respects surpassing the discoveries of Marco Polo and Columbus.

Throughout the 1970s, a series of Soviet and US dual-purpose military space stations were launched into low Earth orbit (LEO). The Soviet Union would be the first to launch its Salyut space station, on 19 April 1971; a total of seven Salyut space stations would be launched in that decade.[2] And the United States' first space station, Skylab, was launched on 14 May 1973.[3] Despite operational problems on both sides, the Soviet Union would continue its space exploration activities by launching sections of the MIR space station in 1986.[4] At this time also, NASA struggled to develop Space Station Freedom, and it would later be amalgamated into the development of the ISS.[5] This was during the time when the Soviet Union became the Russian Federation, and the U.S. was looking to reduce budget costs by involving other nations in the endeavour.

With the end of the Cold War, another awareness had developed on the finiteness of government resources. Whereas the Soviet Union's Salyut space station series, its Mir, and the U.S.'s Skylab were independent attempts at longer duration space exploration, the next stage of space exploration put increased reliance on international cooperation. This approach, sparked long-before the latest financial crisis, has enabled greater and joint access to space for the U.S. and Russia, and

[2] Harland, David M. "Salyut." ENCYCLOPÆDIA BRITANNICA 12 June 2013 http://www.britannica.com/EBchecked/topic/520130/Salyut.

[3] "Skylab." NASA 12 June 2013 http://www.nasa.gov/mission_pages/skylab/.

[4] "Interactive Mir: Orbital construction of the Mir space station." RussianSpaceWeb 12 June 2013 http://www.russianspaceweb.com/mir.html.

[5] "Space Station Freedom." Encyclopedia Astronautica 12 June 2013 http://www.astronautix.com/craft/spaeedom.htm.

empowered other emerging space powers such as Europe, Japan, and Canada; balancing state interests and funding in the shared use of the International Space Station (ISS), and other shared capabilities.

Yet after two decades of development, the completed ISS has already extended beyond its half-life, with operations currently planned until 2020. Thereafter, recertification of the ISS and its spare components until 2028 might be on the horizon; but before a decision will be reached, the station's five partners must first assess the station's potential to function as a springboard for exploration.[6] By 2010, the U.S. had already reduced its spending on life and physical sciences by 60 %, i.e. to $200 million from $500 million in 2002. And current fiscal pressures might result in additional cost reductions and perhaps in shutting-down of one of the ISS operations centres, thereby potentially reinforcing a public perception that the ISS was more about building space infrastructure than about utilising it for the benefit of citizens on Earth. Yet, following its first 3 years of full utilization, some notable examples of the benefits that have already been achieved through the scientific and industrial research on the ISS include: the synthesis of technology used in the ISS's Canadarm2 and Dextre into a spinoff two-armed robot capable of performing brain surgery on Earth; the use of the station's microgravity environment to develop flavour enhancers and stabilisers for food preservation, and the development of ultra-lightweight titanium aluminide turbine blades that should reduce aircraft fuel consumption by half.[7] While under fiscal pressure to demonstrate real-world value and yield rapid results, the ISS – existing as an infrastructure devoted to basic research – should not be expected to incubate and develop technologies at the pace of technologies manufactured on Earth.

8.2 Learning from the Past to Find Our Way Forward

Wisdom is needed when gleaning the future benefits that will arise from space exploration activity, particularly because those benefits are unpredictable to some extent, and might have a longer maturation time than the rapid innovation conducted on Earth. As demonstrated with the ISS, space-based assets play a key role as innovation enablers. Not limited solely to technological advancement, space exploration may lead to other innovations that can impact our lives in ways that have not yet been considered. Take dark matter, first discovered in 1932 by Jan Oort who observed stars moving faster than expected from the gravitation force of the visible mass in the Galaxy. While appearing to move fast enough to escape the galaxy, he hypothesised that an invisible additional mass must exist to explain why

---

[6] Svitak, Amy. "Made In Space: Backers say the $100 billion ISS is not only dressed up, but has somewhere to go." Aviation Week & Space Technology 21 May 2012: 22.

[7] Id.

they stayed bound within the galaxy.[8] In 1933, Fritz Zwicky came to the same conclusion after observing galaxy clusters that moved too fast to be held together from the gravitational attraction created by visible matter alone. Recently, the ISS contributed to the study of dark matter by accommodating the Alpha Magnetic Spectrometer (AMS), which studies cosmic rays prior to entering Earth's atmosphere. While far too early to conclusively rule out other explanations e.g. pulsars, the AMS has observed an excess of antimatter particles in the flux of incoming cosmic rays that could have originated from dark matter particles that collide and annihilate.[9] At first glance, the study of dark matter may not appear to be a path to commercial riches, yet its discovery might bring innovations that lead not only to a better understanding of our universe. Like with development of nuclear energy, predicated on basic science, development of space science will likely enable further normal terrestrial innovations. Quantum computers required quantum mechanics, yet Niels Bohr hardly imagined such a development.

8.3 Business Process Innovation

Learning from the past does not only imply open-mindedness in looking for benefits. In fact, governments and space agencies may be well advised to look also to new approaches to financing when determining how to possibly invest in space exploration. Rather than follow the model established by NASA with its now retired space shuttle, where costs gradually increased due to the shuttle's complexity and reliance on 1970s technology, one could follow in the steps of Elon Musk, founder of SpaceX, whose company achieved space flight within 6 years of incorporation. Musk identified cost reduction as an area ripe for innovation; and in following its long-term goals, the company relied on past experience while focusing on the bigger picture to identify and prioritise opportunities, and later refine the key aspects of the business model in that pursuit.[10] Instead of seeking to cut costs, the company redirected its investments to areas that offered the best prospects for future returns. This farsighted approach, i.e. to identify and prioritise opportunities while being highly cost conscious in areas that are of less priority, could be used in other sectors that could benefit from smaller teams and lower overhead expenses.

Nevertheless, cost reduction only goes so far in space exploration. The key may be a change in attitude, from the highly customised and expensive, to low cost and

---

[8] "B. Hidden Mass" NASA – Goddard Space Flight Center 28 May 2013 http://imagine.gsfc.nasa.gov/docs/teachers/galaxies/imagine/hidden_mass.html.

[9] Spector, Dina. "Scientists Announce First Evidence of Dark Matter." 3 Apr. 2013. Business Insider – Science 28 May 2013 http://www.businessinsider.com/alpha-magnetic-spectrometer-experiment-dark-matter-2013-4.

[10] Agan, Tom. "What SpaceX Can Teach Us About Cost Innovation." 25 Apr. 2013. Harvard Business Review – Blog Network 29 Apr. 2013 http://blogs.hbr.org/cs/2013/04/what_spacex_can_teach_us_about.html.

standardised access to space technology.[11] Here, a paradigm shift can be seen as another innovation that benefits current space exploration activity by making the financial threshold for pursuing activities lower. In other words, the cost-benefit trade-off might be less weighed-down by costs through this route.

8.4 Seeding Future Space Exploration

The recent involvement of emerging space powers and the more economic-driven approach of space exploration have increased the importance and possibilities of international cooperation in future activities in this field. Following a series of discussions among 14 space agencies[12] on global interests in space exploration in 2006, this group drafted its "Global Exploration Strategy (GES) – The Framework for Cooperation" in May 2007.[13] This framework document articulates an essential need to enhance international cooperation, and enables space agencies to exchange information on interests, objectives and plans in space exploration with the goal of strengthening both collective and individual space programmes. Additionally, the GES also gave birth to the International Space Exploration Coordination Group (ISECG), a voluntary and non-binding international coordination mechanism with the purpose of implementing and coordinating the GES. A short while later, the ISECG's lunar architecture strategy inspired its subsequent September 2011 iteration of the "Global Exploration Roadmap" (GER) which articulates the perspectives of the agencies that contributed to GES.[14] This roadmap outlines exploration goals and objectives, mission scenarios, and the coordination of exploration preparatory activities.

8.5 Putting the Needs of Relevant Stakeholders First

In determining the benefits that may develop from future space exploration activities, one should first consider the needs, and assess whether and how space exploration can satisfy those needs. And when considering needs, the foremost

[11] Id.

[12] I.e. ASI (Italy), CNES (France), CNSA (China), CSA (Canada), CSIRO (Australia), DLR (Germany), ESA (European Space Agency), ISRO (India), JAXA (Japan), KARI (Republic of Korea), NASA (United States of America), NSAU (Ukraine), Roscosmos (Russia), and UKSA (United Kingdom).

[13] "International Space Exploration Coordination Group." ISECG 7 May 2013 https://www.globalspaceexploration.org/.

[14] "The ISECG Global Exploration Roadmap." 22 Sept. 2011. ISECG 7 May 2013 https://www.globalspaceexploration.org/documents.

questions to answer are whose needs are the most relevant, and how are those needs prioritised? Where Musk identified the needs of SpaceX on a relatively micro-level, addressing the broader needs of stakeholders and to what extent those needs will be met by space exploration calls for an assessment on a macro-level.

In 2012, the European Space Policy Institute (ESPI) was asked by the European Space Agency (ESA) to make a benefits assessment on such a macro-level. ESPI thus conceived and put into motion a broad-based investigation, through which the needs of the variety of stakeholders relevant to the space exploration were determined. Working in close cooperation with specialists from the European Science Foundation (ESF) and the International Space University (ISU), the study employed sophisticated analysis processes and tapped a network of high-level representatives. Following the completion of the study, with its identification of the benefits relative to the priority and intensity of needs, subsequent studies may focus on a cost-benefit analysis, as the ESPI study only considered needs, not financing.

Identifying the stakeholders was the first step. Whereas in the past, the needs of government stakeholders were pre-eminent, following the switch in priorities toward providing economically sustainable long-term exploration and higher public scrutiny of public spending , new stakeholders in industry, science, and the general public hold an increasing influence in space exploration activity.[15] And as Europe is to be a knowledge-based society; education, the enabler of understanding in society, can be considered as an additional notable stakeholder. Together, these groups of stakeholders, with their incongruous needs, have the potential to impact the manner in which space powers develop their respective space exploration programmes. Yet the influence of these stakeholders might also vary regionally, as their interests may reflect differing political realities, and they may hold non-correlating levels of authority and influence.

Nevertheless, the interests of these stakeholder groups might still align in certain areas although benefits may come to these groups in different ways. For example, suborbital spaceflight can impact industry through space commercialization, while also resulting in space tourism within the broader public sector. In long-term space exploration activity, both political, scientific, and education stakeholders might benefit from the development of an exploration roadmap that is calibrated with the advice and guidance of the scientific community.[16]

---

[15] See generally Ehrenfreund, P., N. Peter, K.U. Schrogl and J.M. Logsdon. "Cross-cultural management supporting global space exploration." Acta Astronautica 66 (2010): 245–256, 250.

[16] C.f. Ehrenfreund, Pascale et al. "Toward a global space exploration program: A stepping stone approach." Advances in Space Research 49 (2012): 2–48: 32.

8.6 Qualifying These Needs

The needs of these stakeholder groups must then be qualified, as their interests may diverge, and have the potential to conflict. Early in its study, ESPI with the aid of the ESF and ISU conducted a benefits workshop involving high-level representatives within these stakeholder groups to assess their primary needs and how they could be captured. The workshop showed that not only were needs different, but the thought process about benefits depended very much on previous experience and the needs themselves. In a sense, the workshop showed that unless tools could be found which could homogenise the heterogeneous there would be a risk of logical circularity, as benefits methodology often appeared premised on the specific benefits and the stakeholder group seeking them. The key in this respect was found to be the ability to compare stakeholder needs against different benefits scenarios, as the comparison in itself would force more generalised reflection.

For the public sector, Manfred Max-Neef's sociological model on human scale development – a standard in measuring public needs – was adopted. In regard to particular factors such as both personal and collective attributes, -objects, -actions, and -settings, the needs that exist in Max-Neef's scale of human development are: subsistence, protection, affection, understanding, participation, idleness (i.e. leisure), creation, identity, and freedom.[17] In view of the financial crisis, it can be assumed that subsistence would be a priority need among the European public, and perhaps that such a need would have to be met through the advancement of space exploration. However, despite the constant discourse on finance the analysis showed a much more differentiated picture, where understanding, creation and protection played major roles, and even affection was found of relevance.

The needs of the science community undoubtedly vary from those found to be fundamental to the public, and they required identification by relevant leaders and authorities from this sector. This collaboration has produced the following need indicators: stability, visibility, infrastructure, communication, education, knowledge creation, science policy, and evaluation. Among these indicators, obviously knowledge creation is the primary goal among science stakeholders. And that need resonates interestingly with the EU's response to the financial crisis by investing in research and development to maintain its position in innovation.

Educators have their own litmus test for needs satisfaction. Whereas the focus for science stakeholders is geared to the advancement of knowledge creation, educators seek to inspire students and advance their understanding and capabilities. Similar to the key persons tapped in the science sector, prominent representatives of European academia at the secondary and tertiary levels were asked to provide need indicators that are relevant to this sector. Their responses emphasised the following needs/enablers: orientation, motivation & inspiration, access to knowledge, learning partnerships, and internships & networking. The priority among these needs is

[17] Max-Neef, Manfred A. Human Scale Development: Conception, Applications and Further Reflections. New York and London: The Apex Press, 32.

not clear cut, as educators might value them on an almost equal basis. Yet, continued development in space exploration may have trickledown applicability to every need within this group.

In the industry sector, stakeholders came up with a number of needs that could harmoniously be condensed within the Excellence Model developed by the EFQM (formerly known as the European Foundation for Quality Management); a model used by over 30,000 organizations in Europe as a comprehensive management framework.[18] With profit as the main goal for industry, this model provides guidance on both business enablers, and their results (i.e. people, customers, society, and overall business) in order to reach that goal. These enablers include: leadership; people; policy & strategies; partnerships & resources; and processes, products & services. It was assumed that in appropriately satisfying the enablers, the results would flow as a natural consequence. Here, it would be interesting to see how human involvement in the areas of leadership and people (i.e. workforce) is relevant to space exploration when compared to the other industry needs.

While space agencies play the largest role in facilitating space exploration activities, they operate under the mandate of their respective national governments. This latter political stakeholder group, with interests spanning across all sectors within their competence, has need indicators that focus both on internal and external levels. Internally, political stakeholders have a need for subsistence, development, and domestic security & safety; externally, international security & stability, and diplomacy & government relations are need indicators that require fulfilment. While development is an unmistakable priority for political stakeholders, issues of domestic security & safety and international security & stability seem to ever be on the rise in importance.

8.7 Establishing Priorities Among Needs

While all needs identified by these stakeholder groups should be regarded as being important; nevertheless, establishing a rank among these needs would help to pinpoint the areas where attention should be directed. This ranking can be done in a variety of configurations; i.e. one way might be through the use of a goals and benefits analysis, where the goals would describe an intended future that would come about after carrying out respective activities, and the benefits would be the observed effect of one or more preceding activities which creates either satisfaction or utility for a particular stakeholder group. The key problem in this approach is that the benefits, both intended and unintended, will likely be discovered ex ante; and forecasting the utility of these benefits without measuring associated quantitative

[18] *See further* "The EFQM Excellence Model." EFQM 2 May 2013 http://www.efqm.org/en/tabid/132/default.aspx.

and qualitative data on an ongoing basis might be as reliable as when forecasting fair weather for a day following a crimson sunset.

Another approach might be to rank the needs on an ordinal scale. While very likely to produce revealing answers on the priority of needs, this approach will inevitably artificially localise these needs indicators in distinct positions that do not reflect exactly where equal priorities exist within a group, as ranking positions could not be shared. Additionally, the preference intensities among the needs is not as easily reflected with the use of equal priority interval separators, i.e. 1st, 2nd, 3rd etc. Moreover, when conducting an ordinal ranking, there is a chance that answers are provided in more of a 'knee-jerk' fashion from a common-sense perspective, rather than by methodically assessing each need against all others.

To resolve these lacunae in the collection of relevant subjective and objective information, a multi-criteria decision making method such as the analytic hierarchy process (AHP), developed by Thomas Saaty, would provide the ability to not only rank needs, but also measure the preference intensity of those needs.[19] Moreover, this pairwise-comparison approach might reveal certain needs that share the same intensity, but for different reasons. This method has stakeholders within respective groups to express judgments on a set of needs and its relevant criteria, and consider benchmark alternatives that are linked to numbers (i.e. 1–9). From these numbers, which express the relationship among these needs, a principal eigenvector of priorities can be determined which can then be used to show the dominance/preference intensity of each need with respect to the other needs.[20]

In its 'Benefits of Space Exploration' study, ESPI developed approaches that could serve as the basis for its analysis. These benchmark alternative approaches, designated as the 'European Roadmap', the 'Contemporary Approach', and 'Robotic Exploration Only', outline the possible scenarios that could be foreseen in the next 20 years.

8.8 Framing a Roadmap to Predict the Way Forward: The Alternatives

8.8.1 The 'European Roadmap' Approach

ESPI's 'European Roadmap'-labelled approach articulates goals that seek to develop robust step-wise technologies and capabilities, and to strengthen existing

[19] *See further* Forman, Ernest H., and Saul I. Gass. "The Analytic Hierarchy Process – An Exposition." Repository of the Croatian Mathematical Society 8 May 2013 http://web5.fer.hr/_download/repository/ahp.pdf.

[20] *See further* Saaty, Thomas L. and Luis G. Vargas. "The Seven Pillars of Analytic Hierarchy Process." Models, Methods, Concepts & Applications of the Analytic Hierarchy Process. New York: Springer Science+Business Media New York (2012): 23–40.

partnerships while also building new ones. In following this approach, a Lunar Lander system would be developed in the pursuit of landing humans on the Moon; and Europe would solidify future human spaceflight cooperation with numerous international space actors by aiding in auxiliary activities involving ATV derivatives (e.g. a cis-lunar tug, a man-tended free flyer, etc.); and supplying a European Support Module (ESM) for NASA's Orion Multi-Purpose Crew Vehicle (MPCV). This path implies a decision to extend the ISS operations beyond 2020; a European launcher scenario; an international partnership scenario; and European political support for human space flight.

8.8.2 Contemporary Approach

ESPI's 'Contemporary Approach' is meant to depict the status quo. Here, bilateral cooperation with a subset of international partners is favoured above full cooperation among all space faring nations, and Europe would support individual partner agencies in projects, sacrificing planning autonomy for mission certainty. European human space exploration would continue, limited specifically to the LEO environment, assuming the ISS is extended beyond 2020. And robotic exploration in this scenario would likely be purely scientifically driven, concentrating on subsequent future missions to Mars. This path implies the decision to extend the ISS operations beyond 2020; a European launcher scenario; an international partnership scenario; and European political support for human space flight.

8.8.3 'Robotic Exploration Only' Approach

This third alternative reference scenario envisions the end of European human space flight post-2020, with missions involving the development of a Lunar Lander system, and ATV derivatives as wholly robotic undertakings. While there would still be room for an international partnership scenario, this approach would channel resources into Europe's robotic technology development, as the development of life science technology would unneeded. And like in the 'Contemporary Approach', robotic exploration would concentrate on subsequent future missions to Mars. Hence, a human lunar lander would no longer be pursued; and in the broader scope, Europe's autonomy in space would be preserved at the expense of sacrificing Europe's presence in long and short duration human space flight.

8.8.4 The Lessons to be Learned

In ESPI's study, a relatively limited number of high-level stakeholders were surveyed in each stakeholder group. One reason for this narrow sampling was the previously untested approach used in its analysis, and the need to determine whether the information provided would create relevant, useful linkages that had not been previously considered (e.g. the need for further outreach in the education sector). The study, similar to the way tech demonstrations are used to prove the feasibility of a space technology, shows that it is possible to gather key information that has a direct impact on space exploration from stakeholders beyond those dedicated to the space sector. And while the majority of benefits can only be determined conclusively ex post, an ongoing analysis of the perceived benefits of space exploration, and of the outcome of those actions would allow all relevant sectors to better estimate the future impact and benefits of these activities, enabling them to make decisions with higher confidence in their long-term benefits.

The value of ESPI's study is probably primarily in the development of the methodology, and the recommendations for how to deploy the methodology on a continuous basis. This has therefore been the focus of this contribution. Previously unrealised common interests among different stakeholder groups might be revealed through the study approach, allowing European and international space agencies to link underappreciated areas. Moreover, the space sector and individual stakeholder groups might be able to reap benefits more effectively by having been made sensitive to other benefits perspectives and other benefits, thus enhancing both cross-fertilisation and serendipity possibilities.

8.8.5 The Reality of Future Cost-Benefit Considerations

In attempting to identify the future benefits of space exploration, the focus should not be strictly limited to perceived long-term technological benefits. Undoubtedly, economic and cultural benefits will be realised through space exploration as well. The ESPI study is flanked by assessments that focused on other sectors within the space industry (e.g. telecom benefits assessments, and studies on the socio-economic benefits of GMES) with each domain high-lighting its own set of relevant factors. While the benefits that come from advances in telecommunication and from GMES are likely to be in the form of innovation developing existing markets with cutting-edge technology the benefits that stem from space exploration are likely to be disruptive innovations that have the potential of opening up completely new markets. Here, the above-referenced stakeholders might find that while their needs would remain unchanged between these various themes, the satisfaction of those same needs might differ substantially. Having been allowed to develop to viability, the benefits that stem from telecommunications and GMES have permeated within every stakeholder group to their specific needs, with increased connectivity and greater ability to respond to our environment. On the other hand, the new markets that space exploration opens is just beginning to come to light among the stakeholder groups, with potential uses ranging from affordable space research opportunities to space tourism, to name a few.

It should be stressed that since this study deliberately focused only on benefits, and not on the required societal resources, the methodology developed by ESPI can only be used for ultimate decision making when married to a proper trade-off of the identified benefits against the cost. This can be done on a whole-sale basis, i.e. can we as a society afford amount X thus harvesting benefits Y; or, it could be done sectorally, i.e. can sector A afford amount X in order to have its needs fulfilled by benefits Y? Doing such an analysis sector by sector (A, B, C, D, etc.), and ultimately adding up all the affordability amounts would then show whether the one or other scenario would be affordable. There would, of course, be a fictitious element to this since almost all funds would ultimately have to come from government, but there would be something very salutary in confronting each stakeholder group not only with the question, which scenario brings you more benefit, but also asking stakeholders how many of the resources available to them they would be ready to devote to a certain scenario. Ideally that would be done by identifying absolute amounts, but that is hardly possible, and hence the question could be asked: how much more, in percentage, would you be ready to pay for scenario H, compared to scenario G, which would meet your needs less well? The summing up would then only be relative, yet it would aid the 'whole-sale' cost-benefit trade-off, by showing not only benefits preferences, but also financing preferences.

At a time of financial crisis everybody is comfortable talking about benefits, and not cost. However, in all political decision-making cost considerations are dominant. The reason the ESPI study, nevertheless, did not address cost was partly methodological, partly because any discussion of cost of exploration scenarios at this time would be wrongly influenced by the current perception that nothing is affordable. The significant expenditures that any exploration scenario would incur would only start to fall due in the medium term, at a time when the financial crisis should be well and truly over. Therefore the space community is well served by looking at benefits scenarios first, and cost-benefits trade-offs only later. Decision–making might be more urgent than we tend to think, but this notwithstanding it is important that bridges are only crossed when necessary. In the context of exploration it is high time to think about benefits and collaboration strategies, but premature to consider what we can afford, because the political gut instinct currently will be: not very much, whereas the right societal answer in a couple of years might well be that we can and should afford to meet needs very extensively.

Chapter 9
Space, Youth Unemployment and Professional Status in the Financial Crisis

Andrea Jaime Albalat

9.1 Introduction

One of the worst repercussions of the financial crisis that has hit Europe is youth unemployment, which may have a lasting negative effect on the future of some countries. This applies not only to Europe: young people are three times more likely to be unemployed than adults, and over 75 million youth worldwide are looking for work.[1] In Europe itself, several studies estimate youth unemployment to be between 40 % and 60 % in countries such as Spain, Greece and Portugal. This is an unacceptable rate for countries that are seeking growth and recovery from this crisis. Moreover, there are several other negative consequences resulting from this tragic situation: brain drain, unbalanced emigration, relative increase of old population, etc. While much data on youth unemployment in general can be found in easily accessible resources, the question this paper addresses is: "what is the actual situation in the space sector in particular?"

The space sector was able to remain relatively untouched at the beginning of the crisis, but it is now starting to affect the space community. Unfortunately, there is not enough data on the current situation of young adults working or starting their careers in the space sector to enable policy-makers to assess the future of the sector and take appropriate decisions.

The so called "Apollo Generation" brought a massive group of brilliant young engineers, scientists, businessmen, and others into the sector. However, there has been a certain stagnation in the community since then, and hiring rates have decreased tremendously in the space agencies and industry. There is indeed, a

[1] "Youth Unemployment" International Labour Organisation. Youth Unemployment 12 Jun. 2013. http://www.ilo.org/global/topics/youth-employment/lang–en/index.htm.

A.J. Albalat (✉)
SGAC (Space Generation Advisory Council), Vienna, Austria
e-mail: andrea.jaime@sgac.gmail.com

C. Al-Ekabi et al. (eds.), *Yearbook on Space Policy 2011/2012*, Yearbook on Space Policy, DOI 10.1007/978-3-7091-1649-4_9, © Springer-Verlag Wien 2014

Table 9.1 Youth unemployment rates forecast in different regions around the world

Youth unemployment rates forecast	2012 (%)	2017 (%)	Tendency
Global	12.7	12.9	↑
Central and South Eastern Europe	16.9	17	↑
Developed economies	17.5	15.6	↓
East Asia	9.5	10.4	↑
Latin America	14.6	14.7	↑
Middle East	26.4	28.4	↑
North Africa	27.5	26.7	↓
South Asia	9.6	9.8	↑
South East Asia & the Pacific	13.1	14.2	↑
Sub-Saharan Africa	12	11.8	↓

Source: International Labour Organization

well-founded fear of a loss of knowledge and transfer of experience to the new generations.

In order to properly assess the future of the sector and be able to make good decisions, policy-makers must look at past and future trends, hence examining the current situation of the young community in the sector is a must in order to ensure robust succession and progression. In order to do this, relevant data and assessments of current needs, status and trends must be analysed. During the 6th ESPI Autumn Conference, it was agreed that there is a lack of this data, and therefore action, in the form of a survey that was prepared and distributed among young professionals, was taken by the SGAC.

It should be noted that this study is not intended to cover the entire sector or to give an exact and accurate picture of the current situation. Rather, it is intended to be more of a starting point to keep attention on the need to keep gathering this data periodically and from a bigger group of people, and extract some relevant conclusions by evaluating the trends (Table 9.1).

9.2 Space Employment

The financial crisis is increasing the number of unemployed people in the countries most affected by it. The sectors most suffering from the effects of the crisis vary from country to country. For example, in Spain the most affected sectors are those linked to the construction and real-estate business. However, the development of the financial crisis and the economic model of many of these countries has meant that other sectors, not initially directly affected by the crisis, have also been affected.

Despite this, particularly during the first years of the financial crisis, the space sector remained relatively robust. However, some space budgets on national level have been tremendously affected by the latest budget cuts that happened in 2012. Nonetheless, due to the late impact of the crisis on the space sector, the space community still can proudly point to statistics that show the strength of the sector, and is positive about its future.

Analysis of some of the developments that may have an impact on the sector during the financial crisis, some interesting findings have been revealed:

- It is known that space-related occupations will grow faster than other sectors, especially in developed and emerging countries.
- STEM Education (Science, Technology, Engineering, Math) graduates are increasing, therefore there will be more professionals prepared to take the lead in the future, thus ensuring continued growth.
- Although budgets appear to be suffering cuts, in general space agencies and industry budgets are actually minimally increasing, and only in some areas, such as exploration, have they been substantially cut.
- Trainee Programmes are ongoing in all agencies and major industries
- The number of available scholarships is increasing, although total expenditure on scholarships is not rising at all.
- The contributions of young professionals and students at the United Nations, space conferences and other bodies historically dominated by senior experts are also increasing tremendously.

The main threat though, as mentioned above, is the aging of the space industry. This is indeed an aging industry that benefits from the wealth of information brought about by the decades of experience of its current leaders.[2] The following charts show the age structure of the staff in three of the main space agencies in the world.

As is shown in Fig. 9.1 the European Space Agency has 70 % of its workforce between the ages of 45 and 60, similar to that at NASA. However, comparing the workforce under 35 in these two agencies, NASA employs only 14.9 %, while ESA has around 24 % of the total. In fact, ESA initiatives such as Young Graduate Trainee programme, allow more than 100 young professionals to enter into the ESA workforce every year. In the same figure, JAXA has been included for comparison. JAXA has a younger profile among its employees.

In Fig. 9.2, NASA civil servant workforce age profiles are shown over the period 1998–2013. Although no similar data could be found specifically for ESA, the pattern can be expected to be similar for historical reasons. The figure very much highlights the aging of the space workforce. In fact, currently in NASA, about 16 % of the workforce is near retirement age, while only 14 % is under 35 years old. There is a need then, to ensure that all the knowledge and expertise is transferred to the new generation and not lost in the transition.

[2] Jaime, Andrea. "Make Way for a new Generation in Space". Space News 8 April 2013: 35 + .

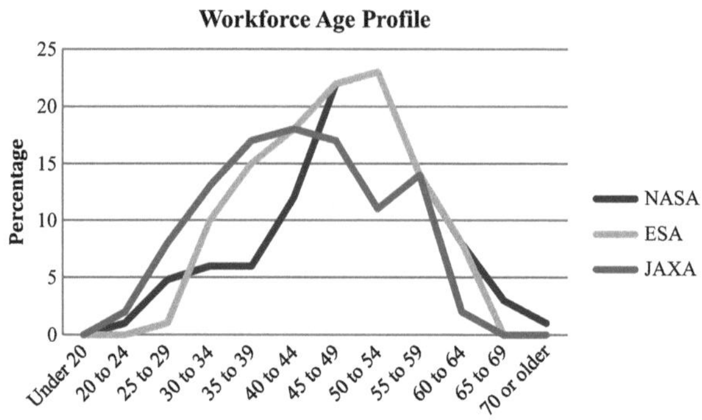

Fig. 9.1 Workforce age profile of NASA, ESA and JAXA

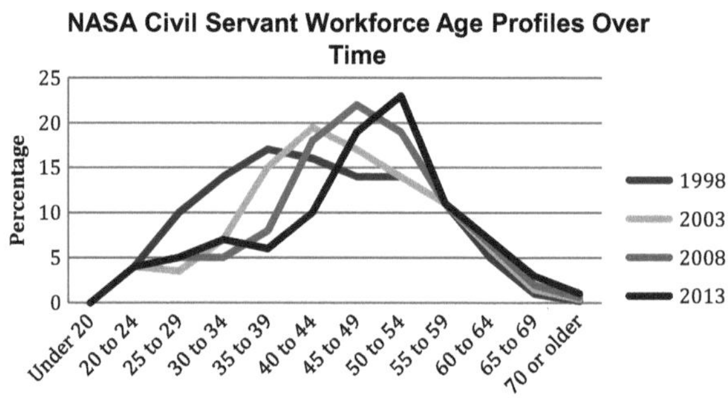

Fig. 9.2 Civil servant workforce age profiles over time (Ibid) (Source: The Space Report 2013. Washington DC: Space Foundation, 2013)

9.3 The Study of the Current Young Workforce in the Space Sector

9.3.1 Introduction

As noted above, there are many studies that refer to general youth unemployment, especially now that it has been aggravated by the financial crisis in Europe. However, detailed and concrete data on the current situation and trend among young space professionals is scarce.

National and regional space agencies such as ESA often publish reports that include their employee demography (usually easily accessible on their websites). Many of these space agencies include education programmes and trainee programmes, specifically to give opportunities to young people to enter the space arena. However, reports of this kind are very general and never address the difficulties and current situation of the younger generation. Especially now, with the financial crisis and budget cuts, young professionals are the group most likely to suffer the immediate consequences. It is for this reason that during the ESPI

Autumn Conference (October 2012), a study was proposed to the European Space Policy Institute, to be carried out through the Space Generation Advisory Council (SGAC), an international non-profit organisation that represents young professionals and university students in the space sector. The proposal was to distribute a survey to young professionals in the space sector, in order to gather data and statistics on the current situation of young professionals working in the space sector.

9.3.2 *Methodology and Demography*

A survey-study was conducted among a group of young professionals in the space sector. The network of SGAC, which includes more than 4,000 people from over 100 countries, was the main source of interviewees. The survey was distributed via SGAC's main social media channels and mailing lists, and also through other organisations such as the Alumni group of the International Space University and the CVA Summer School. The survey was open for 1 month (November-December 2012), during which 183 entries were registered, and it is still available for further consultation.[3]

A total of 38 questions were asked to young professionals in the space sector. The questions covered topics such as salary, internships, type of jobs, difficulties of finding jobs, current situation, etc. Of the 183 responses coming from countries all around the world, this study focuses only on the 126 European entries. Of these, 9 held double nationality, either with other European countries, or with countries such as Australia or the USA. Most responses came from Spain, France, Italy and Germany. This is also a clear indicator of the major space faring countries in Europe with the most professionals in this sector.

The vast majority of respondents had obtained an academic Master's degree (82 %); only a few had a PhD (12 %) or a Bachelor's degree (6 %). These figures correspond to their professional status, with 59 % being Young Professionals, 14 % working as PhD students, and 13 % doing an internship. The rest are unemployed.

It is important to mention that although we define "youth" as those between 18 and 35 years old, the people in this study ranged between 22 and 35 years old, with more than 65 % of them between the ages of 26 and 30. This data also provides an interesting answer to the question of which age students usually enter into space professional life, which is on average, after they are 25 years old.

Another statistic that should not be ignored is gender distribution. It is well known that the space sector has traditionally been dominated by men (for example, of 525 astronauts, only 57 have been women), but this is a trend that has been very much changing in recent years. In this survey the distribution was 35 % female and

[3] Space Generation Advisory Council "Survey: Space Youth Unemployment and Professional Status during Periods of Economic Instability". https://docs.google.com/spreadsheet/viewform?fromEmail=true&formkey=dFNUbU9icTRtVnRfZTcxMERqazF4NlE6MQ.

65 % male. Other SGAC statistics have been consulted as comparison, and they show very similar numbers: in the latest SGAC congresses, about 40 % of the delegates were female.[4]

Exploring now the professional background of the participants in the survey, 72 % of the respondents were engineers and this in very diverse types of engineering. This is an expected result, because many young professionals who work in the space sector decided to work there at an early age, and therefore they had to choose directly space-related courses during their high school and bachelor degrees, for example STEM education. And in fact, although administration is a large part of the space sector (i.e., 21 % of the total employees at ESA),[5] usually the background of these workers is not so focused on a space-related job at earlier stages of education, and many of them have had experience in other sectors before eventually jumping into the space sector later in their professional careers.

In the following section, most of the results of the survey are shown. However, as will be addressed later in the *Recommendations*, there is a lot of data to be managed and much more information can be extracted. The following section then, will show only the most relevant patterns in the European Young Professionals sector.

It is important to note that the results shown here are based on the demography of those who answered the questionnaire, and therefore some of the data here can be not accurate when representing the whole young professionals spectrum in the European space sector.

9.3.3 Findings

Thirty-eight questions were asked covering different areas that affect the employment situation of the young generation, and somehow might affect and define the path of the coming ones, within the space sector.

9.3.3.1 Transition from Student to Professional Life

The first part of the survey covered questions regarding the transition between student life and professional life. The generation between 18 and 35 years old is exposed to big changes in a rather short time (Table 9.2):

Those in between their first job and becoming 35 years old are under particularly high pressure, since this is the time where the young adult must make the right decisions and prove his/her expertise and capabilities. These decisions and the way of work will define the future established professional.

---

[4] "Annual Reports" Space Generation Advisory Council. http://spacegeneration.org/index.php/about-sgac/history-annual-reports.

[5] Space Foundation. The Space Report 2013. Washington DC: Space Foundation, 2013.

Table 9.2 Summary of main threats and challenges of young adults between 18 and 35 years old

Age	Transition	Threats/challenges during the financial crisis
18–19	High school to college	Payment of university fees Temporary Employment
19–23	College	Extra-curricular activities to gain experience Temporary employment
23–28	College to young professional	Internships without payment Lack of jobs Low salary Extra experience is always demanded
28–35	Young professional	Lack of interesting projects Cuts in budgets Risk of losing the job
35–...	Established professional	Having arrived at the desired position

Looking at Table 9.2, the last column shows the main threats experienced, in particular, during the financial crisis, which enhanced the pressure mentioned above. At the early ages, students have to face the payment of university fees, which have increased in many countries during the financial crisis.[6, 7] This forces students to take on temporary and partial jobs during their studies. In addition, while in college, students are requested to join as many extra-curricular activities as possible, since this will differentiate them from each other, and will demonstrate hands-on experience when looking for first jobs and internships.

The first professional experiences are usually very much marked by low salaries, if any, since for many people in this generation, the first contacts with professional jobs are internships, which are most of the time unpaid. In the financial crisis, apart from these challenges, it is important to add the main threat, which is the lack of job opportunities in the mid-term.

The transition from university life to professional life is becoming difficult, and leading to youth unemployment, which is very high in many countries in Europe, especially in the south or periphery.

Participants were asked:

- At what age they finished the studies?
- At what age they had their first job?
- At what age they had their first space job?

Most had finished their studies at the age of 25 years, which indicated that the majority finished their studies in the expected time (average of 5–6 years for bachelor plus master degree; see Fig. 9.3).

[6] Adams, Richard "Universities face more complaints since rise in tuition fees". The Guardian 13 July 2013. http://www.guardian.co.uk/education/2013/jun/13/universities-complaints-tuition-fees.

[7] "Tasas académicas", El Pais. http://elpais.com/tag/tasas_academicas/a/.

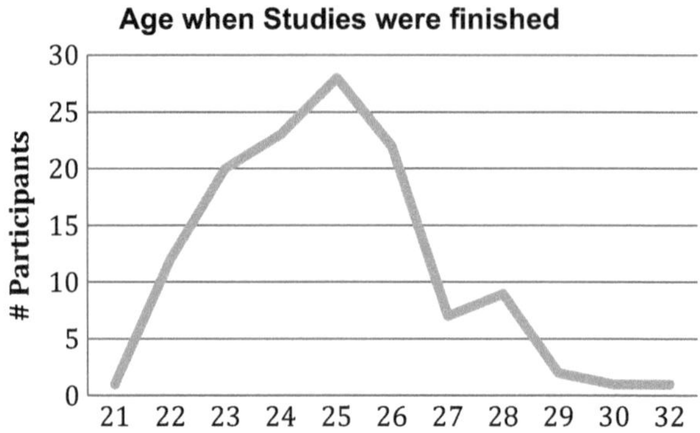

Fig. 9.3 Age when participants finished their studies (Source: SGAC)

Surprisingly, when they were asked about the first time they had a space related job, expecting to have some years of gaining experience before being able to jump into the sector, the results were as follows: 48 % started working in the space sector as soon as they finished their studies, and 21 % got a job in the space sector within 1 year after finishing. But the most surprising result was that a large percentage (21 %) had worked in the space sector even before finishing their studies (internships, partial jobs or job while studying). The mode age at which participants started working in the space sector is around 26 years old.

Another important and relevant piece of information is the typical behaviour of students that pursue a career in space in terms of preparation during their studies for their professional life. Half of the participants (49.2 %) worked during their studies at university, and a large majority (79.37 %) did an internship as part of either their study programme or professional life. Concerning this percentage, and regarding the salary these people earned during their internships, the amounts are very much dispersed. For example 24 % of those who did an internship did not receive any salary, however 20 % also mentioned that they received more than 1,000 euro per month in some of their internships. The average is about 500–750 euros per month.

Continuing with internships, given that they are very important and much valued in the space sector, participants were asked some additional questions. In particular, they were asked whether the internship offer came with a job offer right afterwards, or even if the internship was preparation for a promised future job in the same company. The results are favourable for the hiring companies. Although 76 % of the internships never included a job offer afterwards, it is also true that 78 % of internship offers never promised a job afterwards.

Another reason why internships are very important in the space sector is that space is a very specific area that can nevertheless be accessed from multidisciplinary backgrounds. For many of the jobs available studying a space-related course is not necessary to enter in the space sector. Respondents were asked about the first time they studied or worked in an area related to space. The results can be seen in Table 9.3.

Table 9.3 Education level where space is learned for the first time

First contact with space	Percentage of participants (%)
Bachelor – I studied a space related degree or undertook a space related final project	40
Internships	2
Master – I specialised in outer space studies during my master	35
Professional job – I started my space specialisation when I was already working, but never during my studies	20
Short courses – I took space-related short courses on my own	3

It is clear that there is a trend for students to decide during their bachelor and/or master degree whether they want to work in the space programme, but 20 % still have their first contact with space during their professional careers.

9.3.3.2 Unemployment

Of the total number of respondents, only 14 % were unemployed, mostly coming from France, Italy and Spain. Concerning the length of unemployment, the longer period participants were unemployed was for 6 months or less, and only a few took more than a year to find a new job.

9.3.3.3 Mobility

The open market and the opportunities to travel, work and study in any country within the European Union/Schengen Zone are pushing this new space generation to move from their countries of origin. In fact, 66 % are not currently living in their country of origin. There are many individual reasons behind the mobility of these people, among them more than 50 % moved because a better job was offered in another country, 24 % did so because there were no direct job opportunities in their origin countries, 18 % because the degree they wanted to study was not available in the origin country, and only a small minority of the sample pool (6 %) due to personal reasons.

Table 9.4 shows the mobility pattern of young adults from countries with large numbers of respondents to this survey (Germany, Spain, Italy and France):

Young professionals within the European Union thus appear to be used to be quite mobile, as they are trying professional opportunities outside their country of origin. Countries with ESA centres are those most chosen by respondents, not only to work at ESA centres but to also work in the ecosystem industries generated around them, which includes spin-offs, incubators and contractors.

Table 9.4 Mobility comparison of young adults from countries with major entries in the survey

Country	Stayed in their country	Moved to another
Spain	8	15
Germany	5	9
Italy	5	13
France	6	11

9.3.3.4 Financial Situation

In times of financial crisis, salaries are very much affected, since many companies and organisations try to cut either in their workforce size, or decrease or freeze their salaries. Young professionals, due to their relatively short experience in their companies, are more likely to be affected by these threats and measures. In many cases, the salary of young professionals is very low compared to their senior colleagues, and in some cases they are even paid as if they were still interns. Participants were asked to indicate their gross annual income as employees in the space sector, and to compare it with the salaries of similar aged employees in other sectors in the same country. In addition, they were asked for their number of working hours per day. The results have turned out to be very interesting and at the same time promising.

It is difficult to evaluate the salaries across the whole European Union, because each country has its own "minimum salary" policy and different price levels (PPP). However, since the space sector is very much international, and even more European (most companies have projects with other countries/companies, and Europe is the main example of a regional space programme), it is still appropriate to give interpretations to the results displayed in Fig. 9.4.

In view of the factors just mentioned above, respondents were also asked to compare their salaries with those in a similar position but in other sectors. This would be an indicator of the comparative real value of the salary of young professionals in their country. The results are shown in Fig. 9.5.

On the positive side, it can be noted that 24 % considered the salaries of young professionals in other sectors equal to theirs. However, the same percentage of people thought that those with similar positions in other sectors earned more ('more' for 24 % and 'significantly more' for 8 %). This should not be ignored, particularly because young professionals working in the space sector are highly qualified, and usually work in high technology and science development projects, and the work load and demand for these kinds of projects should not be underestimated as most of the respondents work between 8 and 9 or more hours per day (74 %).

9.3.3.5 Satisfaction

Despite long working hours and a perceived lesser pay than similar positions in other sectors, 71 % of respondents were satisfied with their current job. However,

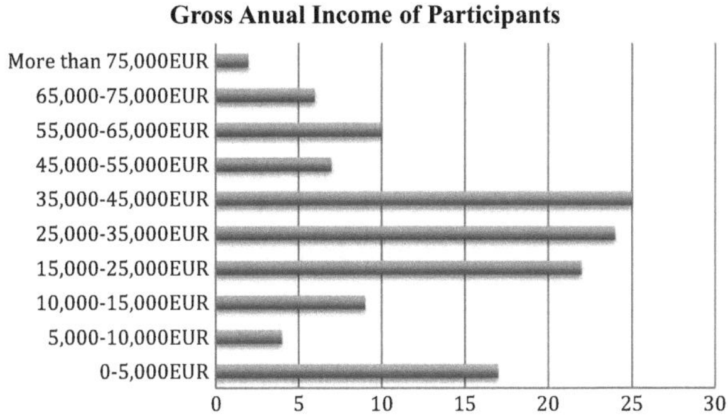

Fig. 9.4 Gross annual income of respondents (Source: SGAC)

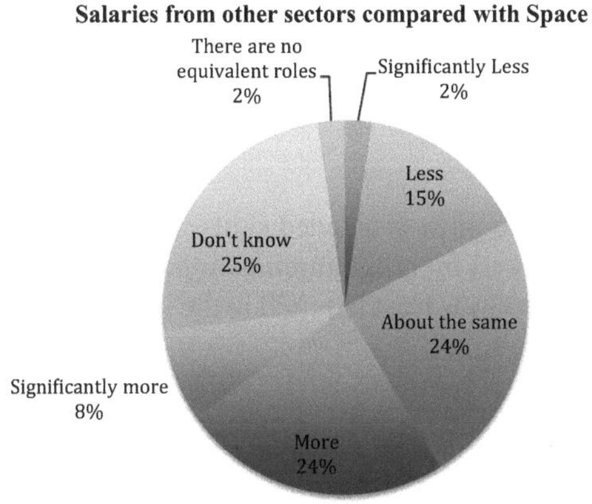

Fig. 9.5 Respondent's comparison of their salaries with other similar positions in other sectors (Source: SGAC)

only 27 % saw themselves as remaining in the same job for more than the coming 2 years. This is an indicator of the ambition of young professionals in the space sector, always looking for a change and better opportunities. In fact, 50 % of the survey respondents thought that they would have more than five different jobs in their lifetime.

9.3.3.6 Current Job Situation in Their Countries of Origin

Young professionals start their professional careers with high expectations, most of them dream of working in Exploration or Human Spaceflight programmes. However, the reality in Europe is slightly different. Although Exploration and Human Spaceflight are included in current national and regional programmes, participants were asked which areas were really in greatest demand when looking for jobs. Sixty-four percent highlighted telecommunications as the area where most job opportunities were available, followed by science programmes (11 %).

An important part of the professional life of a young professional is the search for a job. It is commonly said that "looking for a job is a job itself". The space sector is a very demanding and competitive sector, and although each job vacancy will have its own specific requirements, respondents were asked to rate a list of general skills according to demand (most to least) on the basis of their personal job search experiences.

The resulting list of demanded skills and traits was as follows:

- General space knowledge,
- Programming languages,
- Laboratory or study-case experience,
- Foreign languages,
- Passion,
- Experience in the field,
- Experience in that company/organisation,
- Life experience, and
- International experience.

Results showed that the most demanded skills were experience in the field, and foreign languages, followed by programming languages and international experience, while life experience was least required in the job descriptions and interviews.

9.4 Conclusions

The main conclusion after carrying out the survey is that there is a lack of data and a clear need to improve the survey by having more entries and getting the main space agencies and industries involved.

Summarising, young professionals in the space sector are generally satisfied with their jobs and unemployment is fairly low, since the financial crisis has just recently begun to affect this sector. Most young professionals working in the space sector hold a master's degree, and it is still dominated by men. Typically young professionals become involved in this area during their bachelor's degree and start working at the age of 26. Although not many of them took longer than 6 months to find a job, internships and experience are in strong demand as job requirements. Therefore, many of the participants have done an unpaid or low-paid internship, both during their studies or just after, and in most of the cases this internship has not led to a direct job offer. It is also because of this that mobility is very common among young professionals in the European Union, with 2/3rd currently not living in their country of origin. The profile of these young adults is very competent and ambitious, and therefore they all state that despite their current they would like to change jobs within 2 years.

Salaries for young professionals are still considerably lower than those of senior professionals, however, participants think they do not earn significantly less than other young professionals in similar positions in other sectors, and again satisfaction is quite widespread.

Young professionals currently in the space sector have had good opportunities due to the current trainee programmes in many industries and space agencies. The sector is still enjoying a certain well-being, however there are more professionals ready to retire than young professionals being hired. This will make difficult the transfer of knowledge and raises the threat of losing the great momentum started with the Apollo generation. With this challenge in mind, the most important part of this report is the recommendations in the following section.

9.5 Recommendations and Further Work

The main recommendations presented have been extracted from the results of the study and are in the form of suggestions that should be addressed by policy-makers, space agencies and industry in Europe:

- Most young professionals move away from their country of origin, and most likely never return. Although mobility policies currently exist within the European Union, it is important to look in detail at the pattern and direction of the movement of young professionals in the space sector. A policy to implement jobs of interest in the countries of origin that might attract back young professionals after their training period is needed.
- Although training programmes are offered by many industries and ESA, they are still insufficient to meet the demand. As the survey results show, the majority of young professionals in the space sector have benefitted from such programmes, and indeed they are one of the most demanded skills when looking for a job. Therefore, professional training workshops and courses must be included not only in educational institutions but also in the professional workplace.
- Salaries must be better regulated. Although most young employees are satisfied, space sector salaries are still slightly lower than similar positions in other sectors. The most alarming part is the amount of training that many of the participants did without any remuneration. This forces students to work at the same time as they study, and this limitation of time does not allow them to enrol in training programmes they otherwise would have attended.
- Young professionals must be included in decision-making meetings and policies. It is rather common to have few or no representatives of this sector in higher level meetings where decisions that will affect this demographic group are taken.

The financial crisis has just recently arrived in the space sector, and this is the main reason why young professions employed in the last 2 or 3 years are in a good position, and have found jobs in a relatively easy way. Nevertheless, the statistics might change in the coming 2 years, when the consequences of recent budgets cuts (consider for example the stagnation of the budget at the last ESA Ministerial,

November 2012[8] or the NASA Sequestration[9]) that will affect the latest graduated young generation, or that vast majority who showed willingness to change jobs in the coming 2 years.

Despite the amount of data and the conclusions reached in this study, further work is recommended. The timeline of this study was rather tight, and therefore the preparation and process of gathering data was rather short, and might contain several biases that could have been avoided with longer preparation. Some of the further work suggested is:

- Review the survey to include extra questions that were missing from this one.
- Distribute the survey again to gather not only more entries and data, but also more information from all European countries.
- Conduct a similar survey each year to evaluate trends and be able to prepare a better future estimate.
- Contact the main industries and organisations in Europe to gather more concrete data, and have access to their databases
- Present the outcome of this study to policy makers in Europe

Acknowledgement The author would like to thank the European Space Policy Institute for the opportunity to present these results, and the young professionals of the SGAC network, who took their time to answer the survey and are the main source of this paper.

[8] "European Ministers decide to invest in space to boost Europe's competitiveness and growth" 21 Nov. 2012 European Space Agency 15 Jun. 2013. http://www.esa.int/About_Us/Ministerial_Council_2012/European_Ministers_decide_to_invest_in_space_to_boost_Europe_s_competitiveness_and_growth.

[9] "Consolidated Sequestration Guidance" 4 Apr. 2013 National Aeronautics and Space Administration, 15 Jun. 2013. http://www.nasa.gov/pdf/738826main_consolidated_sequestration_guidance_V1.0_April-1-2013.pdf.

Part III
Facts and Figures

Chapter 10
Chronology: June 2011–May 2012

Cenan Al-Ekabi and Blandina Baranes

10.1 Access to Space

All launch dates are calculated using Greenwich Mean Time (GMT), hence the date at the launch site may differ from the date listed here by 1 day[1].

Europe	Other countries
Launch log	
June 11	
	07 Soyuz – Soyuz TMA-02M (ISS)
	10 Delta II 7320 – SAC-D/Acquarius (R)
	15 Safir 2 – Rasad (R)
	20 Long March 3B – Chinasat 10 (R)[a]
	21 Soyuz – Progress M-11M (ISS 43P) (ISS)
	27 Soyuz U – Cosmos 2472 (I)
	29 Minotaur I – ORS 1 (S)
July 11	
	06 Long March 2C – Shijian 11–03 (S)
	08 Shuttle Atlantis – STS 135 (MPLM, LMC) (MF/ISS)
	11 Long March 3C – Tianlian 1B (C)
	13 Soyuz 2 – Globalstar 2nd Gen. 07, 08, 09, 10, 11, and 12 (C)[a]
	15 Proton M – SES-3 (C)[a], KazSat 2 (C)
	15 PSLV – GSAT 12 (C)

(continued)

[1] Federal Aviation Administration. Commercial Space transportation: 2011 Year in Review. Washington DC: FAA, Jan. 2012: 22+; Federal Aviation Administration. Commercial Space transportation: 2012 Year in Review. Washington, DC: FAA, Jan. 2013: 29+.

C. Al-Ekabi (✉) • B. Baranes
ESPI (European Space Policy Institute), Vienna, Austria
e-mail: Cenan.Al-Ekabi@espi.or.at

C. Al-Ekabi et al. (eds.), *Yearbook on Space Policy 2011/2012*, Yearbook on Space Policy, DOI 10.1007/978-3-7091-1649-4_10, © Springer-Verlag Wien 2014

Europe	Other countries
Launch log	
	16 Delta IV Medium+ (4,2) – Navstar GPS 2F-02 (N) 18 Zenit 3F – Spektr R (S) 26 Long March 3A – Beidou 2-IGS 4 (N) 29 Long March 2C – Shijian 11–02 (S)
August 11	
06 Ariane 5 ECA – Astra IN (C)[a], and BSAT-3c/JCSAT-110R (C)[a]	05 Atlas V 551 – Juno (S) 11 Long March 3B – Paksat 1R (C)[a] 15 Long March 4B – Hai Yang 2A (R) 17 Dnepr M – Sich 2 (R), NX (R), Nigeriasat 2 (R), Edusat (S), RASAT (S), BPA 2 (D), Aprizesat 5 (C)[a], and Aprizesat 6 (C)[a] 17 Proton M – Express AM4 (failure: failure due to placement in the wrong orbit, C)[a] 18 Long March 2C – Shijian 11–04 (failure: failure during initial stages of ascent, S) 24 Soyuz Progress M-12M (ISS 44P) (failure: failed due to misfiring of propulsion system during its third stage, ISS)
September 11	
21 Ariane 5 ECA – Arabsat 5C (C)[a], and SES 2 (C)[a]	10 Delta II 7920H – GRAIL A (S), GRAIL B (S) 18 Long March 3B – Chinasat 1A (C) 20 Proton M – Cosmos 2473 (C) 23 H IIA – IGS-4C (Optical) (I) 24 Zenit 3SL – Atlantic Bird 7 (C)[a] 27 Minotaur IV – TacSat-4 (D) 29 Long March 2F – Tiangong I (D) 29 Proton M – QuetzSat-1 (C)[a]
October 11	
21 Soyuz 2 – Galileo 1 (N), and Galileo 2 (N)	02 Soyuz 2 – Glonass M42 (N) 06 Zenit 3SLB – Intelsat 18 (C)[a] 07 Long March 3B – Eutelsat W3C (C)[a] 12 PSLV – Megha Tropiques (S), Jugnu (R), SRMSAT (S), Vesselsat 1 (C) 19 Proton M – ViaSat 1 (C)[a] 28 Delta II 7920 – NPP (M), RAX 2 (S), M-Cubed (R), AS-1 (S), Explorer-1 (PRIME) Unit 2 (S), DICE 1 (S), and DICE 2 (S) 30 Soyuz – Progress M-13M (ISS 45P) (ISS) 31 Long March 2F – Shenzhou 8 (D)
November 11	
	04 Proton M – Glonass M43, M44, M45 08 Zenit 2M – Phobos-Grunt (failure: failure due to propulsion system malfunction, S), and Yinghuo (failure: failure due to propulsion system malfunction, S)

(continued)

Europe	Other countries
Launch log	
	09 Long March 4B – Yaogan 12 (R), and Tianxun 1 (R) 14 Soyuz – Soyuz TMA-22 (MF/ISS) 20 Long March 2D – Chuang Xing I-03 (C), and Shiyan Wexing 4 (S) 25 Proton M – Asiasat 7 (C)[a] 26 Atlas V 541 – Mars Science Laboratory (S) 28 Soyuz 2 – Glonass M46 (N) 29 Long March 2D – Yaogan 13 (I)
December 11	
16 Soyuz 2 – Pleiades HR I (R), SSOT (R), ELISA 1, 2, 3, and 4 (I)	01 Long March 3A – Beidou 2-IGS 5 (N) 11 Proton M – Luch 5A (C), Amos 5 (C)[a] 12 H IIA – IGS-4B (Radar) (I) 19 Long March 3B – Nigcomsat IR (C)[a] 21 Soyuz – Soyuz TMA-03M (MF/ISS) 22 Long March 4B – Ziyuan I-2C (R) 23 Soyuz 2 – Meridian 5 (failure: failure occurring during third stage firing, C) 28 Soyuz 2 – Globalstar 2nd Gen. 13, 14, 15, 16, 17, and 18 (C)[a]
January 12	
	09 Long March 4B – Ziyuan 3 (R), and Vesselsat 2 (C)[a] 13 Long March 3A – Feng Yun 2F (M) 20 Delta IV Medium +(5, 4) – WGS 4 (C) 25 Soyuz – Progress M-14M (ISS 46P) (ISS)
February 12	
13 Vega – LARES (S), PW-Sat 1 (D), XaTcobeo (D), Robusta (D), e-St@r (D), Goliat (R), ALMASAT (D), MaSat 1 (D), Unicubesat GG (S)	03 Safir 2 – Navid-e Elm-o Sanat (R) 14 Proton M – SES 4 (C)[a] 24 Atlas V 551 – MUOS 1 (C) 24 Long March 3C – Beidou 2-G5 (N)
March 12	
23 Ariane 5 ES-ATV – ATV 3 (ISS)	25 Proton M – Intelsat 22 (C) [a] 30 Proton K – Cosmos 2479 (EW) 31 Long March 3B – APSTAR 7 (C)[a]
April 12	
	03 Delta IV Medium +(5, 2) – NRO L-25 (I) 12 Unha 3 – Kwangmyongsong 3 (failure: presumed to be caused during second stage ignition, R) 20 Soyuz – Progress M-15M (ISS 47P) (ISS) 23 Proton M – Yahsat 1B (C)[a] 26 PSLV XL – Risat 1 (R) 29 Long March 3B – Beidou 2C-M3 (N), and Beidou 2C-M4 (N)

(continued)

Europe	Other countries
Launch log	
May 12	
15 Ariane 5 ECA – JCSAT 13 (C)[a], and Vinasat 2 (C)	04 Atlas V 531 – Advanced EHF 2 (C)
	06 Long March 2D – Tian Hui 1B (R)
	10 Long March 4B – Yaogan 14 (R), and Tiantuo 1 (S)
	15 Soyuz – Soyuz TMA-04M (ISS 30S) (MF/ISS)
	17 Soyuz U – Cosmos 2480 (I)
	17 Proton M – Nimiq 6 (C)[a]
	17 H IIA – GCOM W1 (S), Arirang 3 (R), SDS 4 (D), and Horyu 2 (C)
	22 Falcon 9 – Dragon COTS Demo 2/3 (D)
	23 Safir 2 – Fajr (failure: failure to launch, D)
	26 Long March 3B – Chinasat 2A (C)
	29 Long March 4C – Yaogan 15 (R)

C communications, *D* development, *I* intelligence, *ISS* international space station, *M* meteorological, *MF* manned flight, *N* navigation, *R* remote sensing, *S* scientific, *EW* early warning system

[a]Commercial

10.2 Space Science and Exploration

Europe	Other countries
Earth sciences	
21 June – first map of sea-ice thickness from ESA's CryoSat mission was revealed (ESA, Europe)[2]	*6 July* – Launch of Shijian 11–03 (CAST, China) to conduct various technology and environmental experiments[3]
6 July – ESA's ERS-2 satellite, studying atmosphere, land, oceans and polar ice, begins staggered 2-month guided re-entry into Earth's atmosphere following its retirement (ESA, Europe)[4]	*29 July* – Launch of Shijian 11–02 (CAST, China) to conduct scientific experiments in orbit[5]

(continued)

[2] "NEW ICE THICKNESS MAP OF THE ARCTIC UNVEILED." 21 June 2011. European Space Agency 4 June 2013 http://www.esa.int/Our_Activities/Observing_the_Earth/CryoSat/New_ice_thickness_map_of_the_Arctic_unveiled.

[3] "China launches experimental satellite." 6 July 2011. 17 June 2013 http://news.xinhuanet.com/english2010/china/2011-07/06/c_13969022.htm.

[4] "PIONEERING ERS ENVIRONMENT SATELLITE RETIRES." 5 July 2011. European Space Agency 4 June 2013 http://www.esa.int/Our_Activities/Observing_the_Earth/Pioneering_ERS_environment_satellite_retires.

[5] "China launches another experimental satellite." 29 July 2011. Xinhua, english.news.cn 17 June 2013 http://news.xinhuanet.com/english2010/china/2011-07/29/c_131018243.htm.

Europe	Other countries
12 October – Launch of Megha Tropiques (CNES, France; ISRO, India), a joint French-Indian satellite mission for studying the water cycle and energy exchanges in the tropics[6]	*18 August* – Failed launch of Shijian 11-04 (CAST, China) during initial stages of ascent[7]
13 February – Launch of LARES (ASI, Italy) to measure the frame dragging effect, in regard to general relativity[8]; and Unicubesat GG (University of Rome, Italy)[1]	*12 October* – Launch of SRMSAT (ISRO, India) nanosatellite monitoring greenhouse gases[9]
09 March – first global high-resolution map of the boundary between Earth's crust and mantle produced from data received by ESA's GOCE gravity satellite (ESA, Europe)[10]	*28 October* – Launch of RAX 2[11] (University of Michigan, USA) to study plasma field-aligned irregularities in the ionosphere; AS-1[1] (Auburn University, USA); Explorer-1 (PRIME) Unit 2[1] (Montana State University, USA); and DICE 1 and DICE 2 (Utah State University, USA) to map the geomagnetic SED (Storm Enhanced Density) plasma bulge and plume formations in Earth's ionosphere[12]
08 April – Communication lost with decade-running optical, radar and laser satellite, Envisat (ESA, Europe)[13]	*20 November* – Launch of Shiyan Wexing 4 (CAST, China) to study the environment and test new space technologies[14]
	17 May – Launch of GCOM W1 (JAXA, Japan) to observe water circulation changes[15]

(continued)

[6] "MEGHA-TROPIQUES." CNES 19 June 2013 http://smsc.cnes.fr/MEGHAT/.

[7] Barbosa, Rui C. "China's third launch in a week proves too much as Long March 2C fails." 18 Aug. 2011. NASAspaceflight.com 17 June 2013 http://www.nasaspaceflight.com/2011/08/chinas-third-launch-week-long-march-2c-fails/.

[8] "LARES – Testing of General Relativity." ASI 21 June 2013 http://www.asi.it/en/activity/cosmology/lares.

[9] "SRMSat." Indian Space Research Organisation 19 June 2013 http://www.isro.org/satellites/srmsat.aspx.

[10] "MAPPING THE MOHO WITH GOCE." 9 March 2012. European Space Agency 5 June 2013 http://www.esa.int/Our_Activities/Observing_the_Earth/GOCE/Mapping_the_Moho_with_GOCE.

[11] "RAX-2 (Radio Aurora Explorer-2)." eoPortalDirectory 22 June 2013 https://directory.eoportal.org/web/eoportal/satellite-missions/r/rax-2.

[12] "DICE (Dynamic Ionosphere CubeSat Experiment), DICE-1 and DICE-2." eoPortalDirectory 22 June 2013 https://directory.eoportal.org/web/eoportal/satellite-missions/d/dice.

[13] "INVESTIGATION ON ENVISAT CONTINUES." 20 April 2012. European Space Agency 5 June 2013 http://www.esa.int/Our_Activities/Observing_the_Earth/Envisat/Investigation_on_Envisat_continues.

[14] "Shiyan 4 – Low Earth Orbit Satellite." FindTheData.org 19 June 2013 http://satellites.findthedata.org/l/835/Shiyan-4.

[15] "Global Change Observation Mission 1st – Water | 'SHIZUKU' (GCOM-W1)." JAXA 22 June 2013 http://www.jaxa.jp/projects/sat/gcom_w/index_e.html.

Europe	Other countries
Astronomy	
28 June – NASA/ESA Hubble Space Telescope discovers P4, a new moon circling Pluto[16]	*18 July* – Launch of Spektr R (Roscosmos, Russia) to study the structure and dynamics of radio sources inside and outside the Milky Way galaxy, by operating in conjunction with ground-based antennas to form a gigantic virtual telescope[17]
Exploration	
8 June – Rosetta spacecraft (ESA) put into deep space hibernation for 31 months[18]	*5 August* – Launch of Juno (NASA, USA) to study and reveal the origin and evolution of Jupiter[19]
	10 September – Launch of GRAIL A and GRAIL B (NASA, USA) to map the Moon's gravitational field[20]
	8 November – Failed launch of Phobos-Grunt (Roscosmos, Russia) mission to land on Mars' moon Phobos and return samples of its soil back to Earth, and of the Yinghuo (CAST, China) Mars orbiter. Failure due to propulsion system malfunction; re-entered Earth's atmosphere on 15 January 2012[21]
	26 November – Launch of Mars Science Rover 'Curiosity' (NASA, USA) designed to assess Mars' historic habitability[22]
	10 May – Launch of Tiantuo 1 (PLA, China) microsatellite carrying satellite-based vessel AIS system to conduct optical imaging and various space exploration experiments[23]

(continued)

[16] "New Moon Found At Pluto." 21 July 2011. ESA 6 July 2013 http://www.esa.int/Our_Activities/Space_Science/New_moon_found_at_Pluto.

[17] "Spektr R." RussianSpaceWeb.com 16 June 2013 http://www.russianspaceweb.com/spektr_r.html.

[18] "Rosetta comet probe enters hibernation in deep space." 8 June 2011. ESA 5 July 2013 http://www.esa.int/Our_Activities/Operations/Rosetta_comet_probe_enters_hibernation_in_deep_space.

[19] "Juno – Mission Overview." 24 Aug. 2011. NASA 16 June 2013 http://www.nasa.gov/mission_pages/juno/overview/index.html.

[20] "Grail – Mission Overview." NASA 16 June 2013 http://www.nasa.gov/mission_pages/grail/overview/index.html.

[21] Amos, Jonathan. "Phobos-Grunt: Failed probe 'falls over pacific'." 15 Jan. 2012. BBC News – Science & Environment 16 June 2013 http://www.bbc.co.uk/news/science-environment-16491457.

[22] "Mars Science Laboratory | Curiosity Rover – Mission Overview." NASA 17 June 2013 http://mars.jpl.nasa.gov/msl/mission/overview/.

[23] Hill, Jeffrey. "Chinese Long March Sends Yaogan 14, Tiantuo 1 into Orbit." 11 May 2012. Satellite Today 18 June 2013 http://www.satellitetoday.com/civilspace/headlines/Chinese-Long-March-Sends-Yaogan-14-Tiantuo-1-into-Orbit_38812.html.

Europe	Other countries
Manned spaceflight & cargo transfers	
16 May – 01 June – Roberto Vittori (Italy) participated in STS 134 as mission specialist and the last non-U.S. astronaut to fly on the Space Shuttle[24]	*7 June* – launch of Soyuz TMA-02M (Roscosmos, Russia) with three ISS Expedition 28 & 29 crew
4 November – Mars500, a full-length simulated mission to Mars, ended after 17 months in operation in Moscow[25]	*21 June, 24 August*[a], *30 October, 25 January, 20 April* – launches of 5 Progress (Roscosmos, Russia) cargo transfer vehicles to the ISS (1 launch failure).[1,26]
23 December – 1 July – André Kuipers (Netherlands) participated in ISS Expeditions 30 & 31 as a flight engineer for ESA's PromISSe mission[27]	*8 July* – Launch of STS 135 (NASA, US) on ISS ULF 7 using OV-104 Atlantis. Final mission of Atlantis, and shuttle programme. Four crew, payload Multi-Purpose Logistics Module (MPLM) and the Lightweight Multipurpose Experiment Support Structure Carrier (LMC)[1]
February 18 – Luca Parmitano (Italy) assigned to ISS Expeditions 36/37, to launch in 2013[28]	*29 September* – launch of Tiangong 1 (CAST, China) prototype laboratory, designed for orbital docking tests and space research[29]
23 March – Launch of ATV-3, Edoardo Amaldi (European Space Agency), ISS resupply vehicle[1]	*31 October* – launch of Shenzhou 8 (CAST, China), unmanned spacecraft, designed for orbital docking tests with Tiangong 1[30]
	14 November – Soyuz TMA-22 (Roscosmos, Russia) with three ISS Expedition 29 & 30 crew[1]
	15 November – NASA starts accepting applications for future astronauts through 23 January 2012[31]

(continued)

[24] "NASA – Shuttle Missions – STS-134." 9 June 2011. NASA 31 Oct. 2011 http://www.nasa.gov/mission_pages/shuttle/shuttlemissions/sts134/main/index.html.

[25] "WELCOME BACK AND THANKYOU, MARS500." 4 June 2011. European Space Agency 13 June 2013 http://www.esa.int/Our_Activities/Human_Spaceflight/Mars500/Welcome_back_and_thank_you_Mars500.

[26] Harding, Pete. "Soyuz TMA-04M crew return to Earth following four-month mission to ISS." 16 Sept. 2012. NASAspaceflight.com 13 June 2013 http://www.russianspaceweb.com/progress_m12m.html.

[27] "ESA – PromISSe mission key data." 27 July 2012. European Space Agency 13 June. 2013 http://www.esa.int/Our_Activities/Human_Spaceflight/PromISSe/PromISSe_mission_key_data.

[28] "ESA Portal – ESA astronaut Luca Parmitano assigned to 2013 Space Station mission." 18 Feb. 2011. European Space Agency 31 Oct. 2011 http://www.esa.int/esaCP/SEM5J3PT1KG_index_0.html.

[29] Malik, Tariq. "China's Tiangong 1 Space Lab: Questions & Answers." 29 Sept. 2011. Space.com 13 June 2013 http://www.space.com/13120-china-tiangong-1-space-laboratory-facts-figures.html.

[30] Amos, Jonathan. "China's unmanned Shenzhou 8 capsule returns to Earth." 17 Nov. 2011. BBC News – Science & Environment 13 June 2013 http://www.bbc.co.uk/news/science-environment-15776662.

[31] "NASA Accepting Applications For Future Astronauts." 15 Nov. 2011. NASA 5 July 2013 http://www.nasa.gov/home/hqnews/2011/nov/HQ_11-385_Astro_Apps.html.

Europe	Other countries
	21 December – Launch of Soyuz TMA-03M (Roscosmos, Russia) with three ISS Expedition 30 & 31 crew[1]
	15 May – Launch of Soyuz TMA-04M (ISS 30S) (Roscosmos, Russia) with three ISS Expedition 31 & 32 crew[1]

[a]*24 August* launch of Progress M-12M failed due to misfiring of propulsion system during its third stage

10.3 Applications

Europe	Other countries
Earth observation	
16 December – Launch of Pleiades HR 1 (CNES, France) as part of a two-spacecraft high-resolution optical imaging constellation[32]	*10 June* – Launch of SAC-D/Aquarius (NASA, USA; and CONEA, Argentina) satellite to measure sea surface salinity and provide the global view of salinity variability[33]
13 February – Launch of the Goliat (University of Bucharest, Romania) cubesat[1]	*15 June* – Launch of Rasad 1 (Iranian Space Agency) imaging satellite for environmental monitoring[34]
	15 August – Launch of Hai Yang 2A (CAST, China) to monitor ocean wind fields, sea levels and temperatures, waves, currents, tides, and storms in order to provide disaster and weather forecasting information[35]
	17 August – Launch of Sich 2 (State Space Agency of Ukraine, Ukraine) to monitor agricultural, water, and land resources, as well as disasters and Earth's ionosphere[36];

(continued)

[32] "Pleiades-HR." European Space Agency 19 June 2013 https://earth.esa.int/web/guest/missions/3rd-party-missions/potential-missions/pleiades-hr.

[33] "Aquarias | Sea Surface Salinity from Space." NASA 18 June 2013 http://aquarius.nasa.gov/.

[34] "RASAD 1." Real Time Satellite Tracking 18 June 2013 http://www.n2yo.com/satellite/?s=37675.

[35] Barbosa, Rui C. "China's surge continues with HaiYang-2A launch via Long March 4B." 15 Aug. 2011. NASASpaceFlight 18 June 2013 http://www.nasaspaceflight.com/2011/08/chinas-surge-haiyang-21a-launch-long-march-4b/.

[36] "Sich-2." State Space Agency of Ukraine 19 June 2013 http://www.nkau.gov.ua/nsau/catalogNEW.nsf/systemE/4292067D377BB47CC2256F330055DF6E?OpenDocument&Lang=E.

Europe	Other countries
	along with the NX and Nigeriasat 2 (NASRDA, Nigeria),[37] and RASAT (TÜBİTAK UZAY, Turkey) Earth observation satellites[38]
	12 October – Launch of the Jungu (IIT Kapur, India) nanosatellite Earth observing GPS demonstrator[39]
	28 October – Launch of Suomi NPP (NASA, USA), and M-Cubed (University of Michigan, USA)[1]
	9 November – Launch of Yaogan 12 (CAST, China) and Tianxun 1 (CAST, China) to conduct Earth observation [1]
	16 December – Launch of SSOT (Chilean Space Agency, Chile), to obtain high quality imagery for use in a wide variety of applications both in the defense and civil domains[40]
	22 December – Launch of Ziyuan 1-2C satellite (CAST, China) for Earth observation[1]
	13 January – Launch of Feng Yun 2F (China Meteorological Administration, China) weather satellite[1]
	3 February – Launch of Navid-e Elm-o Sanat (Iranian Space Agency, Iran) satellite for climate monitoring and Earth observation[41]
	12 April – Failed launch of Kwangmyongsong 3 satellite (Korean Committee of Space Technology, North Korea) presumed to be caused during second stage ignition[42]
	26 April – Launch of the Risat-1 (ISRO, India), a Microwave Remote Sensing Satellite carrying Synthetic Aperture Radar (SAR) to

(continued)

[37] "Nigeria launches two satellites." 17 Aug. 2011. BBC News – Science & Environment 19 June 2013 http://www.bbc.co.uk/news/science-environment-14563647.

[38] "First satellite designed and built in Turkey: RASAT." TÜBİTAK UZAY Space Technologies Research Institute 19 June 2013 http://www.uzay.tubitak.gov.tr/tubitakUzay/en/projects/spaceApplications.php.

[39] "Jugnu." Indian Space Research Organization 22 June 2013 http://www.isro.org/satellites/jugnu.aspx.

[40] "SSOT (Sistema Satelital para la Observación de la Tierra)." eoPortal Directory 19 June 2013 https://directory.eoportal.org/web/eoportal/satellite-missions/s/ssot.

[41] "Persian space: Iran launches new satellite." 3 Feb. 2012 RT.com 19 June 2013 http://rt.com/news/iran-satellite-launch-revolution-421/.

[42] "North Korean Flight ends in Failure – Flight Reconstruction." 13 Apr. 2012. Spaceflight101 18 June 2013 http://www.spaceflight101.com/north-korean-kwangmyongsong-3-launch-updates.html.

Europe	Other countries
	observe surface features during night and day in all weather conditions[43]
	6 May – Launch of Tianhui 1–02 (CAST, China) designed for military survey and mapping roles[44]
	10 May – Launch of Yaogan 14[45] (PLA, China) optical military reconnaissance satellite
	17 May – Launch of the Arirang 3 (KARI, South Korea) optical high-resolution observation satellite[46]
	29 May – Launch of Yaogan 15 (PLA, China) to conduct scientific experiments, land resources surveys, crop yield estimates and disaster monitoring purposes, and other reconnaissance applications[47]
Intelligence and early warning	
16 December – Launch of ELISA 1, 2, 3, and 4 (CNES, France) demonstrator for mapping the positions of radar and other transmitters throughout the world and determining their technical characteristics[48]	*27 June* – Launch of Cosmos 2472 (Russian Space Forces, Russia) military imaging reconnaissance satellite[1] *29 June* – Launch of ORS 1 (USAF, USA) to demonstrate the capability to meet emerging and persistent war-fighter needs on operationally relevant timelines[49] *23 September* – Launch of IGS-4C (Optical) (JAXA, Japan)[1] *29 November* – Launch of the Yaogan 13 (PLA, China) satellite[50] *12 December* – Launch of IGS-4B (Radar) (JAXA, Japan)[1]

(continued)

[43] "Earth Observation Satellites | RISAT-1." ISRO 18 June 2013 http://www.isro.org/satellites/risat-1.aspx.

[44] "Tianhui." 8 May 2012. Dragon in Space 18 June 2013 http://www.dragoninspace.com/earth-observation/tianhui.aspx.

[45] Clark, Stephan. "Chinese spy satellite sucessfully launched." 10 May 2012. Spaceflight Now 18 June 2013 http://www.spaceflightnow.com/news/n1205/10longmarch/.

[46] "KOMPSAT-3 (Korea Multi-Purpose Satellite-3)/Arirang-3." eoPortal Directory 22 June 2013 https://directory.eoportal.org/web/eoportal/satellite-missions/k/kompsat-3.

[47] Barbosa, Rui C. "China launches Yaogan 15 via Long March 4C at short notice." 29 May 2012. NASASpaceFlight 19 June 2013 http://www.nasaspaceflight.com/2012/05/china-launches-yaogan-15-long-march-4c-short-notice/.

[48] "Elisa – Mapping radar stations from Space." CNES 22 June 2013 http://www.cnes.fr/web/CNES-en/5940-elisa.php.

[49] "ORS-1 Launch Information." NASA 22 June 2013 http://www.nasa.gov/centers/wallops/missions/ORS.html.

[50] Barbosa, Rui C. "China return Long March 2C to flight with YaoGan Weixing-13 launch." 29 Nov. 2011. NASASpaceFlight.com 22 June 2013 http://www.nasaspaceflight.com/2011/11/china-return-long-march-2c-flight-yaogan-weixing-13-launch/.

Europe	Other countries
	30 March – Launch of the Cosmos 2479 (Russian Space Forces, Russia) missile detection and early warning satellite[51] *3 April* – Launch of NRO L-25 (NRO, U.S.) for intelligence-gathering purposes[1] *17 May* – Launch of Cosmos 2480 (Russian Space Forces, Russia), military imaging reconnaissance satellite[1]
Navigation	
21 October – Launch of Galileo 1 and 2 (ESA, Europe) In-Orbit Validation (IOV) satellites[52]	*16 July* – Launch of the Navstar GPS 2F-02 (USAF, USA) space-based radio positioning, navigation, and time distribution system[53] *26 July* – Launch of Beidou 2-IGS 4 (PLA, China)[1] *2 October* – Launch of Glonass M42 (Russian Space Forces, Russia) satellite[1] *4 November* – Launch of Glonass M39, M40 and M43, M44, and M45 (Russian Space Forces, Russia) satellites[1] *28 November* – Launch of Glonass M46 (Russian Space Forces, Russia) satellite[1] *1 December* – Launch of Beidou 2-IGS 5 (PLA, China) satellite[1] *24 February* – Launch of Beidou 2-G5 (PLA, China)[1] *29 April* – Launch of Beidou 2C-M3 and 2C-M4 (PLA, China)[1]
Telecommunications/broadcasting	
6 August – Launch of Astra 1N (Luxembourg)[1]	*20 June* – Launch of Chinasat 10 (CAST, China)[54]
24 September – Launch of Atlantic Bird 7 (Eutelsat, France)[1]	*11 July* – Launch of the Tianlian 1B (CAST, China) tracking and data relay satellite designed to support near-real-time communications between orbiting spacecraft and the ground control[55]

(continued)

[51] "Ignition! 2012 – Launches to Orbit and Beyond." Zarya 23 June 2013 http://www.zarya.info/Diaries/Launches/Launches.php?year=2012.

[52] "What is Galileo?" European Space Agency 23 June 2013 http://www.esa.int/Our_Activities/Navigation/The_future_-_Galileo/What_is_Galileo.

[53] "Selected Acquisition Report (SAR)." 31 Dec. 2011. US DOD 23 June 2013 http://www.dod.mil/pubs/foi/logistics_material_readiness/acq_bud_fin/SARs/DEC%202011%20SAR/NAVSTAR%20GPS%20-%20SAR%20-%2031%20DEC%202011.pdf.

[54] "SATELLITE RESOURCES: ChinaSat 10(SinoSat-5)." China Satellite Communications Co. Ltd. 27 June 2013 http://www.chinasatcom.com/en/News_Info.aspx?m=20110329113845577061&n=20110705100114623575.

[55] Barbosa, Rui C. "China launches Tian Lian-1B to bolster manned space program network." 11 July 2011. NASASpaceFlight.com 27 June 2013 http://www.nasaspaceflight.com/2011/07/china-launches-tian-lian-1b-bolster-manned-space-network/.

Europe	Other countries
6 October – Launch of Intelsat 18 (Luxembourg)[1]	*13 July* – Launch of Globalstar 2nd Gen 07, 08, 09, 10, 11, and 12 (Globalstar, Inc., USA) mobile communications satellites[1]
7 October – Launch of Eutelsat W3C (Eutelsat, France)[1]	*15 July* – Launch of the SES-3 (SES World Skies) broadcasting satellite, and KazSat 2 (Kazakhstan)[1]
12 October – Launch of Vesselsat 1 (Luxembourg) asset-tracking communications satellite[1]	*15 July* – Launch of GSAT 12 (ISRO, India)[56]
9 January – Launch of Vesselsat 2 (Luxembourg) asset-tracking communications satellite[1]	*6 August* – Launch of BSAT-3c/JCSAT-110R (Sky Perfect JSAT Corp., Japan)[1]
14 February – Launch of SES 4 (Luxembourg) Replacing the NSS-7 to provide enhanced coverage over the Americas, Africa and Europe[1]	*11 August* – Launch of Paksat 1R (India)[1]
25 March – Launch of Intelsat 22 (Luxembourg)[1]	*17 August* – Launch of Aprizesat 5[57] and Aprizesat 6[58] (Argentina) to be used for data collection from small fixed and mobile ground stations, while carrying AIS transponders to track ships at sea
	17 August – Failed launch of Express AM4 (Roscosmos, Russia) due to placement in the wrong orbit; intentional deorbiting conducted on 25 March 2012[59]
	18 September – Launch of Chinasat 1A (CAST, China)[1]
	20 September – Launch of Cosmos 2473 (Roscosmos, Russia)[1]
	21 September – Launch of Arabsat 5C (Arabsat, Saudi Arabia), and SES 2 (USAF, USA) providing commercial coverage over the U.S. and Caribbean, also host to the USAF's experimental Commercially Hosted Infrared Payload (CHIRP)[60,1]
	29 September – Launch of QuetzSat-1 (Mexico)[1]

(continued)

[56] "Geo-Stationary Satellites: GSAT-12." ISRO 27 June 2013 http://www.isro.org/satellites/gsat-12.aspx.

[57] "AprizeSat 5." NASA 27 June 2013 http://nssdc.gsfc.nasa.gov/nmc/spacecraftDisplay.do?id=2011-044E.

[58] "AprizeSat 6." NASA 27 June 2013 http://nssdc.gsfc.nasa.gov/nmc/spacecraftDisplay.do?id=2011-044F.

[59] "Satellite to burn in atmosphere above Pacific." 25 Mar. 2012. RT.com 27 June 2013 http://rt.com/news/express-satellite-pacific-deorbit-411/.

[60] "SES-2." SES 27 June 2013 http://www.ses.com/4628194/ses-2.

Europe	Other countries
	19 October – Launch of ViaSat 1 (Viasat, USA)[1]
	20 November – Launch of Chuang Xing 1–03 (CAST, China) to gather and transmit data for disaster relief and economic development applications[61]
	25 November – Launch of Asiasat 7 (Hong Kong)[1]
	11 December – Launch of Luch 5A (Rocascosmos, Russia) to relay live telemetry and other information from low-orbiting Russian space vehicles, and Amos 5 (Israel)[62, 1]
	19 December – Launch of Nigcomsat 1R (NASRDA, Nigeria)[63]
	23 December – Failed launch of Meridian 5 (Roscosmos, Russia) occurring during third stage firing[64]
	28 December – Launch of Globalstar 2nd Gen 13, 14, 15, 16, 17, 1 (Globalstar, Inc., USA) mobile communications satellites[1]
	20 January – Launch of WGS 4 (USAF, USA) military communications satellite[1]
	24 February – Launch of MUOS 1 (US Navy, USA) tactical satellite communications system[65]
	31 March – Launch of APSTAR 7 (Hong Kong)[1]
	23 April – Launch of Yahsat 1B (Al Yah Satellite Communications Co., Abu Dhabi) providing services to government and commercial customers in the Middle East, Africa, Europe and South West Asia[66]

(continued)

[61] "CHUANG XIN 1–03." 20 Nov. 2011. REAL TIME SATELLITE TRACKING 27 June 2013 http://www.n2yo.com/satellite/?s=37930.

[62] "Luch-5A satellite." RussianSpaceWeb.com 27 June 2013 http://www.russianspaceweb.com/luch5a.html.

[63] Eze, Amaka. "What is NigComSat-1R Bringing to the Table?" 31 Jan. 2013. ThisDayLive 27 June 2013 http://www.thisdaylive.com/articles/what-is-nigcomsat-1r-bringing-to-the-table-/137979.

[64] "Rise or Fall – Meridian 5 Launch Failure." Zarya 27 June 2013 http://www.zarya.info/Galli maufry/Meridian5drop.php.

[65] "US Navy's MUOS-1 Satellite successfully completes on-orbit testing." 19 July 2012. naval-technology.com 27 June 2013 http://www.naval-technology.com/news/newsus-navys-muos-1-sat ellite-successfully-completes-on-orbit-testing.

[66] "Yahsat 1B." SATBEAMS 27 June 2013 http://www.satbeams.com/satellites?norad=38245.

Europe	Other countries
	15 May – Launch of JCSAT 13[67] (Japan) to provide uplink and downlink coverage over Japan, Indonesia, the Middle and Near East, Southwest Asia and Oceania; and Vinasat 2[68] (Vietnam), Vietnam's first communication
	17 May – Launch of Nimiq 6 (Canada) broadcast satellite[1]
	17 May – Launch of Horyu 2 (Kyushu Institute of Technology, Japan)[69]
	26 May – Launch of the ChinaSat 2A (PLA, China) military communications satellite[70]
Technology development	
13 February – Launch of PW-Sat 1 (Warsaw University of Technology, Poland), XaTcobeo (University of Vigo, Spain), Robusta (University of Montpellier II, France), e-St@r (Polytechnic University of Turin, Italy), ALMASAT (University of Bologna, Italy), MaSat 1 (Budapest University of Technology and Economics)[1]	*17 August* – Launch of Edusat[71] (ASI, Italy) as an orbiting testing system for tech qualifications, open to research centres and industry; and BPA 2[72] (Ukraine) intended to test navigation equipment for civil aircraft and satellites, accompanied by an assortment of Earth observation and communications satellites *27 September* – Launch of TacSat-4 (US Navy, USA), an operationally responsive demonstrator satellite[1] *17 May* – Launch of the SDS 4 (JAXA, Japan) demonstration cube satellite carrying an AIS receiver to identify vessels in Japanese waters[73] *22 May* – Launch of unmanned cargo spacecraft Dragon COTS Demo 2/3 (SpaceX, USA)[1]

(continued)

[67] "Lockheed Martin-Built JCSAT-13 Satellite Begins Service for Sky Perfect JSAT of Japan." 10 July 2012. Lockheed Martin 27 June 2013 http://www.lockheedmartin.com/us/news/press-releases/2012/july/0710-ss-jcsat.html.

[68] Bergin, Chris. "Ariane 5 ECA launches JCSAT-13 and VINASAT-2 into orbit." 15 May 2012. NASASpaceFlight.com 27 June 2013 http://www.nasaspaceflight.com/2012/05/ariane-5-eca-jcsat-13-and-vinasat-2-into-orbit/.

[69] "Ignition! 2012 – Launches to Orbit and Beyond." Zarya 27 June 2013 http://www.zarya.info/Diaries/Launches/Launches.php?year=2012.

[70] Barbosa, Rui C. "Long March 3B launches military satellite Chinasat-2A." 26 May 2012. NASASpaceFlight.com 27 June 2013 http://www.nasaspaceflight.com/2012/05/long-march-3b-launches-chinasat-2a/.

[71] "EduSAT project." ASI 21 June 2013 http://www.asi.it/en/educational_en/school/edusat_project.

[72] "Annual Diaries – Events of 2011." Zarya 21 June 2013 http://www.zarya.info/Diaries/2011.php.

[73] "Small Demonstration Satellite-4 (SDS-4) 6 June 2012. JAXA 21 June 2013 http://www.jaxa.jp/projects/sat/sds4/index_e.html.

Europe	Other countries
	23 May – Failure to launch Fajr (Iranian Space Agency, Iran), a student developed solar powered high resolution reconnaissance cubesatellite[1]
Business	
9 June – SES and Qatar's Supreme Council of Information and Communication Technology (ictQatar) signed a cooperative agreement to provide satellite capacity for the joint development of satellite orbital slots[74]	*10 June* – Honeywell made an all-cash $491 million purchase of Atlanta-based EMS Technologies[75]
9 September – RapidEye (Germany, Canada) was contracted by the NGA (US) to provide up to $4.6 million worth of imagery over an 18 month period[76]	*13 June* – EchoStar (US) completed a $2 billion acquisition of satellite broadband provider Hughes Communications[77]
19 September – Astrium Services to provide optical and radar Earth observation data to the EU over a 3-year contract valued at 17 million Euro[78]	*23 June* – SGT wins a NASA contract worth up to $45 million over the course of 4 years to provide geophysics, geodynamics, and space geodesy support[79]
10 October – Thales Alenia to study satellite communications technologies for unmanned aerial vehicles for ESA and the EDA[80]	19 July – NASA awards Navarro Research and Engineering Inc. contract worth up to $80 million to provide environmental management and services[81]
21 October – Norspace (Norway) to be purchased by Kongsberg Defence and Aerospace AS (Norway)[82]	*2 September* – USAF awarded Boeing a $1 billion contract to start building a seventh WGS communication satellite and procure parts for another follow-up satellite[83]
16 February – Telespazio won six contracts valued at a combined total of 112.1 million Euro to provide telemetry, control and telecommunications services to CNES; as well as other activities provided to ESA, the EU Commission, and Brazilian telecommunications operator, Oi[84]	*31 October* – Vietnam contracted to buy two radar imaging satellites from Japan based on its ASNARO remote sensing platform[85]

(continued)

74 "SES and ictQatar Ink Cooperative Agreement." Space News 13 June 2011: 3.

75 De Selding, Peter B. "Honeywell Buying EMS Technologies for $491M." Space News 20 June 2011: 15.

76 "RapidEye Inks Imagery Deal With NGA Worth $4.6 Million." Space News 12 Sept. 2011: 9.

77 "EchoStar Closes $2B Deal For Hughes Communications." Space News 13 June 2011: 8.

78 "Astrium Lands EU Earth Observation Data Contract." Space News 26 Sept. 2011: 8.

79 "SGT Wins $45M Contract To Support Goddard Projects." Space News 4 July 2011: 8.

80 "ESA Taps Thales Alenia Space For Drone Technology Work." Space News 17 Oct. 2011: 8.

81 "Navarro Tapped by NASA For White Sands Contract." Space News 1 August 2011: 8.

82 "Kongsberg Defence To Buy Norspace." Space News 24 Oct. 2011: 3.

83 "U.S. Air Force Gives Boeing $1B for More WGS Satellites." Space News 5 Sept. 2011: 3.

84 "Telespazio Lands Deals Worth $146M." Space News 20 Feb. 2012: 3.

85 Kallender-Umezu, Paul. "Japan, Vietnam Sign Deal for Two Radar Imaging Satellites." Space News 7 Nov. 2011: 1+.

Europe	Other countries
24 February – ESA entered into a $1.8 billion deal with Thales Alenia Space to build six satellites for its METEOSAT Third Generation system[86]	*4 January* – GeoEye signed a multimillion-dollar contract with Russia's ScanEx to complete a national map of Russian real property[87] *13 February* – NASA announced it would withdraw entirely from ExoMars for budgetary reasons, to be replaced by Roscosmos as a main partner with ESA[88] *14 February* – US President Obama signs FAA Air Transportation Modernization and Reform Act (H.R. 658) into law, extending the grace period granted to firms developing private commercial suborbital spacecraft through September 2015[89] *14 May* – LightSquared filed for bankruptcy protection following earlier FCC license restrictions on the proposed mobile satellite-terrestrial broadband network[90] *25 May* – SpaceX eligible to be cleared to begin commercial cargo operations following the successful launch and birthing of its cargo-carrying Dragon capsule to the ISS[91]

10.4 Policy and International Cooperation

Europe	Other countries
General policy	
20 June – In contributing to the development of PPPs in European space programmes, ESA implemented a financial arrangement for a	*20 June* – The Russian space agency announced its plans to improve ISS exploitation and

(continued)

[86] De Selding, Peter B. "ESA Signs $1.8B Deal with Thales Alenia for Six Weather Sats." 24 Feb. 2012. Space News 15 May 2012 http://www.spacenews.com/earth_observation/120224-esa-deal-thales-weather-sats.html

[87] "GeoEye Wins Contract for Mapping Russian Properties." Space News 9 Jan. 2012: 8.

[88] De Selding, Peter. "ExoMars Wins One-month Reprieve." SpaceNews 21 May 2012: 8.

[89] "Obama Signs Bill Extending Commercial Spaceflight Grace Period." Space News 20 Feb. 2012: 9.

[90] "Troubled LightSquared Files for Bankruptcy." Space News 21 May 2012: 3.

[91] Ferster, Warren and Dan Leone. "SpaceX's Successful Mission Boosts Commercial Credibility." SpaceNews 25 May 2012: 1+.

Europe	Other countries
data-relay satellite system over Europe that would have commercial, civil-government and military uses[92]	multiply the scientific experiments conducted onboard[93]
28 October – Germany remains committed to financing a $2 billion upgrade to the Ariane 5 launcher instead of proceeding directly to a new launcher design[94]	*15 September* – NASA announces on its SLS development plans[95]
28 November – French President affirms commitment to maintain France's space budget despite pressure on public spending[96]	*29 December* – China issues a white paper on its five-year space program, with top priorities on developing of three new launch vehicles (including a rapid-response launch system) and mitigating its contribution to space debris[97]
	12 January – United States informally states that it will not adopt the EU's proposed code of conduct for space activity as currently drafted on the grounds that it is too restrictive[98]
	17 January – United States to join Europe and other nations in developing an international code of conduct for space operations so long as the resulting text does not restrict US national security-related activities[99]
General cooperation	
13 October – ESA formally invites Russia to join the U.S.-European ExoMars space exploration mission[100]	*8 July* – Roscosmos became the sole launch provider relied on to transport crew regularly to the ISS
21 October – ESA launches two Galileo positioning satellites on Russian Soyuz rocket from French Guiana Spaceport[101]	*29 January* – US government is seeking a direct line with China, similar to one recently established with Moscow, to prevent

(continued)

[92] De Selding, Peter B. "Europe Knocked for Balking at Public-Private Satellite Venture." Space News, 20 June 2011: 7.

[93] Pyadushkin, Maxim. "Russia To Increase Research On ISS."Aviation Week & Space Technology 20 June 2011: 135.

[94] De Selding, Peter B. "Germany Reaffirms Commitment to $2B Ariane 5 Upgrade." Space News 31 Oct. 2011: 6.

[95] Leone, Dan. "NASA Commits To Building Mandated Heavy-lift Rocket." 19 Sept. 2011. Space News 25 Jan. 2012 http://www.spacenews.com/civil/110919-nasa-commits-heavy-lift.html.

[96] De Selding, Peter B. "Sarkozy Reaffirms Commitment to Space Spending." Space News 28 Nov. 2011: 6.

[97] De Selding, Peter B. "China Outlines Space Priorities: Debris Mitigation, New Rocket." Space News 9 Jan. 2012: 14.

[98] Weisgerber, Marcus. "U.S. Wants Changes to EU Space Code of Conduct." Space News 16 Jan. 2012: 3.

[99] "Clinton: U.S. will Help Draft a Code of Conduct for Space." Space News 23 Jan. 2012: 3.

[100] De Selding, Peter B. "ESA Formally Invites Roscosmos To Join ExoMars Mission as Full-fledged Partner." Space News 17 Oct. 2011: 1.

[101] De Selding, Peter B. "Soyuz Lofts Two Galileo Satellites In Debut from European Spaceport." Space News 24 Oct. 2011: 1 and 14.

Europe	Other countries
	collisions and potentially destabilizing events in space[102]
2 December – ESA ended its efforts to make contact with Russia's Phobos-Grunt Mars spacecraft[103] *9 February* – German-Russian Eurockot Launch Services joint venture announced it will launch two Sentinel satellites aboard Russian Rocket vehicle under contracts with the ESA[105]	*13 February* – NASA announces entire withdraw from ExoMars for budgetary reasons, to be replaced by Roscosmos as a main partner with ESA[104]
7 June – France and Azerbaijan agree to space-based collaboration in telecommunications and the digital economy[106]	
31 October – 17 biological and medical experiments developed by German and Chinese researchers are carried onboard the Chinese Shenzhou-8 capsule[107]	
Space science	
18 October – the UK Space Agency issued a call for proposals for 1.6 million British pounds to award in 2012 to support science associated with Mars exploration[108]	*17 October* – NASA and Japan released an improved version of a digital topographic map produced with measurements from NASA's Terra Earth observing satellite[109]
Applications	
4 October – Astrium Services and ESA plan to develop and launch a European Data Relay System which will use high-speed optical links on two geostationary satellites to relay Earth observation data[110]	*11 July* – China successfully launched its second data relay satellite Tianlian 1–02, as the third nation to build an operational data-relay service[111]

[102] Opall-Rome, Barbara. "U.S. Wants a Space Debris Hotline With China Patterned on the One with Russia." Space News 13 Feb. 2012: A1.

[103] De Selding, Peter B., and Leone, Dan. "ESA Abandons Effort to Contact Russia's Stranded Mars Probe." Space News 5 Dec. 2011: 1+.

[104] De Selding, Peter. "ExoMars Wins One-month Reprieve." SpaceNews 21 May 2012: 8.

[105] "Rockot to Launch 2 ESA Sentinel Craft." Space News 13 Feb. 2012: 3.

[106] SpaceNews Staff. "France-Azerbaijan Accord Emphasizes Space Cooperation." 8 June 2011 SpaceNews 5 July 2013 http://www.spacenews.com/article/france-azerbaijan-accord-emphasizes-space-cooperation.

[107] De Selding, Peter B. "German Experiments Onboard Chinese Shenzhou-8 Capsule." Space News 7 Nov. 2011: 12.

[108] "UK Space Agency Solicits Mars Research Grant Bids." Space News 24 Oct. 2011: 8.

[109] "NASA, Japan Release Improved Topographic Map of the Earth." Space News 24 Oct. 2011: 9.

[110] De Selding, Peter B. "Astrium, ESA Sign Deal for High-speed Data Relay System." Space News 10 Oct. 2011: 10.

[111] "2nd Chinese Data-Relay Satellite Reaches Orbit." Space News 18 July 2011: 9.

10.5 Country Profiles

Austria

Population[112]	8.44 million
GDP[113]	300.7 billion Euro
Responsibility[114]	The Austrian Space Program is funded by the Federal Ministry for Transport, Innovation and Technology and managed by the Agency for Aeronautics and Space of the Austrian Research Promotion Agency
Activities[114]	In addition to ESA programmes, three main national programmes, along with an Austrian initiative: the Austrian Space Applications Programme (ASAP) and the Austrian Radionavigation Technology and Integrated Satnav Services and Products Testbed (ARTIST), an impulse programme meant to strengthen the research-, technology- and co-operations competence of Austrian players in the aeronautical sector (TAKE OFF), along with an Austrian initiative in Nanoscale Sciences and Nanotechnologies (NANO)
Budget[115 116]	2012 ESA contribution of 52.2 million Euro; and 2012 National expenditure of 20 million Euro
Staff[117]	ALR: 12
Direct employment in space manufacturing industry[118]	328

[112] "Eurostat – Tables, Graphs and Maps Interface (TGM) table: Population." 6 July 2013 http://epp.eurostat.ec.europa.eu/tgm/refreshTableAction.do?tab=table&plugin=1&pcode=tps00001&language=en.

[113] "Eurostat – Tables, Graphs and Maps Interface (TGM) table: GDP." 6 July 2013 http://epp.eurostat.ec.europa.eu/tgm/refreshTableAction.do?tab=table&plugin=1&pcode=tec00001&language=en.

[114] "Die Österreichische Forschungsförderungsgesellschaft FFG | FFG." 6 July 2013 http://www.ffg.at/.

[115] "ESA Budget for 2012." European Space Agency 25 June 2013 http://download.esa.int/docs/corporate/pies_final_final.ppt.

[116] European Space Directory 2013. 28th Edition. Paris: ESD Partners, 2013.

[117] "Agentur für Luft- und Raumfahrt – MitarbeiterInnen-Verzeichnis." 25 June 2013 http://www.ffg.at/team#Agentur für Luft- und Raumfahrt.

[118] Lionnet, Pierre. "Eurospace Facts and Figures." European Space Directory 2013. 28th Edition. Paris: ESD Partners, 2013. 52–58.

Belgium

Population[112]	11.09 million
GDP[113]	369.98 billion Euro
Responsibility[119]	The "Space Research and Applications" directorate, overseen by the Belgian Federal Science Policy Office, is responsible for managing Belgium's participation in the programmes and activities of the European Space Agency and of other national or international organisations which are responsible for research and scientific public service
Activities[120]	With ESA, the directorate manages all the space programmes, i.e. space sciences, earth observation, navigation, telecommunications, international space station, exploration, launchers, technology and science programmes. Cooperation on the bilateral level extends to France, Russia, and Argentina; and the directorate is also responsible for the Belgian state's participation in Airbus aeronautical programmes
Budget[115,116]	2012 ESA contribution of 169.8 million Euro; and 2012 National expenditure of 13.9 million Euro
Staff[121]	Department for space research and applications: around 20
Direct employment in space manufacturing industry[118]	1,495

Czech Republic

Population[112]	10.50 million
GDP[113]	155.49 billion Euro
Responsibility[122]	The Ministry of Education, Youth and Sports is the main governmental body responsible for space activities in the Czech Republic. Additionally, the Czech Space Office maintains direct communication with space-related experts and institutions in science, R&D and business sectors inside the

(continued)

[119] "belspo." belspo 1 July 2013 http://www.belspo.be/belspo/index_en.stm .

[120] "Space Research and Applications." belspo 1 July 2013 http://www.belspo.be/belspo/space/intro_en.stm.

[121] "European space agencies." Czech Republic Ministry of Transport: Space Technologies and Satellite Systems Department 6 July 2013 http://www.spacedepartment.cz/en/european-space-agencies-1/.

[122] "Welcome to Czech Space." Czech Space Office 1 July 2013 http://www.czechspace.cz/en/welcome-to-czech-space.

	country and at the same time develops relationships and programmes supporting international cooperation
Activities[122]	The key sectors include astronomical studies, magnetospheric, ionospheric and atmospheric research, remote sensing, Earth observation, microgravity research experiments, and competitive industry production, including small satellites construction
Budget[115]	2012 ESA contribution of 11.5 million Euro
Staff[123,124]	CSO: 14 Members: 19

Denmark | DTU Space Institut for Rumforskning og -teknologi

Population[112]	5.57 million
GDP[113]	240.45 billion Euro
Responsibility[125]	The National Space Institute (NSI) at the Technical University of Denmark (DTU) is the main governmental body responsible for space activities
Activities[126]	In addition to ESA programmes, key research areas include astronomical studies, Earth's physics and geodesy, climate and environment, and technological instruments, systems and methods in application to space activity
Budget[115,116]	2012 ESA contribution of 27.8 million Euro; and 2012 National expenditure of 5 million Euro
Staff[127]	DTU: around 112
Direct employment in space manufacturing industry[118]	210

[123] "Organisation structure of the CSO." Czech Space Office 1 July 2013 http://www.czechspace.cz/en/cso/organisation-structure-of-the-cso.

[124] "Members." Czech Space Office 1 July 2013 http://www.czechspace.cz/en/cso/members.

[125] "About DTU Space." DTU Space | National Space Institute 1 July 2013 http://www.space.dtu.dk/english/About_NSI.

[126] "Research." DTU Space | National Space Institute 1 July 2013 http://www.space.dtu.dk/english/Research.

[127] "Research Divisions." DTU Space | National Space Institute 2 July 2013 http://www.space.dtu.dk/english/Research/Research_divisions.

Finland Tekes

Population[112]	5.40 million
GDP[113]	189.49 billion Euro
Responsibility[128]	The Finnish Space Committee has representatives from all stakeholder groups and is responsible for the coordination of all of the activities. Tekes is the executive body for space activities and, together with the Academy of Finland for basic research, manages the Finnish participation within ESA programmes and other international projects
Activities[129]	In addition to its national space technology programme; Finland is closely involved in The EU Seventh Framework Programme for research and technological development, ERA-NET, EUREKA, COST, TAFTIE, and participation in ESA programmes
Budget[115,116]	2012 ESA contribution of 19.4 million Euro; and 2012 National expenditure of 15 million Euro
Staff[130]	Approximately 400 people in Finland and abroad (90 work in regional Centres for Economic Development, Transport and the Environment (ELY Centres)). Twenty programme coordinators from outside Tekes support the work of Tekes' own managers and provide an important contribution
Direct employment in space manufacturing industry[118]	173

France cnes

Population[112]	65.33 million
GDP[113]	2,001.40 billion Euro
Responsibility[131, 132]	The Centre National d'Etudes Spatiales (CNES) is responsible for the French space activities. It is under the shared responsibility of the Ministry of Education and Research

(continued)

[128] "Tekes." Tekes 2 July 2013 http://www.tekes.fi/en/community/Home/351/Home/473.

[129] "European cooperation." Tekes 2 July 2013 http://www.tekes.fi/en/community/European_cooperation/568/European_cooperation/1439.

[130] "Tekes organisation." 2 July 2013 http://www.tekes.fi/en/community/Organisation/344/Organisation/1288.

[131] "Organisation | National Centre for Space Studies (CNES)." 11 Dec. 2012. erawatch 2 July 2013 http://erawatch.jrc.ec.europa.eu/erawatch/opencms/information/country_pages/fr/organisation/organisation_mig_0019.

[132] "ONERA | The French Aerospace Lab." ONERA 2 July 2013 http://www.onera.fr/en.

	and of the Ministry of Defence. The Office Nationale d'Etudes et de Recherches Aérospatiales (ONERA) is also responsible for space related research
Activities[133]	In addition to ESA programmes, civil, military and science programmes are undertaken by France (i.e. Pleiades, SPOT, ELISA), as well as bilateral cooperation with the U.S. (CALIPSO, Jason 2 and 3) and India (Saral, Oceansat 3, Altika-Argos, Megha Tropiques)
Budget[115,116,134]	Overall 2011 budget of approximately 2,150 million Euro. 2012 ESA contribution of 718.8 million Euro; and 2012 National expenditure of 881.2 million Euro
Staff[135]	CNES: app. 2,400
Direct employment in space manufacturing industry[118]	12,869

Germany

Population[112]	81.84 million
GDP[113]	2,592.6 billion Euro
Responsibility[136]	The DLR is the national research centre responsible for space and aerospace activities in Germany. It is under the responsibility of the Ministry of Economics and Technology (BMWi)
Activities[137]	In addition to ESA programmes, Germany has national civil and commercial programmes in Earth observation (RapidEye, TerraSAR-X, TanDEM-X, EnMAP), Human space flight (ISS, Microgravity experiments), launch services (Eurockot, OHB-Cosmos), associated ground systems and space technologies (such as intersatellite links). Germany is involved in bilateral cooperation with the U.S. (GRACE, Dawn, Sofia) and its military programs include remote sensing satellites (SAR-Lupe radar satellites) and satcoms (Satcom BW)

(continued)

133 "Alphabetical index of the programs." CNES 2 July 2013 http://www.cnes.fr/web/CNES-en/461-cnes-programmes-alphabetical-index.php.

134 "Organisation | National Centre for Space Studies (CNES)." 11 Dec. 2012. erawatch 2 July 2013 http://erawatch.jrc.ec.europa.eu/erawatch/opencms/information/country_pages/fr/organisation/organisation_mig_0019.

135 "About CNES." CNES 2 July 2013 http://www.cnes.fr/web/CNES-en/3773-about-cnes.php.

136 "DLR at a glance" DLR 2 July 2013 http://www.dlr.de/dlr/en/desktopdefault.aspx/tabid-10443/637_read-251/.

137 "DLR." DLR 2 July 2013 http://www.dlr.de/dlr/desktopdefault.aspx/tabid-10002.

Budget[115,116]	2012 ESA contribution of 750.5 million Euro; and 2012 National expenditure of 460.0 million Euro
Staff[138]	DLR for space activities: app. 2,000
Direct employment in space manufacturing industry[118]	5,702

Greece

Population[112]	11.29 million
GDP[113]	208.53 billion Euro
Responsibility[139]	The General Secretariat for Research and Technology (GSRT) of the Ministry of Education is responsible for Greek space activities
Activities[139]	The Greek space activities cover mainly the fields of space physics, ionospheric physics, Earth observation and telecommunications
Budget[115]	2012 ESA contribution of 8.6 million Euro
Staff	For space: <5

Hungary

Population[112]	9.93 million
GDP[113]	99.82 billion Euro
Responsibility[140]	The Hungarian Space Office, recently administratively integrated into the body of the Ministry of National Development, manages, coordinates and represents Hungarian space activities
Activities[140]	Participation in microgravity, Earth observation, life and material sciences and optional GTSP programmes of ESA
Budget[141]	In 2011, Hungary spent about 4.5 million Euro (US$ 6 million) as its overall expenditure in civil space activities
Staff[142]	About 300 personnel working in over 30 different research groups

[138] "DLR Portal." DLR 28 Oct. 2011 http://www.dlr.de/dlr/desktopdefault.aspx/tabid-10002.

[139] "ςρχική Σελίδα." ΓΕΝΙΚΉ ΓΡςΜΜςΤΕΊς ΈΡΕΥΝςΣ ΚςΙ ΤΕΧΝΟΛΟΓΊςΣ 3 July 2013 http://www.gsrt.gr/central.aspx?sId=119I428I1089I646I488772.

[140] "The structure of the Hungarian space activity." A magyar űrkutatási hirei 3 July 2013 http://www.hso.hu/page.php?page=215.

[141] EuroConsult Report 2012. Overview – Global Dynamics in Government Space Programs: 8.

[142] European Space Directory 2013. 28th Edition. Paris: ESD Partners, 2013.

Ireland ENTERPRISE IRELAND

Population[112]	4.58 million
GDP[113]	158.99 billion Euro
Responsibility[143]	Enterprise Ireland, in association with the Office of Science and Technology of the Department of Enterprise, Trade and Employment, manages and coordinates space activities in Ireland
Activities[143]	Irish space activities are in the fields of software systems and services, precision mechanical components, advanced materials, electronics/microelectronics and telecommunications systems and service engineering
Budget[115]	2012 ESA contribution of 15.6 million Euro
Staff	For space: <5
Direct employment in space manufacturing industry[118]	46

Italy

Population[112]	59.39 million
GDP[113]	1,578.50 billion Euro
Responsibility[144]	The Italian Space Agency (ASI), under the Ministry of Education, University and Research, manages Italian space activities
Activities[144]	Italian civil space activities include ESA programmes and a national programme. National activities include scientific missions (AGILE, AMS, LARES, MIOSAT, PRISMA), dual-use Earth observation satellites (Cosmos-Skymed) and military telecommunication (Sicral), etc.
Budget[115,116]	2012 ESA contribution of 350.5 million Euro; and 2012 National expenditure of 400.0 million Euro
Staff	ASI: app. 200
Direct employment in space manufacturing industry[118]	5,134

[143] "Home." Enterprise Ireland 3 July 2013 http://www.enterprise-ireland.com/en/.

[144] "Agenzia Spaziale Italiana." ASI 3 July 2013 http://www.asi.it/en.

Luxembourg

Population[112]	524.85 thousand
GDP[113]	42.63 billion Euro
Responsibility[145]	Luxinnovation|National Agency for Innovation and Research, under the responsibility for the Ministry of Culture, Higher Education and Research, coordinates space activities
Activities[145]	Luxembourg focuses mainly on telecommunications, through SES Astra commercial satellite services provider
Budget[115]	2012 ESA contribution of 15.0 million Euro
Staff[146]	35 staff members are listed in the 2011 Annual Report, not all of whom work in space
Direct employment in space manufacturing industry[118]	31

Netherlands

Population[112]	16.73 million
GDP[113]	601.97 billion Euro
Responsibility[147]	The Netherlands Space Office (NSO) acts as the Dutch agency for space affairs, and is responsible for coordinating international space activities with ESA and other international space actors/. The Dutch Space Research Organization (SRON) is responsible for national and multilateral space science programmes
Activities[147]	In addition to its contributions to ESA programmes, such as Gaia, Exo-Mars, and GMES, NSO conducts bilateral activities with Japan (SPICA-SAFARI) and is responsible for planning the Dutch National Space Programme. SRON conducts space research in the fields of Earth observation, microgravity and planetary exploration
Budget[115,116]	2012 ESA contribution of 60.3 million Euro; and 2012 National expenditure of 25.0 million Euro

(continued)

[145] "Luxinnovation | National Agency for Innovation and Research." Luxinnovation 3 July 2013 http://en.luxinnovation.lu/Accueil#.

[146] "Annual Report 2011 (UK)." Luxinnovation 3 July 2013 http://en.luxinnovation.lu/Publications/Luxinnovation/Annual-Report-2011-UK : 32.

[147] "Home." Netherlands Space Office 3 July 2013 http://www.spaceoffice.nl/.

Staff[148]	NSO: 26, SRON: app. 200
Direct employment in space manufacturing industry[118]	767

Norway

Norsk Romsenter
NORWEGIAN SPACE CENTRE

Population[112]	4.99 million
GDP[113]	352.86 billion Euro
Responsibility[149]	The Norwegian Space Centre (NSC), under the Ministry of Trade and Industry, manages Norwegian space activities
Activities[149]	In addition to ESA programmes (in particular Earth observation, telecommunications and launchers), Norway has national support programmes and commercial activities (Telenor). Moreover, Norway operates the Andøya rocket range and the Svalbard ground station. Norway has also a bilateral agreement with Canada on the use of Radarsat 2 data
Budget[115,116]	2012 ESA contribution of 63.1 million Euro; and 2012 National expenditure of 8.0 million Euro
Staff[150]	NSC: 36
Direct employment in space manufacturing industry[118]	297

Poland

Polskie Biuro do spraw
Przestrzeni Kosmicznej

Population[112]	38.54 million
GDP[113]	370.85 billion Euro
Responsibility	While transitioning into a member state of ESA, responsibility for Polish space activities has been split among several bodies, in descending primacy: i.e. the

(continued)

[148] "Jobs | Vacatures" SRON: Netherlands Institute for Space Research 4 July 2013 http://www.sron.nl/jobs-mainmenu-2480.html.

[149] "The Norwegian Space Centre." Norsk Romsenter | Norwegian Space Centre 4 July 2013 http://web.spacecentre.no/eng.

[150] "Staff." Norsk Romsenter | Norwegian Space Centre 4 July 2013 http://web.spacecentre.no/eng/About/Staff.

	Ministry of Economy; the CBK; and the Polish Agency for Enterprise Development (PARP)
Activities[151, 152]	Polish space activities focus on the fields of space science, navigation and Earth observation applications. Poland is participating in several ESA science missions, such as Integral, Rosetta, BepiColombo and Solar Orbiter, and in Earth observation activities with Envisat and GMES
Budget[116]	2012 space expenditure of around 35 million Euro
Staff[153]	Space Research Centre: 199

Portugal

Population[112]	10.54 million
GDP[113]	171.05 billion Euro
Responsibility[154]	Portuguese space activities are coordinated by the Portuguese Space Office within the Foundation for Science and Technology (FCT), which is accountable to the Ministry of Science, Technology and Higher Education (MCTES)
Activities[154]	Mainly participation in ESA programs (telecommunications systems, technology developments, Earth observation, exploration), as well as space outreach programmes
Budget[115]	2012 ESA contribution of 15.8 million Euro
Staff	App. 6
Direct employment in space manufacturing industry[118]	131

Romania

Population[112]	21.36 million
GDP[113]	131.33 billion Euro

(continued)

[151] "Centrum Badań Kosmicznych." CBK 4 July 2013 http://www.cbk.waw.pl/.

[152] "Polskie Biuro do spraw Przestrzeni Kosmicznej." 4 July 2013 http://www.kosmos.gov.pl/.

[153] "Pracownicy etatowi CBK." CBK 4 July 2013 http://www.cbk.waw.pl/index.php?option=com_contact&view=category&catid=49&Itemid=82.

[154] "FCT Space Office." FCT 4 July 2013 http://www.fct.pt/apoios/cooptrans/espaco/index.phtml.en.

Responsibility[155]	The Romanian Space Agency (ROSA), under the responsibility of the Ministry of Education and Research, is managing Romanian space activities
Activities[155]	Romania has national activities covering space science (space physics and astronomy), space systems (construction of nanosatellites and microgravity experiments) and space applications (telemedicine, Earth observation and navigation space-based services); and its participation with ESA includes ESA's microgravity and exploration programme and technology activities
Budget[115,156]	In 2011, Romania spent about 25.5 million Euro (US$ 34 million) as its overall expenditure in civil space activities; and made a 2012 ESA contribution of 7.6 million Euro
Staff	ROSA: app. 36

Spain

Population[112]	46.20 million
GDP[113]	1,063.36 billion Euro
Responsibility[157]	The Centre for the Development of Industrial Technology (CDTI), under the Ministry of Science and Innovation, is funding and coordinating Spanish space activities
Activities[157]	In addition to ESA programmes, Spain has a national space programme including governmental and commercial programmes, especially in civil and military telecommunications (Hispasat, Spainsat), and Earth observation (SEOSAT/INGENIO, SEOSAR/PAZ, INTAuSAT 1) and it is involved in bilateral cooperation projects with France, Canada, the U.S. and Russia. Also, Spain manages national, ESA, and NASA ground facilities
Budget[115,116]	2012 ESA contribution of 184.0 million Euro; and 2012 National expenditure of 250.0 million Euro
Staff[157]	CDTI: app. 300
Direct employment in space manufacturing industry[118]	2,491

[155] "Romanian Space Agency." ROSA 4 July 2013 http://www.rosa.ro/.

[156] EuroConsult Report 2012. Overview – Global Dynamics in Government Space Programs: 8.

[157] "What is CDTI | Presentation." Centro para el Desarrollo Tecnológico Industrial 4 July 2013 http://www.cdti.es/index.asp?MP=14&MS=59&MN=1.

Sweden RYMDSTYRELSEN Swedish National Space Board

Population[112]	9.48 million
GDP[113]	387.60 billion Euro
Responsibility[158]	The Swedish National Space Board (SNSB) is under the Ministry of Education and Research, and is responsible for national and international activities relating to space and remote sensing, primarily with research and development
Activities[159]	Conducting magnetospheric and ionospheric research, astronomical studies, remote sensing, observation of the Earth's atmosphere and environment and specialized industrial competence (technology development), including development of small cost-efficient satellites. Promoting R&D is the main task of SNSB; and special focus is put on *inter alia* on-board computers, telecom equipment, nano-technology and formation flying platforms. Bilateral cooperation is mainly with France. Low-cost, short-term profile projects are represented by the Swedish-multilateral Viking, Freja and Odin scientific satellites
Budget[115,116]	2012 ESA contribution of 65.3 million Euro; and 2012 National expenditure of 16.0 million Euro
Staff[160]	17
Direct employment in space manufacturing industry[118]	759

Switzerland Schweizerische Eidgenossenschaft Confédération suisse Confederazione Svizzera Confederaziun svizra

Population[112]	7.95 million
GDP[113]	476.05 billion Euro
Responsibility[161]	The Swiss Space Office of the State Secretariat for Education and Research of the Federal Department of Home Affairs is responsible for Swiss space activities and cooperates

(continued)

[158] "Swedish National Space Board." RYMDSTYRELSEN 4 July 2013 http://www.snsb.se/en/Home/Home/.

[159] "Space Activities in Sweden." 14 Sept. 2012. RYMDSTYRELSEN 4 July 2013 http://www.snsb.se/en/Home/Space-Activities-in-Sweden/.

[160] "About SNSB." 30 Nov. 2010 RYMDSTYRELSEN 4 July 2013 http://www.snsb.se/en/Home/About-SNSB/.

[161] "State Secretariat for Education, Research and Innovation." SERI 4 July 2013 http://www.sbfi.admin.ch/index.html?lang=en.

	closely with the Swiss Department of Foreign Affairs in that field. The Federal Commission for Space Affairs (CFAS) is developing Swiss space policy. And the interdepartmental coordination committee for space (IKAR) is responsible for coordinating activities
Activities[161]	Most of the Swiss activities are undertaken within ESA programmes (space science, human spaceflight, launchers, Earth observation, Prodex and navigation)
Budget[115,116]	2012 ESA contribution of 105.6 million Euro; and 2012 National expenditure of 2.0 million Euro
Staff[162]	SNSB: 16
Direct employment in space manufacturing industry[118]	737

United Kingdom

UK SPACE AGENCY

Population[112]	63.26 million
GDP[113]	1,746.59 billion Euro
Responsibility[163]	The UK Space Agency, an executive agency of the Department for Business, Innovation and Skills (BIS), and reporting to the Minister of State for Universities and Science, is responsible for the strategic decisions on UK civil space activities
Activities[164]	UK space activities include bilateral cooperation with JAXA, India and the U.S. and within ESA in Earth observation and space exploration (Cassini-Huygens, James Webb Space Telescope, Herschel, and Planck missions)
Budget[115,116,165]	The UK Space Agency's net expenditure for the financial year 2011/12 amounted to £256M, while other sources list UK civil space expenditure to consist of a 2012 ESA contribution of 240.0 million Euro; and 2012 National expenditure of 80.0 million Euro
Staff	UKSA: 16
Direct employment in space manufacturing industry[118]	3,413

[162] "Staatssekretariat für Bildung und Forschung SBF." SBF 28 Oct. 2011 http://www.sbf.admin.ch/htm/themen/weltraum_de.html.

[163] "Who we are." UK Space Agency 4 July 2013 http://www.bis.gov.uk/ukspaceagency/who-we-are.

[164] "Home." UK Space Agency 4 July 2013 http://www.bis.gov.uk/ukspaceagency.

[165] "UK Space Funding." UK Space Agency 4 July 2013 http://www.bis.gov.uk/ukspaceagency/who-we-are/how-we-work/uk-space-funding.

Chapter 11
Bibliography of Space Policy Publications July 2011–June 2012

Blandina Baranes

11.1 Monographs

Abeyratne R (2012) Aeronomics and law. Fixing anomalies. Springer, New York

Alber I (2012) Aerospace engineering on the back of an envelope. Springer Praxis, New York

Albrecht M (2011) Falling back to Earth: a first hand account of the great space race and the end of the cold war. New Media Books, Lexington

Al-Rodhan NRF (2012) Meta-geopolitics of outer space: an analysis of space power, security and governance. Palgrave Macmillan, Basingstoke

Andersen T, Enmark A (2011) Integrated modeling of telescopes. Springer, New York

Arnould J (2011) Icarus' second chance: the basis and perspectives of space ethics. Springer, Wien/NewYork/Vienna

Asphaug V, Sorensen E (eds) (2011) Global positioning systems. Nova Science, Hauppauge

Badescu V (ed) (2012) Moon. Prospective energy and material resources. Springer, New York

Beysens D et al (eds) (2011) Laboratory science with space data. Springer, New York

Bignami GF (2012) We are the Martians. Connecting cosmology with biology. Springer, New York

Bizony P (2011) The space shuttle: celebrating thirty years of NASA's first space plane. Zenith Press, Minneapolis

Braude SY et al (eds) (2012) A brief history of radio astronomy in the USSR. A collection of scientific essays. Springer, New York

B. Baranes (✉)
ESPI (European Space Policy Institute), Vienna, Austria
e-mail: blandina.baranes@espi.or.at

C. Al-Ekabi et al. (eds.), *Yearbook on Space Policy 2011/2012*, Yearbook on Space Policy, DOI 10.1007/978-3-7091-1649-4_11, © Springer-Verlag Wien 2014

Brünner C, Soucek A (eds) (2011) Outer space in society, politics and law. ESPI studies in space policy. Springer, Vienna

Burgess C (2011) Selecting the Mercury seven: the search for America's first astronauts. Springer Praxis Books, New York

Chow M, Schilling G (2011) Tweeting the universe: tiny explanations of very big ideas. Faber and Faber, London

Clément G (2012) Fundamentals of space medicine. Springer, New York

Cox B, Forshaw J (2012) The quantum universe (and why anything that can happen, does). Da Capo Press, Jackson

Damon TD (2011) Introduction to space: the science of spaceflight. Krieger, Malabar

Dator JA (2012) Social foundations of human space exploration. Springer, New York

de Montluc B (2011) A new international strategic context for space policies. Space science exploration and policies. Nova Science, Hauppauge

Doran J, Bizony P (2011) Starman: the truth behind the legend of Yuri Gagarin. Walker & Company, New York

Eickhoff J (2012) Onboard computers, onboard software and satellite operations. An introduction. Springer, New York

Emslie AG et al (eds) (2012) High-energy aspects of solar flares. Springer, New York

Evans B (2012) At home in space. The late seventies into the eighties. Springer Praxis, New York

Evans B (2012) Tragedy and triumph in orbit. The eighties and early nineties. Springer Praxis, New York

Farquhar RW (2011) Fifty years on the space frontier: halo orbits, comets, asteroids, and more. Outskirts Press, Parker

Fischer MM, Wang J (2011) Spatial data analysis. Models, methods and techniques. Springer, New York

Flournoy DM (2012) Solar power satellites. Springer, New York

Gargaud MA (editor in chief) et al (eds) (2011) Encyclopedia of astrobiology. Springer, Berlin

Gendler R, Christensen LL, Malin D (2012) Treasures of the southern sky. Springer, New York

Genta G (2012) Introduction to the mechanics of space robots. Springer, New York

Geppert ACT (ed) (2012) Imagining outer space: European astroculture in the twentieth century. Palgrave Macmillan, Basingstoke

Gompert DC, Saunders PC (2011) The paradox of power: Sino-American restraint in an age of vulnerability. NDU Press, Washington, DC

Gonzalez W, Burch JL (eds) (2012) Key processes in solar-terrestrial physics. Springer, New York

Grego P (2012) Mars and how to observe it. Springer, New York

Haapala I (ed) (2012) From the Earth's core to outer space. Springer, New York

Handy R et al (2012) Sketching the Moon. An astronomical artist's guide. Springer, New York

Hays PL (2011) Space and security: a reference handbook. ABC-CLIO, Santa Barbara

Hobe S, Hofmannová M, Wouters J (eds) (2011) A coherent European procurement law and policy for the space sector. Towards a third way. LIT Verlag, Berlin

Huang B (ed) (2011) Satellite data compression. Springer, New York

Hubbart S (2012) Exploring Mars. Chronicles from a decade of discovery. University of Arizona Press, Tucson

Huntress WT Jr, Wesley T, Marov MY (2011) Soviet robots in the solar system: mission technologies and discoveries. Springer Praxis, New York

Jakhu RS, Sgobba T, Dempsey PS (eds) (2011) The need for an integrated regulatory regime for aviation and space. ICAO for space? ESPI studies in space policy. Springer, Vienna

Kaufman M (2011) First contact: scientific breakthroughs in the hunt for life beyond Earth. Simon & Schuster, New York

Kennedy LA (2012) One-shot color astronomical imaging. In less time, for less money! Springer, New York

Kessler A (2011) Martian summer: robot arms, cowboy spacemen, and my 90 days with the Phoenix Mars Mission. Pegasus, New York

Khorram S et al (2012) Remote sensing. Springer, New York

King C (2011) Blast off! A guide to world of space exploration, including global spaceflight programs, astronauts, space accidents, space tourism, and more. Webster's Digital Services, Baldwin City

Krauss LM (2012) A Universe from nothing: why there is something rather than nothing. Free Press, New York

Krukin J (2011) New space nation: America's emerging entrepreneurial space industry, 2nd edn. Kindle Edition, NN

Lafferranderie G, Marchisio S (eds) (2011) The astronauts and rescue agreement – lessons learned. The European Centre for Space Law, Paris

Lane H et al (eds) (2011) Wings in orbit: scientific and engineering legacies of the Space Shuttle, 1971–2010. Progressive Management, San Francisco

Lasota J-P (ed) (2011) Astronomy at the frontiers of science. Springer, New York

Launay F (2012) The astronomer Jules Janssen. A globetrotter of celestial physics. Springer, New York

Lee R (2012) Law and regulation of commercial mining of minerals in outer space. Springer, New York

Lefebvre J-L (2011) Stratégie spatiale. Penser la guerre des étoiles: une vision française. L'Esprit du livre editions, Sceaux

Lemmens M (2011) Geo-information. Technologies, applications and the environment. Springer, New York

Léna P et al (2012) Observational astrophysics. Springer, New York

Logsdon J (2011) John F. Kennedy and the race to the Moon. Palgrave Macmillan, New York

Long KF (2012) Deep space propulsion. A roadmap to interstellar flight. Springer, New York

Lutes CD et al (eds) (2011) Toward a theory of spacepower: selected essays. National Defense University Press, Washington, DC

Lyall F (2012) International telecommunications: the International Telecommunication Union and the Universal Postal Union. Ashgate, Farnham

Mahmoud MS, Xia Y (2012) Applied control systems design. Springer, New York

Marboe I (ed) (2012) Soft law in outer space: the function of non-binding norms in international space law. Böhlau, Wien

McGill University (2012) Active debris removal – an essential mechanism for ensuring the safety and sustainability of outer space. McGill University, Montreal

Menshikov VA, Perminov AN, Urlichich YM (2012) Global aerospace monitoring and disaster management. Springer, New York

Mineiro, MC (2011) Space technology export controls international cooperation in outer space. Springer, Berlin

Moltz JC (2011) The politics of space security: strategic restraint and the pursuit of national interests, 2nd edn. Stanford University Press, Stanford

Moltz JC (2012) Asia's space race: national motivations, regional rivalries, and international risks. Columbia University Press, New York

Orchiston W, Nakamura T, Strom RG (eds) (2011) Highlighting the history of astronomy in the Asia-Pacific region. Springer, New York

Pelton JN (2012) Satellite communications. Springer, New York

Petersen MB (2011) Missiles for the fatherland: Peenemunde, national socialism, and the V-2 missile. Cambridge University Press, Cambridge

Phillips RW (2012) Grappling with gravity. How will life adapt to living in space? Springer, New York

Plaxco KW, Gross M (2011) Astrobiology: a brief introduction, 2nd edn. Johns Hopkins University Press, Baltimore

Portree DSF (2012) Humans to Mars. Fifty years of mission planning, 1950–2000. CreateSpace Independent Publishing Platform, Washington, DC

Reddy F (2012) Celestial delights. The best astronomical events through 2012. Springer, New York

Samy I, Gu D-W (2011) Fault detection and flight data measurement. Demonstrated on unmanned air vehicles using neural networks. Springer, New York

Sanford, JF, Potkonjak M, Slijepcevic S (2012) Localization in wireless networks. Foundations and applications. Springer, New York

Scheeres DJ (2012) Orbital motion in strongly perturbed environments. Springer Praxis, New York

Seedhouse, E (2012) Astronauts for hire: the emergence of a commercial astronaut corps. Springer Praxis, New York

Simpson P (2012) Guidebook to the constellations. Telescopic sights, tales and myths. Springer, New York

Sing P (2012) Applications of spaceflights including Earth observation satellites, space exploration, space tourism, satellite navigation, space architecture, and space colonization. Webster's Digital Services, Baldwin City

Smith CM, Davies ET (2012) Emigrating beyond Earth. Human adaption and space colonization. Springer Praxis, New York

Smith LJ, Baumann I (2011) Contracting for space: contract practice in the European space sector. Ashgate, Surrey

Spudis PD (2011) Blogging the Moon: the once & future Moon collection. Apogee Prime, Burlington

Stüber GL (2012) Principles of mobile communication. Springer, New York

Sun J et al (eds) (2012) China satellite navigation conference (CSNC) 2012 proceedings, vols 159–161. Springer, Berlin/Heidelberg

Tewari A (2011) Automatic control of atmospheric and space flight vehicles. Springer, New York

Thakur JK et al (eds) (2012) Geospatial techniques for managing environmental resources. Springer, New York

Thomas AR, Thomarisos PN (2012) The final journey of the Saturn V. University of Akron Press, Akron

Thomas KS, McMann HJ (2012) U.S. spacesuits. Springer, New York

Troshichev O, Janzhura A (2012) Space weather monitoring by ground-based means. Springer Praxis, New York

Tyson de Grasse N (2012) Space chronicles: facing the ultimate frontier. WW Norton, New York

Vakoch DA (ed) (2011) Psychology of space exploration: contemporary research in historical perspective. NASA Publications, Washington, DC

Vakoch DA, Harrison AA (eds) (2011) Civilisations beyond Earth: extraterrestrial life and society. Berghahn Books, New York

Valavanis KP (ed) (2012) Recent developments in unmanned aircraft systems. Springer, New York

van Pelt M (2012) Rocketing into the future. The history and technology of rocket planes. Springer Praxis, New York

Venet C (2011) Key trends in the European earth observation sector. IFRI, Paris

Venet C (2012) Launcher policy in Europe: from symbol to market? IFRI, Paris

Venet C (2012) The European space industry, an endangered species? IFRI, Paris

Venet C, Nardon L (2011) The use of space for civil protection in Europe. IFRI, Paris

Walter U (2012) Astronautics: the physics of space flight, 2nd edn. Wiley-VCH, Weinheim

Wang W-Q (2011) Near-space remote sensing. Springer, New York

Wilkinson J (2012) New eyes on the sun. A guide to satellite images and amateur observation. Springer, New York

Woods WD (2011) How Apollo flew to the Moon. Springer Praxis, New York

Worden A, French F (2011) Falling to Earth: an Apollo 15 astronaut's journey to the Moon. Smithsonian Books, Washington, DC

Wotring Virginia E (2012) Space pharmacology. Springer, New York

Zalevsky Z (ed) (2012) Super-resolved imaging. Geometrical and diffraction approaches. Springer, New York

11.2 Articles

Abeyratne R (2012) The law and practice relating to aeronautical treaties and agreements. Ann Air Space Law XXXVII

Acevedo R et al (2011) Space activities in the Bolivarian Republic of Venezuela. Space Policy 27:174–179

Almár I (2011) SETI and astrobiology: the Rio scale and the London scale. Acta Astronaut 69:899–904

Anitha CS, Sharma RK (2011) Decay of high eccentricity satellite orbits with air drag using uniformly regular KS canonical elements. Acta Astronaut 68:1865–1871

Ansdell M (2012) Language protocols in international human spaceflight: time for a common tongue? Space Policy 28:2–6

Ansdell M, Ehrenfreund P, McKay C (2011) Stepping stones toward global space exploration. Acta Astronautica 68:2098–2113

Anselmo L, Pardini C (2011) Orbital evolution of the first upper stages used for the new European and Chinese navigation satellite systems. Acta Astronaut 68:2066–2079

Aoki S (2012) Japanese space activities act in the making. ZLW Zeitschrift für Luft-und Weltraumrecht (German J Air Space Law) 61:111–128

Argoun MB (2012) Recent design and utilization trends of small satellites in developing countries. Acta Astronaut 71:119–128

Atri D, DeMarines J, Haqq-Misra J (2011) A protocol for messaging to extraterrestrial intelligence. Space Policy 27:165–169

Bannova O, Bell L (2011) Space architecture education as a part of aerospace engineering curriculum. Acta Astronaut 69: 1143–1147

Baseley-Walker B (2012) Outer space, Geneva and the conference on disarmament: future directions. Space Policy 28:45–49

Baum SD, Haqq-Misra JD, Domagal-Goldmans D (2011) Would contact with extraterrestrials benefit or harm humanity? A scenario analysis. Acta Astronaut 68:2114–2129

Beck T et al (2011) Thermal analysis of a reflective baffle designed for space applications. Acta Astronaut 69:323–334

Cabrières B, Alby F, Cazaux C (2012) Satellite end of life constraints: technical and organisational solutions. Acta Astronaut 73:212–220

Cameron BG, Seher T, Crawley EF (2011) Goals for space exploration based on stakeholder value network considerations. Acta Astronaut 68:2088–2097

Carbajal Smith JD (2012) Mexico: new developments in space law and policy. ZLW Zeitschrift für Luft-und Weltraumrecht (German J Air Space Law) 61:95–98

Chengzhi L (2011) Agronomy in space – China's crop breeding program. Space Policy 27:157–164

Clark TK et al (2011) Numerical simulation of human orientation perception during lunar landing. Acta Astronaut 69:420–428

Cole TJ et al (2012) The challenges of designing a lightweight spacecraft structure for landing on the lunar surface. Acta Astronaut 71:83–91

Cornell A (2011) Five key turning points in the American space industry in the past 20 years: structure, innovation, and globalisation shifts in the space sector. Acta Astronaut 69:1123–1131

Darnopykh VV, Malyshev VV (2012) Operative planning of functional sessions for multisatellite observation and communication systems. Acta Astronaut 73:193–205

de Leon P, Peña Jr MAS (2011) Miguel Sánchez Peña (1925–2009) organizer of the space activities in Argentina. Acta Astronaut 69:892–898

Delgado-López LM (2012) Sino – Latin American space cooperation: a smart move. Space Policy 28:7–14

Doda LN et al (2011) Earthquakes forecasts following space- and ground-based monitoring. Acta Astronaut 69:18–23

Dunk von der FG (2011) Space tourism, private spaceflight and the law: key aspects. Space Policy 27:146–152

Elkin K et al (2011) Effectiveness of long-term program of scientific and applied experiments on the ISS Russian segment. Acta Astronaut 69:667–675

El-Madany HT et al (2011) Spacecraft power system controller based on neural network. Acta Astronaut 69:650–657

Fairblum SM (2011) Designing transformations: schools of excellence. Acta Astronaut 69:1132–1142

Freeland S (2011) For better or for worse? The use of 'soft law' within the international legal regulation of outer space. Ann Air and Space Law XXXVI

Freeland S (2012) Reshaping Australia's space policy and regulation – recent developments. ZLW Zeitschrift für Luft-und Weltraumrecht (German J Air Space Law) 61:99–110

From PJ, Ytterstad Pettersen K, Gravdahl JT (2011) Singularity-free dynamic equations of spacecraft-manipulator systems. Acta Astronaut 69:1057–1065

Gibbs G (2012) An analysis of the space policies of the major space faring nations and selected emerging space faring nations. Ann Air Space Law XXXVII

Gushin V (2012) Some psychophysiological and behavioral aspects of adaption to simulated autonomous Mission to Mars. Acta Astronaut 70:52–57

Haga RA, Saleh JH (2011) Epidemiology of satellite anomalies and failures: a subsystem-centric approach. Acta Astronaut 69:676–690

Hansen R, Wouters J (2012) Towards an EU industrial policy for the space sector – lessons from Galileo. Space Policy 28:94–101

Harris R, Miller L (2011) Earth observation and public good. Space Policy 27:194–201

He Y, Liang B, Xu W (2011) Study on the stability of Tethered satellite system. Acta Astronaut 68:1964–1972

Hobe S (2012) The permanent court of arbitration adopts optional rules for arbitration of disputes relating to outer space. ZLW Zeitschrift für Luft-und Weltraumrecht (German J Air Space Law) 61:4–26

Hoerber T (2012) New horizons for Europe – a European studies perspective on European space policy. Space Policy 28:77–80

Horri NM, Palmer P, Roberts M (2011) Design and validation of inverse optimisation software for the attitude control of microsatellites. Acta Astronaut 69:997–1006

Howe S et al (2011) The Mars hopper: an impulse driven, long range, long-lived mobile platform utilizing in-situ Martian resources. Acta Astronaut 69:1050–1056

Hubert G et al (2011) Multi-scale modeling to investigate the single event effects for space missions. Acta Astronaut 69:526–536

Inamori T, Sako N, Nakasuka S (2011) Magnetic dipole moment estimation and compensation for an accurate attitude control in nano-satellite missions. Acta Astronaut 68:2038–2045

Jaffer G, Nader R, Koudelka O (2011) Internet-to-orbit gateway and virtual ground station: a tool for space research and scientific outreach. Acta Astronaut 69:658–666

Jakhu RS (2012) McGill declaration on active space debris removal and on-orbit satellite servicing, 12 Nov 2011. Air Space Law 37:277–280

Jakhu RS, Doldirina C, Nyampong YOM (2012) Findings of an independent review of Canada's remote sensing space systems act of 2005. Ann Air Space Law XXXVII

Kaiser SA (2012) Satellite navigation systems – the impact of interoperability. Ann Air Space Law XXXVII

Kanas N et al (2011) High versus low crewmember autonomy during a 105-day Mars simulation mission. Acta Astronaut 69:240–244

Keil R (2011) An attitude model for the spacecraft of the ESA mission Gaia. Acta Astronaut 69:869–881

Kessler H, Peeters W (2011) Man and space – the Isle of man, the IISC and the new space economy. Space Policy 27:222–226

Kharytonov OM, Kiforenko BM (2011) Finite-thrust optimization of interplanetary transfers of space vehicles with bimodal nuclear thermal propulsion. Acta Astronaut 69:223–233

Kisdi A, Tatnall ARL (2011) Future robotic exploration using honeybee search strategy: examples search for caves on Mars. Acta Astronaut 68:1790–1799

Kojima H, Sugimoto Y, Furukawa Y (2011) Experimental study on dynamic and control of tethered satellite systems with climber. Acta Astronaut 69:89–108

Konstantinov MS, Petukhov VG (2012) The analysis of manned Mars mission with duration of 1000 days. Acta Astronaut 73:122–136

Langston SM (2011) Suborbital flights: a comparative analysis of national and international law. J Space Law 37:299

Larsen MB, Smith RS, Blanke M (2011) Modeling of tethered satellite formations using graph theory. Acta Astronaut 69:470–479

Launius PD (2012) Planning the post-Apollo space program: are there lessons for the present? Space Policy 28:38–44

Launius RD (2012) Why go to the moon? The many faces of lunar policy. Acta Astronaut 70:165–175

Lee J, Jo HJ, Choi JD (2011) The organizations for space education and outreach programs in the Republic of Korea. Acta Astronaut 69:730–736

Lee JJ, Chung S (2011) Space policy for later comer countries: a case study of South Korea. Space Policy 27:227–233

Levin E, Pearson J, Carroll J (2012) Wholesale debris removal from LEO. Acta Astronaut 73:100–108

Liu X, Kumar KD (2011) Input-to-state stability of model-based spacecraft formation control systems with communication constraints. Acta Astronaut 68:1847–1859

Logsdon JM (2011) Analyzing the new Kennedy tape. Space Policy 27:153–156

Machay M (2011) Contemporary Czech space policy and its future prospects. Space Policy 27:170–173

Machay M (2012) A brief analysis of US space employment. Space Policy 28:125–129

Marboe I (2012) The new Austrian outer space act. ZLW Zeitschrift für Luft-und Weltraumrecht (German J Air Space Law) 61:26–61

Martinez P (2012) The African leadership conference on space science and technology for sustainable development. Space Policy 28:33–37

Masutti A (2012) Legal problems arising from the installation of the Galileo and EGNOS ground stations in Non–EU countries. Air Space Law 37:65–79

Mathers N, Pakakis M, Christie I (2011) Mars mission program for primary students: building student and teacher skills in science, technology, engineering and mathematics. Acta Astronaut 69:722–729

Mayorova V (2011) Integration of educational and scientific-technological areas during the process of education of aerospace engineers. Acta Astronaut 69:737–743

Meulenberg A, Karthik PS (2011) The LEO archipelago: a system of earth rings for communications, mass-transport to space, solar power, and control of global warming. Acta Astronaut 68:1931–1946

Milligan T (2011) Property rights and the duty to extend human life. Space Policy 27:190–193

Mineiro MC (2011) An inconvenient regulatory truth: Divergence in US and EU satellite export control policies on China. Space Policy 27:213–221

Montluc de B (2012) What is the state of play in European governance of space policy? Space Policy 28:74–76

Murtazin R, Petrov N, Ulybyshev Y (2011) Launch strategy for manned spacecraft: improving safety or increasing of launch mass? Acta Astronaut 69:644–649

Mutschler MM, Venet C (2012) The European Union as an emerging actor in space security? Space Policy 28:118–124

Obara T, Matsumoto H, Koga K (2012) Space environment measurements by JAXA satellites and ISS/JEM. Acta Astronaut 71:1–10

Oikonomu I (2012) The European Defence Agency and EU military space policy: whose space odyssey? Space Policy 28:102–109

Ollongren A (2012) Processes in *Lingua Cosmica*. Acta Astronaut 71:170–172

Ormston T et al (2011) An ordinary camera in an extraordinary location: outreach with the Mars Webcam. Acta Astronaut 69:703–713

Osterlund J, Lawrence B (2012) Virtual reality: Avatars in human spaceflight training. Acta Astronaut 71:139–150

Pace S (2011) How far – if at all – should the USA cooperate with China in space? Space Policy 27:127–130

Pagkratis S (2011) Space activities in 2009/2010. Acta Astronaut 69:335–345

Paolozzi A, Ciufolini I, Vendittozzi C (2011) Engineering and scientific aspects of LARES satellite. Acta Astronaut 69:127–134

Peters L (2012) A regime of third party liability of global navigation satellite system (GNSS) services. Ann Air Space Law XXXVII

Petroni G, Santini S (2012) Innovation and change? The evolution of Europe's small satellite manufacturers. Space Policy 28:25–32

Pezzella G, Viviani A (2011) Aerodynamic analysis of a Mars exploration manned capsule. Acta Astronaut 69:975–986

Polkowska M (2011) State sovereignty in the airspace: an old notion in a new reality. Ann Air Space Law XXXVI

Purdy R (2011) Attitudes of UK and Australian farmers towards monitoring activity with satellite technologies: lessons to be learnt. Space Policy 27:202–212

Ram S, Doldirina C, Nyampong YOM (2012) Findings of an independent review of Canada's remote sensing space systems act of 2005. Ann Air Space Law XXXVII

Rees D (2012) We think we know where we are, but can anyone tell us where we are going? Has the European Union lost its way? Space Policy 28:81–85

Robinson GS (2011) Space law: addressing the legal status of evolving 'envoys of mankind'. Ann Air Space Law XXXVI

Rößner MR et al (2012) Broadband light source for fiber-optic measurement system in spaceborne applications. Acta Astronaut 70:95–99

Ryan S, Christiansen EL (2011) A ballistic limit analysis programme for shielding against micrometeoroids and orbital debris. Acta Astronaut 69:245–247

Sadowski R (2012) Insuring commercial space travel: seeking solutions to a futuristic problem. ZLW Zeitschrift für Luft-und Weltraumrecht (German J Air Space Law) 61:79–94

Saito H et al (2011) Small satellite REIMEI for auroral observations. Acta Astronaut 69:499–513

Salotti JM (2011) Simplified scenario for manned Mars missions. Acta Astronaut 69:266–279

Sandal GM, Bye HH, van de Vijver FJR (2011) Personal values and crew compatibility: results from a 105 days simulated space mission. Acta Astronaut 69:141–149

Schidler K et al (2011) PANIC – a surface science package for the in situ characterization of near-Earth asteroid. Acta Astronaut 68:1800–1810

Sedelnikov AV, Kireeva AA (2011) Alternative solutions to increase the duration of microgravity calm period on board the space laboratory. Acta Astronaut 69:480–484

Selva D, Krejci D (2012) A survey ad assessment of the capabilities of Cubesats for Earth observation. Acta Astronaut 74:50–68

Sherwood B (2011) Comparing future options for human space flight. Acta Astronaut 69:346–353

Sigalas E (2012) The role of the European Parliament in the development of a European union space policy. Space Policy 28:110–117

Smith LJ (2011) Business innovation symposium 'At what price? IP-related thoughts on new business models for space information' Acta Astronaut 69:714–721

Smith LJ (2012) Legislating for outer space: the example of Germany: considerations on a national space law. ZLW Zeitschrift für Luft-und Weltraumrecht (German J Air Space Law) 61:62–78

Steinberg A (2011) Space policy responsiveness: the relation between public opinion and NASA funding. Space Policy 27:240–246

Stephenson P (2012) Talking space: the European Commission's changing frames in defining Galileo. Space Policy 28:86–93

Streeper T et al (2011) Development of an integrated countermeasure device for use a long-duration spaceflight. Acta Astronaut 68:2019–2037

Suedfeld P, Brcic J (2011) Resolution of psychological crisis associated with flying in space. Acta Astronaut 69:24–29

Szajnfarber Z et al (2011) Moon first versus flexible path exploration strategies: considering international contributions. Space Policy 27:131–145

Takeichi N (2012) Geostationary station keeping control of a space elevator during initial cable deployment. Acta Astronaut 70: 85–94

Whitman Cobb WN (2011) Who's supporting space activities? An 'issue public' for US space policy. Space Policy 27:234–239

Wood D, Weigel A (2012) A framework for evaluating national space activity. Acta Astronaut 73:221–236

Wood D, Weigel A (2012) Charting the evolution of satellite programs in developing countries – the space technology ladder. Space Policy 28:15–24

Xuehua L et al (2011) Reachable domain for satellite with two kinds of thrust. Acta Astronaut 68:1860–1864

About the Authors and Editors

Andrea Jaime Albalat joined the Space Generation Advisory Council (SGAC) in 2011, and became the new Executive Director of the organisation in 2012. Prior to that, Andrea worked at the European Space Agency (ESA) as a Young Graduate Trainee. Based at ESTEC in the Netherlands she worked for the Human Spaceflight and Operations Directorate. In 2007 she obtained a double undergraduate degree in both Aeronautical and Mechanical Engineering from the Universitat Politecnica de Catalunya (UPC), in Barcelona. During these studies, she spent 1 year at the North East of Wales Institute (NEWI) in the United Kingdom. In 2010, Andrea obtained her master's degree in Aerospace Science and Technology, also from UPC in Barcelona (2010). Additionally, Andrea is an alumni of the International Space University's (ISU's) Summer Studies Program (SSP) 2009 which was hosted by the NASA Ames Research Center in California, USA. Andrea also completed the CVA (Communauté des Villes Ariane) Summer School, hosted by Rome in 2011, and has completed several short training courses on space related matters.

Cenan Al-Ekabi joined the European Space Policy Institute in Vienna in 2011, functioning as a project manager from 2012, and a Resident Fellow from 2013. Prior to that, he obtained two Advanced LL.M. degrees in Air & Space Law, and European & International Business Law from Leiden University in the Netherlands. He also holds a U.S. JD with concentration in studies in international law from the Thomas M. Cooley law school, and a bachelor's degree in Political Science from McMaster University in Canada.

Blandina Baranes joined the European Space Policy Institute (ESPI) in Vienna in February 2005 and currently holds the position of ESPI Communications Manager. Prior to this, she was the chief librarian of the Jewish Studies Department of the University of Vienna. During the past years she has also worked as a documentalist and librarian for different institutions, such as the Austrian Broadcasting Corporation (ORF), the Der Spiegel magazine and others. She conducted her studies and research in Austria and Israel and graduated with a Masters Degree from Vienna University, Faculty of Philosophy, Department of Social and Cultural

C. Al-Ekabi et al. (eds.), *Yearbook on Space Policy 2011/2012*, Yearbook on Space Policy, DOI 10.1007/978-3-7091-1649-4, © Springer-Verlag Wien 2014

Anthropology. She is co-editor of the past three volumes of ESPI's "Yearbook on Space Policy", published several articles and participated as a speaker in a number of space related events on the topic of "Space and Interdisciplinarity".

Jean Bruston is a senior Administrator at ESA in the Director's General European Union Relations Office since 2010 where he is dealing with institutional matters. During his previous career at ESA, he was seconded for 5 years as Secretary General of Eurisy, an organisation representing most European space agencies, providing support to European civil society for its appropriation of satellite based data and services, for economic, social and environmental purposes. From 2000 to 2002, he was scientific payload manager for the Herschel space Telescope instruments at ESA ESTEC in the Netherlands, at which time his main duty was the coordination of the scientific and industry consortia. This position followed his 6 years at NASA – Jet Propulsion Laboratory in the USA, where he held several positions from Research to project management and programme coordination. Jean Bruston also worked for 2 years as Director in a Consulting firm accompanying CEOs and leaders to devise international development strategies. Jean Bruston has a Ph.D. in Physics from the University of Paris, a Master degree in International Relations from the Institut de Sciences Politiques in Paris, and a Master degree in Humanities from the Ecole Normale Supérieure, also in Paris.

John Eyre has been Head of Satellite Applications at the Met Office (UK) since 1995. He leads research and development to improve exploitation of satellite data in numerical weather prediction and other applications in weather and climate. He has chaired activities with both EUMETSAT and the World Meteorological Organisation concerning the development of future observing systems. He was born in 1954 and received a doctorate in Atmospheric Physics from the University of Oxford in 1978.

Stéphane Hallegatte is senior economist with the World Bank, in the office of the chief economist of the Sustainable Development Network. His work includes green growth and climate change mitigation strategies, urban economics, and climate change adaptation and disaster risk management. He is lead author of the IPCC for its fifth assessment report. He also co-led the World Bank report "Inclusive Green Growth" in 2012, and is a core writing team member for the 2014 World Development Report "Managing Risks for Development." Stephane Hallegatte holds an engineering degree from the Ecole Polytechnique and a Ph.D. in economics from the Ecole des Hautes Etudes en Sciences Sociales in Paris.

Peter Hulsroj pursued his legal studies at the University of Copenhagen from 1974 to 1979. After military service, Peter Hulsroj did post-graduate studies at Harvard Law School in the USA, where he earned a Master's Degree in 1981. Returning from the USA, Mr. Hulsroj joined a large commercial law firm in Copenhagen, where he became a full member of the Danish Bar. After 3½ years of private practice in Copenhagen, Mr. Hulsroj became a Contracts Officer in ESTEC, and thereafter served for almost 14 years as the Head of Contracts and Legal Affairs at Eumetsat in Darmstadt, Germany. Before rejoining ESA, Mr. Hulsroj was the Legal Adviser of the Preparatory Commission for the Comprehensive Nuclear-Test-Ban Treaty Organization in Vienna. Peter Hulsroj was appointed ESA Director of Legal Affairs and External Relations (D/LEX) in April 2008, comprising four specialist areas: the Legal Department, the International Relations Department, the Communication Department and the Education and Knowledge Management Office at ESA from 2008 to 2011. Since 2011 Peter Hulsroj is Director of ESPI.

Robert Husband gained an engineering degree from Manchester University in 1978 and a Doctorate in Stability Theory from Oxford University in 1982. Thereafter he has been involved predominantly in the space sector working for British Aerospace, the European Space Operations Centre, British Satellite Broadcasting and EUMETSAT. After a period as Head of Operations and then Director of Operations at EUMETSAT, he left to form a consultancy company (MSYS Ltd) in 2001, and now provides consultancy support to a variety of organisations in the space sector.

Arne Lahcen joined the European Space Policy Institute in Vienna in 2011, functioning as a project manager from 2012, and a Resident Fellow from 2013. Prior to joining ESPI, he obtained an Advanced Master's degree in Space Studies with a specialisation in space law, policy, business and management at the Faculty of Science of the Catholic University of Leuven, Belgium. He also holds a Bachelor's and Master's degree in Social-Economic Sciences pursued at the Faculty of Applied Economics, University of Antwerp, Belgium. In addition to ESPI research activities, he is editor-in-chief of the ESPI Perspectives series and co-responsible for the follow-up of the ESPI Autumn Conference and the editing of the ESPI Yearbook.

Roland Potthast is head of the division for "Data Assimilation" at Deutscher Wetterdienst, Germany, since 2010, and Professor for Applied Mathematics at the University of Reading, UK, where he works since 2006. He has been Professor at the University of Göttingen, Germany, 2001–2006. Professor Potthast has had visiting positions at the University of Delaware, USA, and at the University de Rennes, France. As partner and director, he has been involved in several start-up activities at the Universities of Göttingen and Reading. He is strongly involved in international activities in inverse problems and data assimilation, including the management committee of various COST actions, as organiser of the "International Symposium on Data Assimilation" and winter and summer school activities in the field.

Per Tégner is a Swedish citizen born 1944 and pursued his studies at the Stockholm School of Economics. Between 1985 and 1998 he was Director at the Swedish Ministry of Enterprise. Subsequently, between 1998 and 2009, he was Director General of the Swedish National Space Board and Head of the Swedish delegation to ESA Council. From 2005 to 2006 and from 2007 to 2008, Per Tegner was Chair of the ESA Council. In addition, between 2005 and 2007 he held the position of vice chair of the ESA Council. Since 2012 Mister Tegner is Chair of the ESA Procurement Review Board.

Jean-Jacques Tortora currently serves as the Secretary General of ASD-Eurospace, which is the trade association of the European Space Industry. From 2004 to mid-2007, he was head of the French Space Agency (CNES) office in North America and the Attaché for Space and Aeronautics at the Embassy of France in Washington, D.C. Previously he was Deputy Director for Strategy and Programs, responsible for the Industrial Strategy of CNES, the French Space Agency. Prior to that position, he was France's representative in the ESA Industrial Policy Committee and Joint Communication Board. From 1998 to 2000, Mr. Tortora was adviser to the French Ministry of Research for Industrial Policy Funds management, aiming at industry competitiveness support and new space applications and services development and promotion. From 1996 to 1998, he was appointed by Arianespace in Evry, France, to lead a risk mitigation and cost reduction plan covering to the end of operational life of the Ariane 4 launcher. From 1990 to 1996, Mr. Tortora was based in Kourou, French Guiana, as Head of Arianespace Operations Quality Department. There, he led the Quality Management of the Ariane 4 launchers final integration and of the launch pad operations and maintenance. He supervised the industrialization and the implementation of the associated control plans. Mr. Tortora started his career in 1984 at the French military procurement agency, DGA, as a naval weapons integration and test engineer, expert in signal processing and warships acoustic discretion.

Index

C. Al-Ekabi et al. (eds.), *Yearbook on Space Policy 2011/2012*, Yearbook on Space Policy, DOI 10.1007/978-3-7091-1649-4, © Springer-Verlag Wien 2014

MIX
Papier aus verantwortungsvollen Quellen
Paper from responsible sources
FSC® C105338

If you have any concerns about our products,
you can contact us on
ProductSafety@springernature.com

In case Publisher is established outside the EU,
the EU authorized representative is:
Springer Nature Customer Service Center GmbH
Europaplatz 3, 69115 Heidelberg, Germany

Printed by Libri Plureos GmbH
in Hamburg, Germany